LEARNING
SUPPORT
SERVICES

Please return
on or before
the last date
stamped below

 City College
NORWICH

26. SEP. 2003

24 JUL 09

-9 NOV 2012

A FINE WILL BE CHARGED FOR OVERDUE ITEMS

120 187

 D1449149

Second Edition

TOOL DESIGN

Herman W. Pollack

Professor Emeritus
Orange County Community College
Middletown, NY.

PRENTICE HALL, Englewood Cliffs, N.J. 07632

Library of Congress Cataloging-in-Publication Data

POLLACK, HERMAN W.
    Tool design, 2/E.

    Includes index.
    1. Metal-cutting tools—Design and construction. 2. Jigs and fixtures—Design and construction. 3. Dies (Metal-working)—Design and construction. I. Title. II. Title: Tool design, Two/E.
TJ1186.P585 1988   621.9   87–25885
ISBN   0–13–925181–2

Editorial/production supervision
    and interior design: Eileen M. O'Sullivan
Cover design: Diane Saxe
Manufacturing buyer: Peter Havens

> *Dedicated To Betté,*
> *My Lovely Daughter And Friend*

Printed in the United States of America
10   9   8   7   6   5   4

ISBN   0-13-925181-2   025

PRENTICE-HALL INTERNATIONAL (UK) LIMITED, *London*
PRENTICE-HALL OF AUSTRALIA PTY. LIMITED, *Sydney*
PRENTICE-HALL CANADA INC., *Toronto*
PRENTICE-HALL HISPANOAMERICANA, S.A., *Mexico*
PRENTICE-HALL OF INDIA PRIVATE LIMITED, *New Delhi*
PRENTICE-HALL OF JAPAN, INC., *Tokyo*
SIMON & SCHUSTER ASIA PTE. LTD., *Singapore*
EDITORA PRENTICE-HALL DO BRASIL, LTDA., *Rio de Janeiro*

Contents

question 2

Contents

Preface

It is the intention of this edition of Tool Design to present as complete a picture as space permits of the knowledge and skills necessary for the effective design of tools and fixtures. With this in mind, this text includes presentations of tool and fixture design with background materials such as metallurgy, materials, manufacturing processes, and tool room drafting.

Because the manufacture of components and assemblies must address a competitive marketplace, the beginning chapters deal with processes, systems, processing, and cost analysis. This is followed with a presentation of components available to the designer, the performance that can be expected from them, and their uses in the design of tools, jigs, and fixtures as they relate to production.

The chapters that follow start with single point tool design and progress in complexity into a discussion of multiple point tool design. The latter is preparation for a discussion of jigs and fixture design, dies and numerical control.

A major addition to the second edition is Geometric Control. This material has surfaced since the publication of the first edition. Therefore a new chapter on this subject has been added. It should be understood that enough of this material has been added to give students a firm grounding in this topic so that they may research more complicated aspects of Geometric Control.

Also included in this edition is a discussion of some of the new tool materials available to the tool designer that may be used in high speed cutting and longer tool life. The latter is a necessity for economic production when using computer numerical control machines, or flexible manufacturing systems.

Many tables and graphs are included throughout this text. It should be evident that it is not possible to include in one text all the data, statistics, and other information needed to design tools, jigs and fixtures. The tool designer, as well as the student, must have access to tables, handbooks and literature to be a completely effective designer. The materials in this text will provide the foundations for the student so that effective search and research may take place.

It is assumed that the student will have the necessary knowledge and skills in drafting and will have had, or is concurrently taking, a course in Manufacturing Processes. It is assumed that a tool design course is taken in the second year and that the student will have had at least a one semester college level mathematics course. Such a course should include algebra and trigonometry.

This text includes many illustrative examples, project type problems, theory problems, tables and graphs.

I wish to express my thanks to Ed Moura, Acquisitions Editor, for his support of this effort and to Eileen O'Sullivan, Senior Editor, for her patience with me and her dedication to her professionalism.

Herman W. Pollack

Chapter 1

Structure and Properties of Metals

1.1 PRODUCTION OF FERROUS METALS

The flowchart shown in Fig. 1.1 is a graphic representation of the processing of the raw iron ore material from the time it is put into the blast furnace until the time it is shipped to be processed into castings, sheet, or bar stock. In this book the emphasis is on the tools needed to process these ferrous and nonferrous raw materials. Other materials, such as plastics, are included to complete the picture.

As shown in Fig. 1.1, the end product of the blast furnace is *pig iron*. The pig iron may then be processed into either cast iron or steel. In either case the impurities are carefully controlled. Any element added to pure iron is classified as an *impurity*. The control of the addition of impurities to the base material—in this case iron—is the first step needed to ensure the desired physical properties of the end product at that point of manufacture.

If the pig iron is fed into a cupola and processed, the end product is *cast iron*. Cast iron is composed of carbon, silicon, manganese, sulfur, and phosphorus in combination with iron. The percentages of each of these components are: approximately 2 to 4% carbon, 1 to 3% silicon, 1 to 2.5% manganese, 0.04% sulfur, 0.06 to 3% phosphorus, and about 90% iron. To achieve special properties, other elements may be added.

Cast iron is used in the main sections of jigs and fixtures in many instances because it can be cast into the desired shape and therefore requires a minimum amount of machining. Cast iron is a very stable material and it has suitable compressive strength for use as jig and fixture bodies or for die sets.

1

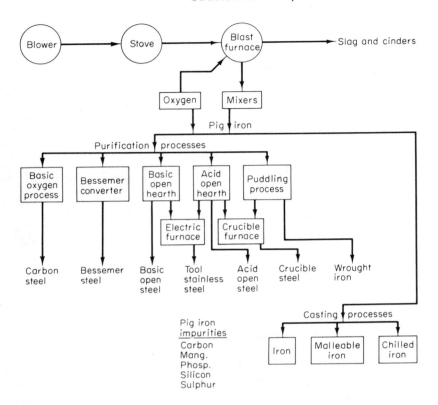

Figure 1.1

1.2 PLAIN CARBON STEELS

Plain carbon steels with carbon contents of up to 0.30% are called *low-carbon steels*. Structural cold- and hot-rolled flat steel is used in the construction of large or small jigs and fixtures. These steels need not have high strength or wear-resistance properties. They may be carburized and case hardened if better mechanical properties are desired. Cold-rolled steel has a good finish and therefore need not be machined, whereas hot-rolled steel has an oxide scale surface and must be machined if a smooth surface is required. It should be noted that rolling stresses are locked into the surface of cold-rolled steel. These stresses warp it when only one face is machined or if this material is welded. Hot-rolled steel does not have these stresses and since it generally costs less, it is used in welded structures.

Plain carbon steels with carbon contents of between 0.30 and 0.70% are called *medium carbon steels*. Their strength characteristics are better than those of the low-carbon steels. They will also respond to heating and quenching to produce better mechanical properties.

Plain carbon steels that have carbon contents over 0.70% are called *high-carbon steels*. These are the steels that have high strength and high hardness. They have properties that permit their use in cutting tools.

1.3 ALLOYS AND THEIR EFFECT ON TOOL STEELS

Plain carbon steels lack certain necessary characteristics, such as red hardness, hot and impact strength, and abrasive resistance, when used as cutting tools. When alloys are added to carbon steels, many of these shortcomings are overcome. The addition of alloys to produce tool steel is for the purposes of:

1. Improving toughness
2. Improving tensile and impact strength without greatly affecting toughness
3. Increasing hardenability
4. Improving wear resistance
5. Improving corrosion resistance
6. Increasing high temperature (red hardness)
7. Controlling the grain size in the material

The elements discussed below are for the purpose of producing one or more of these characteristics. These properties are discussed in Sec. 1.5.

Chromium. This alloy increases the hardenability, abrasion and wear resistance, and the impact strength of steel. It also increases its resistance to corrosion and oxidation. The addition of chromium also raises its critical temperature. This has the effect of causing distortion in the steel when it is water-quenched, and therefore great care must be exercised when designing parts made from this type of steel. As a carbide former its effectiveness is generally to produce properties between tungsten and manganese. Therefore, its response to tempering is good.

Manganese. This material, when added to steel as an alloy, increases the depth to which the steel will harden. When approximately 1.5% manganese in combination with approximately 1% carbon is added, the critical temperature of the steel is lowered so that it will harden when quenched in oil. When added to steel, its ability to form carbides is greater than that of ferrite, but not as great as chromium. Manganese as an alloy also counteracts the embrittling effect that sulfur may have. Its use as an alloy has very little effect on the tempering characteristics of steel.

Molybdenum. This alloy, when used with manganese, chromium, and silicon, increases the hardness, strength, and toughness of steel. It increases the hardenability of alloy steels to a much greater degree than does chromium. When alloyed together with chromium and vanadium, it greatly increases the hot strength and red hardness of the steel. Because of its strong carbide-forming effect, it promotes secondary hardening during heat treating, which in turn has the effect of making tempering more difficult. Since molybdenum, when used as an alloy, has the effect of raising the grain-coarsening temperature of the austenite phase, its use makes it easier to achieve fine-grain room-temperature structure in all steels quenched from the austenite phase.

Vanadium. This element greatly increases the hardenability of steel when it is used as an alloy. It is a very strong carbide former and as such resists the tempering operation because of the secondary hardening that is present. It does raise the grain-coarsening temperature, which makes it easier to produce a fine-grained structure in the steel when it is heat treated.

Silicon. When silicon is used with molybdenum, manganese, and chromium in percentages of up to 2%, it intensifies the effects which these three alloys have on the properties of steel. It increases the hardenability of steel, but not to the same degree as manganese. Its carbide-forming effect is less than that of ferrite and therefore also less than that of manganese. It does have the effect of increasing the resistance of steel to corrosion. In percentages of less than 0.25%, it has no effect as an alloy in steel.

Phosphorus. This alloy increases the hardenability of steel about the same as managanese. It has practically no effect as a carbide former and therefore has no effect on the tempering operation. It increases the corrosion resistance of steels. It also increases the machinability of steel. Sulfur and phosphorus are usually held to a minimum in tool steel.

Nickel. This alloy tends to aid the retention of austenite in high-carbon steel. In so doing, it has an effect on hardenability. If the steel has a high chromium content, the retention of austenite during heat treating is increased. Its addition has no effect on the tempering operation because it is a weak carbide-former.

Cobalt. When cobalt is used as an alloy with vanadium, chromium, or tungsten, it greatly increases the red hardness of the material. It has about the same carbide-forming tendencies as ferrite. When used as an alloy, cobalt is one of the few elements that decreases hardenability.

Tungsten. When used in low- or medium-carbon steels in small quantities, tungsten increases the hardenability only slightly. When approximately 4% is used as an alloy, it imparts hardness and wear resistance to the steel. If 18% tungsten is used as an alloy, the steel develops red hardness and hot strength. It also makes tool steel abrasion resistant. It is a strong carbide former, which, because of secondary hardening, makes it oppose softening during the tempering operation.

Titanium.   This element increases the austenite hardenability. In medium-chromium steel, it reduces the effect of hardening that results from martensitic formation.

Aluminum.   As an alloy in steel, aluminum acts as a deoxidizer. It also restricts grain growth at elevated temperatures. When alloyed with chromium, it forms complex nitrides under special heat-treating conditions. This produces a very hard surface on the steel. This process is called *nitriding*.

1.4 CLASSIFICATION OF TOOL STEEL

Tool steels may be classified as plain high-carbon, low-alloy, intermediate-alloy, and high-alloy steels. In general, alloys are used in steel to increase its hardenability and wear resistance by forming hard carbides and to reduce its tendency to softening through tempering (see Fig. 1.2).

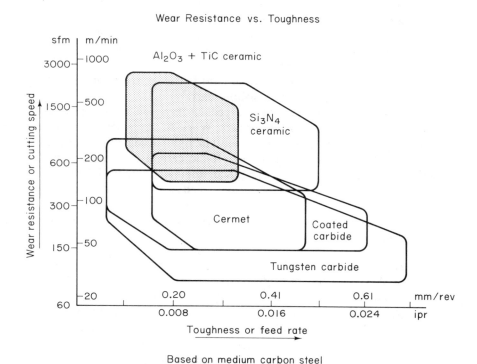

Figure 1.2   Reproduced with permission of Society of Manufaturing Engineers form Volume III of the Tool and Manufacturing Engineers Handbook, 4th ed.

Plain carbon tool steels are those which possess:

Carbon:	0.50 to 1.50%
Manganese:	0.12 to 0.40%
Silicon:	0.10 to 0.30%
Sulfur and phosphorus:	up to 0.50%

Carbon is the most important impurity in steel. Steel that is heated shows a marked increase in hardness for carbon contents of up to 0.85%. Above 0.85% carbon, the hardness increases only slightly. The effect of carbon content above 0.85% is to increase the wear resistance and the cutting properties of steel.

Manganese, when used with carbon, forms carbides (Mn_3C) and this together with iron carbide (Fe_3C) increase the wear resistance, tensile strength, and hardness of the material. As indicated in Sec. 1.3, when 1.5% manganese is alloyed in combination with 1% carbon, the steel must be quenched in oil. If quenched in water, the steel will crack. In quantities of 0.30% manganese, alloyed with about 0.80% carbon, the alloy may be quenched in water without cracking.

The quantities of silicon, sulfur, and phosphorus should not be greater than the percentages shown above. They have little effect on the properties of plain carbon tool steels.

Low-alloy tool steels harden deeper, faster, and with less distortion than do plain carbon tool steels. The quenching medium need not be severe, and therefore the danger of cracking is decreased.

Medium-alloy tool steels form hard, wear-resistant carbides. They are used in cutting tools for taking light finishing cuts.

High-speed tool steels have especially high wear-resistant qualities because of the high amounts of carbides formed. When heated, they produce secondary hardness. This makes them especially useful in high-speed cutting. If the carbon content is low, the steels are impact resistant. If the carbon content is high, they have high abrasion-resistant qualities. Because of their extremely high hardenability characteristics, they may be oil- or air-hardened without fear of the steel cracking or distorting.

Tool steels have been classified according to some special purpose. The symbols used in this classification are:

High-speed tool steel:
  M molybdenum type
  T tungsten
Hot-working tool steel:
  H H1-H19 chromium type
      H20-H39 tungsten type
      H40-H59 molybdenum type
Cold-working tool steel:
  D high-carbon, high-chromium
      type

Shock-resisting tool steels:
  S shock-resisting type
Mold steel:
  P mold type
Water-hardening tool steels:
  W water-hardening type
Special-purpose tool steels:
  L low-alloy type
  F carbon-tungsten type

TABLE 1.1   AISI IDENTIFICATION AND TYPE CLASSIFICATION OF TOOL STEELS

Type	C	Mn	Si	W	Mo	Cr	V	Co	Ni	Al
High-Speed Tool Steel (M = molybdenum type)										
M1	0.80	0.20	0.20	1.55	8.50	4.00	1.00			
M4	1.30	0.20	0.20	5.50	4.50	4.25	4.00			
M6	0.75	0.20	0.20	4.00	5.00	4.00	1.25	12.00		
M10	0.90	0.20	0.20		8.00	4.00	2.00			
M30	0.80	0.20	0.20	1.50	8.25	4.00	1.10	5.00		
M41	1.10	0.20	0.20	6.75	3.75	4.25	2.00	5.00		
M43	1.25	0.20	0.20	1.75	8.75	3.75	2.00	8.25		
High-Speed Tool Steel (T = tungsten type)										
T2	0.80	0.20	0.20	18.00	0.60	4.00	2.00			
T6	0.80	0.20	0.20	20.00	0.80	4.50	1.60	12.00		
T8	0.75	0.20	0.20	14.00	0.75	4.00	2.00	5.00		
T9	1.25	0.20	0.20	18.00	0.75	4.00	4.00			
T15	1.50	0.20	0.20	13.00	0.50	4.50	5.00	5.00		
Hot-Working Tool Steel (H = chromium type)										
H10	0.40	0.60	1.00		2.50	3.25	0.40			
H14	0.40	0.30	1.00	5.00	0.25	5.00	0.25	0.50		
H19	0.40	0.30	0.30	4.25	0.45	4.25	2.20	4.25		
Hot-Working Tool Steel (H = tungsten type)										
H20	0.35	0.20	0.40	9.00	9.50	2.00	0.50			
H23	0.30	0.20	0.50	12.00		12.00	1.00			
H26	0.60	0.20	0.20	18.00		4.00	1.00			
Hot-Working Tool Steel (H = molybdenum type)										
H41	0.65	0.20	0.20	1.60	8.75	4.00	1.00			
H43	0.55	0.20	0.20		8.00	4.00	2.00			
Cold-Working Tool Steel (D = carbon chrome type)										
D1	1.00	0.30	0.25		0.80	12.00	0.50			
D2	1.50	0.30	0.25		0.75	12.00	0.60			
D3	2.25	0.30	0.25		0.80	12.00	0.60			
D7	2.35	0.40	0.40		1.00	12.50	4.00			
Cold-Working Tool Steel (A = air-hardening type)										
A2	1.00	0.60	0.25		1.00	5.00	0.30			
A3	1.25	0.50	0.25		1.00	5.00	1.00			
A4	1.00	2.00	0.25		1.00	1.00				
A7	2.25	0.55	0.55	1.20	1.00	5.25	4.75			
A8	0.55	0.25	1.00	1.25	1.50	5.00	0.25			
A9	0.50	0.40	1.00		1.50	5.00	1.00		1.50	

(*continued*)

TABLE 1.1 (*Continued*)

Type	C	Mn	Si	W	Mo	Cr	V	Co	Ni	Al
			Cold-Working Tool Steel (O = oil-hardening type)							
O1	0.90	1.25	0.25	0.50		0.50	0.25			
O2	0.90	1.60	0.25		0.30	0.25	0.20			
O6	1.45	0.65	1.00		0.25	0.20				
O7	1.20	0.25	0.25	1.55	0.25	0.50	0.25			
			Shock-Resisting Tool Steel (S = shock-resisting type)							
S1	0.50	0.25	0.25	2.50		1.50	0.20			
S2	0.55	0.50	1.00		0.50		0.20			
S4	0.55	0.85	2.00			0.20	0.20			
S6	0.45	1.40	2.25		0.40	1.25	0.30			
			Weld Steels (P)							
P2	0.07				0.20	2.00			0.50	
P5	0.10					2.25				
P21	0.20								4.00	1.20
			Special-Purpose Tool Steels (L = low-alloy type)							
L1	1.00					1.25				
L3	1.00					0.25	0.75		1.50	
L7	1.00	0.35				0.40	1.40			
			Water-Hardening Tool Steels (W = water-hardening type)							
W1	0.60–1.4									
W2	0.60–1.4						0.25			
W4	0.60–1.4					0.25				
W5	1.10					0.50				
			Special-Purpose Tool Steels (F = carbon-tungsten type)							
F1	1.00			1.25						
F3	1.00			3.50	0.75					

Source: AISI Product Manual, Tool Steels, April, 1963.

A   medium-alloy,   air-hard-
    ening type
O   oil-hardening type

Table 1.1 (partial list) shows the classification adopted by the American Iron and Steel Institute (AISI). Table 1.2 shows a proposed classification from the American Society for Metals (ASM).

Table 1.3 shows several practices used in the heat treating of tool steels. A complete listing may be found in the handbook *Tool Steels* (ASM). The reader should understand that the temperatures and temperature ranges are appropriate under ideal

TABLE 1.2 TOOL STEEL

Class	AISI	Tool steel
110	W1	Carbon
120	W2	Carbon-vanadium
130	W4, 5	Carbon-chromium
140		Carbon-chromium-vanadium
210	L1, 2, 7	Chromium up to 3% and carbon greater than 0.65%
220	L2	Chromium up to 3% and carbon less than 0.65%
230	L6	Nickel and carbon greater than 0.65%
240		Nickel and carbon less than 0.65%
310	S	Silicon
320	S1	Tungsten chisel
330		Nontempering chisel
340	F2, 3	Tungsten finishing
350	F1	High-carbon low-alloy
360		
370	P	Mold steels carburized cavities
380	P20, 21	Mold steels for machined cavities
390	O and A10	Graphic tool
410	O	Oil-hardening cold-working die
420	A	Air-hardening cold-working die
430	D	High-carbon high-chromium cold-working die
440	A7, 0	Special wear-resistant cold-working die
510	S7	3 to 4% chromium die steel (hot working)
520	H, A8–9	Chromium-molybdenum hog-working die
530	H-14	Chromium-tungsten hot-working die
540	H-20 series	Tungsten hot-working die
550	H15–H4 D series	Molybdenum hot-working die
610	T1, 9, 7	Tungsten
620	T4, 5, 6, 8, 15	Tungsten-colbalt
630	M1, 10, 7	Molybdenum
640	M30	Molybdenum-cobalt
650	M2, 3, 4	Tungsten-molybdenum
660	M6, 15, 35, 36	Tungsten-molybdenum-cobalt
710		Wortle die and self-hardening steel

Roberts, Hamaker, and Johnson, *Tool Steel,* 3rd ed. (Metals Park, Ohio: American Society for Metals, 1962).

conditions. Since the heat treating of tool steels requires very precise procedures, great care should be exercised when using data.

Example 1.1

Given a W-steel, list (a) the hardening, (b) the tempering, and (c) the annealing temperatures. (d) What type of tool steel is this? Use Table 1.3.

Solution

(a) The hardening data show that this material need not be preheated. The temperature is raised to between 1400 and 1450°F and held there for 10 to 30 minutes to permit the

TABLE 1.3 Heat Treating of Tool Steel

AISI Series	ASM Class[a]	Hardening, H					Tempering	Annealing		Normalizing	Type of Tool Steel
		Preheat Temp. (°F)	Aust. Temp. (°F)	Holding Time (min)	Quench. Med.[b]	Hardness, R_c	Temp. (°F)	Temp. (°F)	Hardness, BHN	Temp. (°F)	
M-	630–660	1425	2150–2275	2–5	O, A, S	63–66	1000–1100	1600–1650	210–295	None	High-speed molybdenum
T-	520	1550	2200–2375	2–5	O, A, S	63–67	1000–1100	1600–1650	220–295	None	High-speed tungsten
H10	520	1500	1850–1900	15–40	A	56–60	1050–1250	1550–1650	192–229	None	Hot-working chromium
H16	530	1500	2050–2150	2–5	O, A	55–58	1050–1250	1600–1650	212–241	None	
H19	530	1550	2000–2200	2–5	A, O	52–55	1050–1250	1600–1650	207–241	None	
H20	540	1500	2000–2200	2–5	A, O	53–55	1100–1250	1600–1650	207–235	None	Hot-working tungsten
H23	530	1500	2200–2300	2–5	O	33–35		1600–1650	212–255	None	
H26	540	1600	2150–2300	2–5	A, O, S	63–64	1050–1200	1600–1650	217–241	None	
H41	550	1350	2000–2175	2–5	A, O, S	64–66	1050–1200	1500–1600	207–215	None	Hot-working molybdenum
H43	550	1350	2000–2175	2–5	A, O, S	54–58	1050–1200	1500–1600	207–235	None	
D-	430	1500	1750–2000	15–45	A	61–65	400–1000	1600–1650	200–260	None	Cold-working carbon-chromium
D3	430	1500	1700–1800	15–45	O	64	400–1000	1600–1650	215–250		

Type											Description
A- A_{10}	420 390	1200 1200	1450–1850 1450–1500	15–90 30–60	A A	60–65 62–64	300–1000 300–400	1550–1600 1410–1460	200–260 235–270	None 1450	Air hardening
O	410	1200	1400–1650	5–30	O	63–65	300–500	1375–1500	180–215	1600	Cold-working, oil-hardening
S- $S_{2,4}$ S_7	310–320 310 513	1300 1200 1250	1600–1750 1550–1700 1700–1750	5–45 5–20 15–45	O B, W A, O	55–60 60–63 60–61	300–1200 300–800 400–1000	1400–1550 1400–1450 1500–1550	187–230 192–230 187–223	None None None	Shock-resisting
P P_{20}	370–380 380		1450–1600 1450–1600	15 15	O, W A, O	62–65 60–62	300–500 300–500	1350–1650 1400–1450	100–135 150–180		Mold steel
L $L_{1,2,3}$	210–230 210	1200 1200	1450–1700 1450–1500	10–30 10–30	O O, W	60–65 63–65	300–600 300–600	1400–1500 1400–1450	160–215 160–200	1650 1650	Special-purpose low-alloy
F F_3	340–350 340	1200 1200	1450–1600 1450–1600	15 15	W, B W, B, O	65–67 65	300–500 300–500	1400–1500 1450–1500	180–240 210–250	1650 1650	Special-purpose tungsten
W	110–120 130–140		1400–1450	10–30	B, W	50–65	300–600	1350–1450	160–200		Water-hardening

[a] Proposed.

[b] W, Water; O, oil; B, brine; S, fused salts; A, air.

Source: ASM, *Metals Handbook*, 8th ed., Vol. 2. (Metals Park, Ohio: American Society for Metals, 1964). For specific values see Roberts, Hamaker, and Johnson *Tool Steel* 3rd ed. (Metals Park, Ohio: American Society for Metals, 1962).

austenite transformation to take place. The length of time depends on the thickness of the
workpiece. The higher temperature produces larger grains. After the allotted time, the
workpiece is quenched in brine or water. The workpiece should Rockwell between 50
and 65 R_c.

(b) Tempering, after hardening, is accomplished by reheating the workpiece to 300 to 600°F.

(c) Should annealing be required, the workpiece should be reheated to between 1350 and
1450°F.

(d) This is a water-hardening steel.

1.5 MECHANICAL PROPERTIES OF TOOL STEEL

Steels are used extensively in the making of tools, jigs, and fixtures. The
process of making steel from pig iron is essentially one of refining. The impurities are
removed and then reintroduced in amounts in the form of alloys to achieve a desired
result. The end product is steel. This steel may be rolled hot or cold into bars, strips,
sheets, and structural shapes. If it is rolled hot it will be relatively stress-free but will
have a scale on its outer surface. This may require extensive machining. Cold-rolled
steel has a smooth surface and may not need to be machined. However, the rolling,
when done cold, sets up stresses in the material that could cause the material to warp
when being machined or welded.

The operating parts of the fixtures or tools that do the cutting (drills), are made
from tool and die steel. These tool and die steels should have physical properties that
suit the job at hand. These properties may be: heat resistance (red hardness), abrasion
or wear resistance, shock resistance, deformation, and machinability. Because no tool
steel possesses all of these properties to an optimum degree, many tool steels have
been developed to service a wide range of needs.

Heat resistance.   Cutting tools must operate generally above 900°F. A material that has the ability to operate above 900°F is said to possess *red hardness*. Plain
high-carbon steels will operate at temperatures of up to 800°F. They will soften if this
temperature is exceeded. The addition of tungsten, molybdenum, and chromium will
produce tool steels that possess red hardness. Table 1.4 shows a graded list of tool
steels, from best to poor.

TABLE 1.4 HEAT RESISTANCE

Type	T	H20-; H40; M	D; H10	S, A	W; O; L; F; P
	Best	Very good	Good	Fair	Poor

Abrasion or wear resistance (Table 1.5).   Among other things, this property is related to hardness. It is caused by either the *scratching* of one material by
another, or by metal-to-metal contact as the materials move over each other while
they are in contact. The first is generally overcome by selecting the appropriate
lubrication or a combination of suitably different metals. Scratching may be overcome by selecting the proper hardness or chemical composition of the materials used.

TABLE 1.5 RESISTANCE TO ABRASION

Type	D	A; T; M; F	O; H20-; H40; L; P	W; S; H10-
	Best	Very good	Good	Fair

Shock resistance, or impact strength (Table 1.6).   Toughness is that property of a material which enables it to bend before it breaks when subjected to repeated or continuous loads. Shock resistance is therefore one aspect of toughness. Tools that have good shock resistance should therefore have a balance between toughness and hardness since the latter two properties are opposites. A correct combination of the silicon content with other elements, such as molybdenum or manganese, will yield optimum shock resistance.

TABLE 1.6 SHOCK RESISTANCE

Type	S	W	H10-; H20-; P	O; A; L; H40-	D; T; M; F
	Best		Good	Fair	Poor

Deformation (Table 1.7).   This property is extremely important. Steels will go through substantial changes in volume when they are heated and quenched during the heat treatment process. Any abrupt changes in the shape of the material will cause an uneven rate of change of volume, or set up lines or areas of stress concentration that will cause distortion or cracking. Many tools are designed so that they are first shaped within dimensional tolerances and then heat treated. In all cases the deformation of the material used must be held to a minimum. To some extent this can be controlled by the addition of manganese or high chromium.

TABLE 1.7 NONDEFORMING PROPERTIES

Type	A; D	O; D3; H11-	F; H; T; M	S	L; W; P
	Best	Very good	Good	Fair	Poor

Machinability.   This property refers to the ease with which a material may be machined relative to water-hardening alloy steels. The cutting ability of a tool is related to the type of material in the tool and in the workpiece. When referring to the tool it is better to think about its cutting ability as related to the material that is being cut. Of all the alloy tool steels, water-hardening tool steels have the best machinability characteristics. Therefore, in Table 1.8 the water-hardening tool steels are

TABLE 1.8 MACHINABILITY OF ALLOY TOOL STEELS (AVERAGE VALUES)

Type			P			H(W)	H(W)			
		L	A	F			H(W)			
	W	O	S	H(Cr)			H(Mo)	T	M	D
	100	90	80	70	60		50			40

TABLE 1.9 MACHINABILITY OF CARBON STEELS

AISI	Rating	AISI	Rating
B1112	100	A6145	50
C1010	55	A8620	60
C1015	50	A8650	50
C1118	80	A9140	60
C1020	65	Malleable	120
C1045	60	Cast iron	
C1070	45	Soft	80
B1113	130	Medium	65
A3120	60	Hard	50
A3130	55	Wrought iron	50
A3140	55	Stainless (12% Cr)	70
A3145	50	Stainless 18–8 (aust.)	25
A4130	65	Cast steel (0.35% C)	70
A4145	55	Manganese oil hard	30
A4150	50	Tool steel	
A4340	45	Low W, Cr, C	30
A5120	65	High C, Cr	25
A5140	60	High-speed steel	30
A5150	55		

arbitrarily assigned the value 100. Table 1.9 shows the machinability rating for several carbon alloy steels.

The greater the amount of carbon and of alloying elements in the steel, the more difficult it is to machine. This is especially true when the alloying elements form carbides. Such is the case when carbon is alloyed with tungsten, vanadium, molybdenum, or chromium.

Depth of hardening. Hardenability is a term that refers to the depth to which hardening takes place. Manganese, molybdenum, tungsten, and chromium when used as alloys have a strong influence on the depth to which a steel will harden. Vanadium, silicon, nickel, and copper have the same effect, but to a lesser degree. Coarse-grained steels have better hardenability characteristics than do fine-grained steels because the coarse grains tend to retain austenite for longer periods of time. In so doing the austenite-to-martensite transformation has time to materialize. Table 1.10 shows the hardenability characteristics of various tool steels.

TABLE 1.10 DEPTH OF HARDNESS

Type	A; D; H; T; M	S; O; L	W; F; P
	Deep	Medium	Shallow

Decarburization. Since carbon is probably the most important element in steel, any change in its content will change the physical properties of the steel during heat treating. If machining is to take place after heat treating, it is important to know

TABLE 1.11 RESISTANCE TO DECARBURIZATION

Type	W	O; P; L; F	A; M; H; T	S; A; D; T	M; H40-; S
	Best	Very good	Good	Fair	Poor

the depth to which decarburization has taken place. If it is not to be machined after heat treating, it is equally important to know if there has been any surface decarburization. Table 1.11 shows the type of steel and its resistance to decarburization.

The physical properties discussed in this section are summarized in Table 1.12.

TABLE 1.12 PHYSICAL PROPERTIES[a]

Class	Type	Quench	Non-def.	Shock-Resist.	Wear-Resist.	Heat-Resist.	Decarb.-Resist.	Depth-Hard.	Mach. Rate Av.
M	High-speed Mo	A O S	F G F	P	V.G.	V.G.	G-P	D	52
T	High-speed W	A O S	F G F	P	V.G.	B	G-F	D	48
H_{10}	Hot-working Cr	A O	G F	G	F	G	G	D	75
H_{20}	Hot-working W	A O	G F	G	G	V.G.	G	D	55
H_{40}	Hot-working Mo	A O S	F G F	F	G	V.G.	P	D	55
D	Cold-working C-Cr	A O	B V.G.	P	B	G	F	D	45
A	Air-hardening	A	B	F	V.G.	F	G-F	D	85
O	Cold-worked, oil-hardening	O	V.G.	F	G	P	V.G.	M	90
A	Shock-resisting	O W A	F P G	B	F	F	F-P	M	85
F	Special purpose W-	W B O	F G G	P	V.G.	P	V.G.	S	75

(continued)

TABLE 1.12 PHYSICAL PROPERTIES[a] (*Continued*)

Class	Type	Quench	Non-def.	Shock-Resist.	Wear-Resist.	Heat-Resist.	Decarb.-Resist.	Depth-Hard.	Mach. Rate Av.
L	Special purpose low alloy	O W	F P	F	G	P	V.G.	M	80
W	Water-hardening	W B	P P	G	F	P	B	S	100
P	Mold steel	O W A	P	G	G	P	V.G.	S	90–75

[a]B, Brine; W, water; O, oil; S, molten salt.
B, Best; V.G., very good; G, good; F, fair; P, poor.
D, Deep; M, medium; S, shallow.

1.6 TOOL STEELS: TYPES

High-speed tool steels (M- and T-types).   These tool steels are deep-hardening and they retain that hardness at elevated temperatures, as well as having a high resistance to abrasion and wear. The M-type tool steels have high percentages of molybdenum in combination with chromium or vanadium. The T-type tool steels have high tungsten percentages in combination with chromium, vanadium, and sometimes cobalt. They are used for blanking and hot-forming dies, lathe centers, cutting tools, drills, and taps.

The M-type high-speed tool steels were made by replacing tungsten in the T-type steels with molybdenum. The typical molybdenum (M-) tool steel has about 4 to 8.5% molybdenum, 4% chromium, 1.5 to 6% tungsten, and about 2% vanadium.

The T-type high speed tool steels' hardness are controlled by adjusting the carbon content. Those materials that have a high carbon content have high wear resistance and are very hard. Those with a low carbon content are tougher but not as hard as those with a high carbon content. The most widely used combination of tungsten, chromium, and vanadium seems to be the 18–4–1 tool. As the percent of tungsten increases, the toughness decreases. The 4% chromium is used to control the hardenability. As the percent of chromium decreases, the tool loses those qualities that make it acceptable as a high-speed material. The percent of vanadium affects the abrasion-resisting qualities of the steel and therefore the cutting qualities of the tool. When cobalt is added as an alloy, the carbon content must be increased, because cobalt reduces hardenability.

Hot-working die steels (H-types).   These steels are separated into three groups: H-chromium, H-tungsten, H-molybdenum types, or combinations of these elements.

The H-chromium-type die steels (H10 series) are resistant to thermal shock and cracking which results from high-temperature use. Thus they must be able to withstand distortion when used in hot-work dies. They are used in aluminum, magnesium, or zinc extrusion dies and die casting, hot shear blades, plastic molds, header dies, punches and dies for piercing shells, and hot press dies.

The H-chromium-molybdenum hot-working die steels (H10 series) are the same as the H-chromium type series except that the molybdenum quantities are increased. These steels are more widely used than the H-chromium series because they are less subject to cracking with temperature change and because they have greater resistance to heat checking. They have good resistance to shock loading at normal or elevated temperatures, and excellent shock resistance at room temperatures. They are used in die castings and forging dies.

The H-chromium-tungsten-type hot-working die tool steels (H16) have high red hardness and wear resistance, which increases as the tungsten is increased up to 7%. As the carbon content increases, the hardenability increases. They are considered deep-hardening steels. As the chromium content is increased, their resistance to shock loading decreases. Because of this, they are not necessarily considered to be shock-resistant steels. They are suitable for use in hot dies, for forging or extruding brass, die casting of aluminum, in permanent molds for molding brass, hot blanking or forming punches and dies, and piercing points or mandrels used in shell molding.

The H-tungsten-type hot-working die steels (H-20) uses 9 to 18% tungsten in combination with 2 to 4% chromium and about 0.50% vanadium. These steels have good shock-resisting qualities and have the best red-hardness properties of any of the hot-working steels. The risk of cracking these steels, when cooled rapidly, is great because they have poor resistance to thermal shock. They are used essentially for forging, extruding, and die casting dies for nonferrous materials, and in hot-blanking forming, hot or cold headers, thread-rolling, hot trimming, or shearing dies.

The H-molybdenum-type hot-working die steels (H-40) replace the tungsten component with molybdenum. They have good red hardness, fair wear-resistant qualities, and fair toughness. They are used for essentially the same operations as the H-tungsten steels.

Cold-working tool steels (D-, A-, and O-types).   The classifications in this group are the high-carbon, high-chromium D-steels; the medium alloy in combination with medium- to high-carbon air-hardening A-steels; and the oil-hardening O-steels. The D- and S-steels are deep hardening. The O-steels have medium hardenability. They all have excellent resistance to deformation, average red hardness, and high wear-resistant properties.

The D and A groups, because of their excellent resistance to wear and deformation, are used for gages, blanking, drawing and piercing dies, shears, forming and banding rolls, wear plates, lathe centers, mandrels, and breaches. The O group is used for making reamers, taps, threading dies, blanking, formdies, gages, plastic molds, cold trimming and coining dies, and burnishing and knurling tools.

Shock-resisting tool steels (S-type). These steels have up to 0.5% carbon in combination with silicon, nickel, and chromium. These classes of steel have very good shock-resistant qualities and excellent toughness. They possess medium-hardenability characteristics, but low red hardness and resistance to wear.

They are used in form tools, chisels, punches, cutting blades, springs, crimping, trimming, and swaging dies. When tungsten is added to S-steels, they develop red hardness. The latter development permits their use in cold- or hot-working tools, such as concrete and rock drills, bolt and pipe cutters, and hot or cold header dies.

Mold tool steels (P-type). These types of steel are classified into two general types: (1) those that are machined by hubbing (pressing a form into the material), and then carburizing and hardening the surface; and (2) those that are machined, or hubbed and machined. The first group has a low carbon content. The latter group has a high carbon content. The low-carbon group has low red hardness. The high-carbon group has poor-to-medium red hardness. Both groups possess high toughness qualities.

Special-purpose tool steels (L-and F-types). The L-type steels are low-alloy steels with a chromium content that makes them a good low-cost substitute for cold-working steels. These steels are used for gages, drills, taps, broaches, threading dies, ball and roller bearings, knurls, and files.

The F-type steels have a high carbon and tungsten content. They have poor-to-medium red hardness with medium hardenability characteristics. They have good toughness and high wear resistance. They are used in finishing tools because they will maintain a sharp cutting edge. For the same reason they are used in shearing dies.

Water-hardening tool steels (W-type). These tool steels have a carbon content of 0.60 to 1.4%. They also contain small amounts of manganese and silicon. In addition, one group has vanadium added, another group has chromium added, and a third group combines all of these alloys and 1.00% carbon. They have average toughness and wear resistance. They have shallow hardenability and poor red hardness. They may be used in cutting tools but only where the heat generated will be low. They are used in wood-working tools of all kinds, such as chisels, punches, hammers, files, and reamers.

1.7 SPECIAL MATERIALS

There are three special types of materials that are of interest to tool designers: cast alloy, carbide, and ceramic materials.

Cast materials. Cast materials are suitable for use as cutting tools because they exhibit high hardness, excellent red hardness, and excellent wear- and corrosion-resistant properties. Also, when used to cut steel, they have a low coefficient of friction. The alloys used in these tools are 1 to 3% carbon, 10 to 20% tungsten, and 25 to 30% chromium. Cobalt makes up the remainder of the alloy material.

They are cast to desired size and shape, and may be used at higher machine speeds than the high-speed steels. Heat treating does not affect their strength or hardness. They are generally not as hard as high-speed steels. Because they are cast, they are not as tough as high-speed steel. They are used in all types of cutting tools, in hard-facing dies, gages, cams, in crushing and grinding machinery, and in excavating equipment.

Cemented carbides.   These materials are made by blending the carbides of tungsten, titanium, or tantalum with a binder that has a lower melting point than the carbide. Such materials are cobalt, iron, or nickel. Special effects may be achieved by adding the carbides of vanadium, columbium, chromium, zirconium, or molybdenum to tungsten.

These powder mixes are compacted into a desired shape by pressures of up to 30 tons per square inch. They are then placed in furnaces and sintered at temperatures of 2500 to 2900°F for about 1 hour. In some instances the compacting and heating are done at the same time. At these pressures and temperatures the cobalt changes phase and forms a eutectic with the hard component (carbide). Once cooled, further heating will not affect the structure. These materials are used to machine cast iron, nonferrous, and nonmetallic materials.

These materials have Rockwell hardnesses about $70R_c$ and have compressive strengths generally greater than 7×10^5 psi. It is because of this that carbide is very brittle, lacks toughness, and has low tensile properties. It has very high red hardness.

When tantalum carbide is added to tungsten carbide, the coefficient of friction between the tool and the material being cut is greatly reduced. When titanium carbide is added to tungsten carbide, the ability of the tool to cut steel is greatly improved.

Besides their use as cutting tools in tool bits, reamers, milling cutters, and in dies to machine ferrous materials and tool steels, they are used in the machining of such nonferrous metals as aluminium, brass, and copper. Small form tools are shaped in a mold. Larger tools use carbides as inserts, slugs, or disposable tips.

Table 1.13 shows 10 groups of carbide materials. Groups 1 through 3 are straight tungsten carbide. Groups 4 through 8 contain tantalum and titanium carbide, and are used to machine steel. Groups 9 and 10 contain tantalum primarily and are used where high-heat-resistant cutting materials are needed.

One company* reports the hard phase as either titanium carbide, tungsten carbide, or titanium-tungsten double-carbides compacted and vacuum sintered with a suitable matrix to produce the desired results. The control of the matrix is related directly to the physical properties achieved, while the abrasion and wear resistance is related to the hardness of the hardest component. The report therefore stresses the use of the various matrices to achieve different results. Thus a medium-tool-steel matrix

*M. K. Mal and S. E. Tarkan, (Chromalloy American Corp., Sintercast Div.), *Manuf. Eng. Manage.*, Nov. 1972, p. 26.

TABLE 1.13 CARBIDES

Carbide Group	Composition (% Remaining WC)		Rockwell Hardness R_A
	Cobalt	TaC + TiC	
	Straight Tungsten Carbide		
1	2.5–6.5	0–3	93–91
2	6.5–15	0–2	92–88
3	15–30	0–5	88–85
	TiC Predominantly Added		
4	3–7	20–42	93.5–92.0
5	7–10	10–22	92.5–90.0
6	10–12	8–15	92.0–89.0
	TaC Predominantly Added		
7	4.5–8	16–25	93.0–91.0
8	8–10	12–20	92.0–90.0
	TaC Added Exclusively		
9	5.5–16	18–30	91.5–84.0
10	12–16	25–30	84.0–86.0

Source: Metals Handbook, 8th ed., p. 660. Prepared by American Society for Metals, Metal Park, Ohio, 1964.

as a binder produces a general-purpose carbide material. A high-chromium or stainless steel matrix produces a better-than-average heat-resisting steel in the former and a corrosion-resistant steel in the latter. A nickel-based alloy as a matrix produces a carbide that will withstand temperatures of 2000°F. In addition, other carbides are made with nickel-aluminum, copper-nickel, and high-speed matrices.

Cermet materials. These are a class of materials that result when metallic and nonmetallic materials are combined. When Al_2O_3 and nickel are combined, they produce a material that has high heat-resisting properties. If Teflon and iron are combined, the resulting material is self-lubricating and may be used in bearings.

Still another class of cermets is the group of materials that results when an oxide material (aluminum, titanium, or chromium oxide) is combined with a binder material. After compacting and sintering, the material is rolled or shaped as desired. Since the oxides are extremely hard, these cermets have great potential as cutting tools. Table 1.14 shows the relative hardness of various materials.

These tools have characteristics which make their potential as high-production cutting tools very promising. The need is great for rigid setups using machine tools with practically no play in the bearings and moving parts, and greater power and cutting speeds. They exhibit high red hardness, high abrasive resistance, resistance to cratering, galling, or welding, low heat conductivity, and high compressive strength. They have very low resistance to bending, tension, or shock loading.

At present they are used to cut nonmetallic and nonferrous materials such as fiberglass, plastics, rubber, graphite, aluminum, copper, and brass. They are also

TABLE 1.14

Material	Knoop Hardness No.
Tungsten carbide and cobalt	1050–1500
Topaz	1250
Aluminum oxide	1625–1680
Silicon carbide	2130–2140
Boron carbide	2250–2260
Sapphire	1600–2200
Diamond	6000–6500

Source: ASM, *Metals Handbook*, 8th ed., p. 600. (Metals Park, Ohio: American Society for Metals, 1964).

used to machine cast iron (noninterrupted cuts) and low-carbon and low-alloy steel at high speeds. This class of materials is usually referred to as ''white ceramics.''

The ceramic materials (white ceramics) described above are cold pressed. Hot pressed ceramics (black ceramics) are a composite of aluminum oxide and titanium or tungsten carbide. They may be used to machine hard cast iron, 200 BHN and above, or steel as high as $65R_c$. They can be used with spindle speeds of up to five to six times that of carbide. Rupture strength and shock resistant qualities are also higher than other cutting materials.

The latest development of these materials is a cutting tool made with silicon nitride which produces a dramatic decrease in the horsepower utilization during the machining operation. This results from a dramatic decrease in the force-to-speed ratio.

Yet another recently developed ceramic composite is a ''reenforced silicon nitride'' ceramic.* This material promises to have a positive effect on the fracture toughness and the resistance to thermal shock. Figure 1.2 shows the relationship for the various tool materials between wear resistance and toughness.

1.8 TOOL MATERIAL DEVELOPMENT

The history of the development of cutting materials is shown in Fig. 1.3. Materials used to cut, or shape objects have been with us since man had a need to extend his capabilities beyond the use of his hands. The development of tool materials, as we know them today, started at about the middle of the 18th century.

As the need for new and better tool materials became necessary various materials were developed to meet these needs. With the advent of the computer, a radical change took place in the need for better tool materials. Machine tools were built with greater efficiency, and capable of higher cutting speeds which in turn required cutting materials that would not limit these capabilities.

*Greenleaf Corp., Saegertown, Pa., WG-300 material.

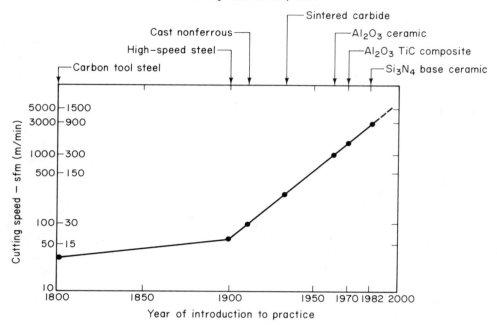

Figure 1.3 Reproduced with permission of Society of Manufacturing Engineers from Volume III of the Tool and Manufacturing Engineers Handbook, 4th, ed.

As a result carbide materials were improved to withstand the greater heat generated, to possess greater toughness requirements, and to better wear resistance. Single, double or triple coatings (See Fig. 1.4) of titanium carbide, aluminum oxide and titanium nitride were added to tool materials to reduce friction between the tool material cutting surface and the chip produced. Cubic boron nitride was developed to coat tungsten carbide tool inserts. This is a very hard material and when used as a coating, permits cutting speed of 4 to 5 times the cutting speeds of that used with an uncoated carbide tool.

Ceramic, cermet and polychrystalline diamonds were developed to provide other options to the machine tool industry.

Micro-structures were examined which resulted in "whisker" structure materials that made possible very tough cutting tools. These whiskers are very hard microcomponents that are free from grain boundaries. Thus, a crack, or a strain, will not propagate past this whisker. It must either displace it, or go around it. The effect is a tougher cutting tool material. In addition, the high heat generated while producing the chip, creates an oxidizing condition that produces a powder which is easily disposed.

High Strain Micrograin carbides have a cobalt base. This material creates a structure that has high transverse resistance to rupture, hardness and high edge strength. The latter permits the use of positive rake angles with carbide tools.

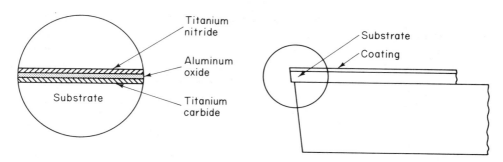

Figure 1.4

1.9 NONFERROUS MATERIALS

There are many nonferrous materials which are used in tool, jig, and fixture design. Such materials include aluminum, magnesium, zinc-base alloys, bismuth alloys, lead-base alloys, and beryllium. These materials may be used where lightness of weight is a factor, for temporary dies, limited production runs, or for some special purposes.

Aluminum-bronze and duraluminum are two of the more widely used aluminum alloys. The former group is lightweight and have strengths which permit their use as dies for forming or drawing stainless steel. The latter may be used for lightweight fixture bodies. In several of the alloys of aluminum the tensile strength may be as high as 100,000 psi, which rates with many of the steels. Its low density (about one-third that of steel) and corrosion resistance and machinability make it a desirable material in tool and fixture design.

Magnesium is another of the lightweight (two-thirds that of aluminum) materials which may be used in large fixtures. Like aluminum, it is machinable and may be welded. It is corrosion resistant in most atmospheres, but needs surface protection when in saltwater atmospheres. It is also used as facing material on forming blocks for short-run production.

Zinc-based materials may be cast into shapes quickly for the purpose of being used as short-run punches and dies for either short-run production or for experimental short runs, when punching thin sheet aluminum parts. When cast into tools they require very little machining. Since the material is comparatively cheap and is easy to cast, it is an ideal material for short, quick production runs.

One of the better known zinc-based die materials is Kirksite. It is cast into complicated shapes, polished, and used in blanking or drawing dies. Since it is cast as a dense smooth surface, it needs only to be polished. It has been used as the sole material in a die, as the hard component in combination with the matting component made from a softer material, or as part of a die in combination with steel components. Another advantage of Kirksite is the fact that the material may be remelted and used over again.

Bismuth alloys are low-melting alloys which expand upon solidification. This is especially true of the high-bismuth alloys. This characteristic of bismuth makes it possible to use this material for duplicating mold configurations which would otherwise require many hours to reproduce. The material is suitable for making tracing or engraving patterns that have intricate shapes, and for making forms used in stretch forming. It has been used effectively to anchor components in punches and dies.

Lead-based alloys, when one of the alloys is antimony, have been used with Kirksite (a zinc alloy) dies.

When alloyed with copper, beryllium produces a material that has characteristics similar to those of aluminum bronze. When only 2% beryllium is alloyed with copper and heat treated, it develops properties that make it suitable for chisels and other cutting tools. It develops tensile strengths of up to 1.7×10^5 psi and a hardness of 40 R_c.

1.10 NONMETALLIC MATERIALS

In addition to the ferrous and nonferrous materials, there are a wide variety of nonmetallic materials which are important to the tool designer. Several of these are discussed below.

Diamonds. Diamonds used in industry are the hardest substance known. They may be natural or synthetic. In either case, the diamond has high strength, high heat conductivity, high melting point, high modulus of elasticity, and high abrasion resistance. It is also chemically inert. As a tooling material it may be powdered and used in polishing or in grinding in either its powdered form or as the hard component in a wheel. In its solid form whole diamonds are used as inserts in turning tools, in drawing dies, and in wheel dressers.

Diamond cutting tools are used to cut nonferrous or nonmetallic materials that require a high finish. They are used to cut babbit or silver bearing material, oilless materials such as Oilite, carbon, plastics, precious metals, or difficult-to-machine materials. The procedure is to take very light cuts at high cutting speeds. Any type of impact or interrupted cutting should be avoided. Inclusions in materials, such as sand holes or hard spots, may destroy these tools.

Plastics. The thermosetting plastics, such as the phenolic, epoxy, urethane, and polyester plastics, are used widely as tooling materials. Where the base plastic is not completely suitable as a tooling material, it may be impregnated with metallic or other abrasive powders. In some instances, steel wear plates are inserted in the plastic. They are currently used in many operations, such as drawing and forming dies, stretching dies, drill jigs, assembly, machining, and inspection fixtures. These materials have the advantage over many other materials in that they are resistant to moisture, chemicals, and temperature. They are generally easy to machine, therefore saving labor costs. Related to this are the advantages of ease of machine repairs, rework, and design changes.

Rubber. A rubber pad confined in a container is used, in several processes, as the female die. The desired form is contained on the solid punch. The formed punch pushes the metal into the rubber, which takes on the shape of the punch and in so doing causes the metal to be formed. It should be noted that if the punch has sharp edges, and the rubber pad the appropriate hardness, the metal workpiece will be cut or blanked. Operations such as forming, blanking, bulging, and drawing may be done with rubber as one of the components of a die set.

Wood. Densified wood is made by impregnating wood with phenolic resin and compressing it to about one-half its original thickness. These woods may then be used in dies for drawing or forming soft materials.

Hardboard. Hardboard, such as Masonite, made from compressed wood material, is used to form or draw thin-gage metals. It may also be used as the solid form when rubber forming materials, or as the solid form when stretching materials. When used for punching or blanking, steel inserts are used as the cutting edges. Special high-tensile-strength masonite may be purchased in the desired thickness.

PROBLEMS

1.1. Sketch a flowchart for the processing of iron ore. Explain each point in the chart as to the product produced at that point.

1.2. How are the ingredient percentages controlled in a particular material?

1.3. (a) What is the end product of the cupola?
(b) What is its composition?

1.4. What are some of the advantages of cast iron that make it suitable as a tool material? How is it used?

1.5. (a) What is the surface structural difference of hot- and cold-rolled steel?
(b) List and explain the advantages and disadvantages of each as a tool material.

1.6. List the range of carbon percentages for
(a) low-carbon steel
(b) medium-carbon steel, and
(c) high-carbon steel.
(d) What are some of the advantages of one over the other?

1.7. List some of the characteristics needed in cutting tools that are lacking in plain carbon steels.

1.8. List the seven characteristics that are affected when alloys are added to steel. Explain each with a short statement.

1.9. List the effects of chromium when it is added to steel as an alloy.

1.10. Given an O-steel, list the
(a) hardening,
(b) tempering,
(c) annealing, and

      (d) normalizing data.

      (e) What type of tool steel is this? Use Table 1.3.

1.11. What effects does the addition of manganese have on steel when used as an alloy?

1.12. Repeat Prob. 1.11 for molybdenum as an alloy.

1.13. Repeat Prob. 1.11 for the alloy vanadium.

1.14. Repeat Prob. 1.11 for the alloy silicon.

1.15. Repeat Prob. 1.11 for the alloy phosphorus.

1.16. Repeat Prob. 1.11 for the alloy nickel.

1.17. Repeat Prob. 1.11 for the alloy cobalt.

1.18. Repeat Prob. 1.11 for the alloy tungsten.

1.19. Repeat Prob. 1.11 for the alloy titanium.

1.20. Repeat Prob. 1.11 for the alloy aluminum.

1.21. State the physical properties for each of the tool steels discussed in Sec. 1.4.

1.22. **(a)** List the symbols used for the various tool steels and their classification in the AISI system.

      (b) What type of steel does each letter represent?

1.23. Given an A_{10} tool steel, list the

      (a) hardening,

      (b) tempering,

      (c) annealing, and

      (d) normalizing data.

      (e) What type of tool steel does A_{10} represent? Use Table 1.3.

1.24. Repeat Prob. 1.23 for an F-tool steel.

1.25. **(a)** Discuss the physical property red hardness as it relates to tool steels.

      (b) Which type of tool steel has the best red-hardness property?

      (c) Which has the worst?

1.26. Repeat Prob. 1.25 for abrasion or wear resistance.

1.27. Repeat Prob. 1.25 for shock resistance.

1.28. Repeat Prob. 1.25 for deformation.

1.29. Repeat Prob. 1.25 for machinability.

1.30. Repeat Prob. 1.25 for depth of hardening.

1.31. Repeat Prob. 1.25 for resistance to decarburization.

1.32. What effect does the amount of carbon have on machinability? Explain.

1.33. **(a)** Discuss the M- and T-type steels as tool materials.

      (b) What is the effect of changing the percentages of the various alloys?

1.34. List the physical properties of the H-chromium and H-chromium-molybdenum tool steels.

1.35. What are the physical properties of

      (a) the H-chromium-tungsten tool steels and

      (b) the H-tungsten tool steels?

      (c) How does the H-molybdenum tool steel differ from the H-tungsten type?

1.36. **(a)** List the cold-working tool steels and their physical properties.

      (b) List some of their uses.

1.37. Repeat Prob. 1.36 for shock-resisting tool steel.

1.38. Discuss the two types of mold tool steel as to their characteristics and uses.

1.39. List the physical properties and uses of L- and F-type tool steels.

1.40. List the three major categories of water-hardening tool steel and state their uses.

1.41. Discuss the cast tool materials and their uses.

1.42. (a) What components could be used in carbide tool materials?
(b) How are these carbides manufactured?
(c) How does the addition of tantalum or titanium carbide affect tungsten carbide?
(d) List other uses for carbide besides as a cutting tool.

1.43. List the 10 groups of carbide available and give the major constituents.

1.44. (a) What is the function of the matrix in the carbide compact?
(b) List the materials that may be used as matrices.

1.45. (a) What is cermet material?
(b) There are two groupings of cermet materials. List them.
(c) Compare the hardness of cermet with other materials.

1.46. (a) What are the characteristics of cermet materials?
(b) How are they used in cutting tools?

1.47. Describe some of the machining advantages of the so called "black ceramics" over the "white ceramic" materials.

1.48. Trace the history of the development of cutting tool materials.

1.49. Discuss the relationship between wear resistance and toughness for cutting tool materials. Refer to Fig. 1.2.

1.50. How does the structure of polychrystalline diamond differ from commercial diamond?

1.51. List two characteristics of high strain micrograin carbide.

1.52. (a) What is "whisker" cutting tool material?
(b) Describe its structure and why this makes it desirable as a cutting tool material.
(c) What kind of a chip does this cutting tool produce?
(d) Why is this chip desirable?

1.53. What are the advantages over steel which may be achieved when using aluminum or magnesium alloys as a tool material?

1.54. (a) Discuss the use of zinc-based alloys as tool materials.
(b) What is Kirksite?
(c) What are its advantages?
(d) How is it used?

1.55. Discuss the characteristics of bismuth as a tool material.

1.56. Explain the use of lead and beryllium as tool materials.

1.57. (a) How are diamonds used in tool design?
(b) What are the physical properties which make it desirable as a tool material?

1.58. (a) Discuss the uses of plastic as a tool material.
(b) What are the characteristics of this material which make it suitable as a tool material?

1.59. How is rubber used in punches and dies for forming or shearing metal parts?

1.60. Discuss densified wood as a tool material.

1.61. Discuss hardboard as a tool material.

1.62. Trace the historical development of the cutting tool.

Chapter 2

Manufacturing Processes

2.1 METHODS OF FABRICATING MATERIALS

This chapter is intended to review the processes used to fabricate metal parts. In Chapter 3 we group them into modules to form systems. The actual design of cutting tools is treated in a subsequent chapter.

There are various methods of manufacturing parts. This results in parts that may be used as they are fabricated by a primary process, or parts that need further processing. Thus pig iron may be processed further by refining and controlling the impurities in the open hearth, Bessemer, or electric furnace, and then processed into usable parts, by one of the fabricating processes. Pig iron may be melted in a cupola, cast into desired shapes, and used with or without machining. It should be noted that other materials, such as aluminum, brass, magnesium, and plastics, should not be excluded in any discussion of the methods of fabrication.

The fabricating processes that a tool designer might be called upon to consider are: sand, die, and mold casting; forging; extruding; shearing or drawing presswork; welding; and machining. Quality control is essential in all of these processes and may require the design of very intricate pieces of equipment.

2.2 CASTING

One of the most commonly used processes for making parts is casting. The general term *casting* covers sand, shell mold, investment, plaster mold, centrifugal, permanent mold, die, continuous, electroform, and powder metallurgy casting. One need only consider this impressive list of processes to realize that we can do no more than acquaint the reader with each of these processes.

Sand casting.   This process relies on the moldmaker creating a cavity in a block of sand. This cavity must have the desired configuration so that when molten metal is poured into it and allowed to solidify, the solid metal will take on the desired shape and have dimensions that match, or can be machined to match, a finished parts print.

The process of making a sand mold requires that a pattern be made. This pattern may be made from solid wood, laminated wood, metal, or a series of shapes (sweep pattern). It must take into account dimensional requirements, which may differ from the finish part print because of shrinkage as the material solidifies, extra material needed when surfaces must be machined, allowances needed for draft (taper which permits pulling the pattern out of the sand mold), or warpage. This pattern is therefore used to make the cavity in the sand block. The mold must provide some means of removing the pattern so that the cavity is free to accept the molten metal. The process also requires that there be an opening in the mold into which the molten metal can be poured. The mold must also allow for escaping gases.

In addition, should a casting require that it have a cast hole, sand cores are used inside the cavity. They are located inside the mold cavity, and in many instances their construction is as complicated as the castings themselves.

Shell molding.   This casting process uses a solid metal pattern in the shape of the workpiece to be cast. A very fine grade of washed sand and a thermosetting plastic powder is throughly mixed. The pattern is heated to about 400°F, after which the mixture of sand and plastic powder is heaped on the pattern. The heated pattern causes the plastic to cure partially, forming a thin shell that clings to the outline of the pattern. This shell is removed and placed in a furnace, where the curing is completed. If needed, the shell may be completed by gluing other sections together, or by attaching pouring basins, risers, and gates to it. The mold is placed in a container and surrounded with some type of supporting material. It is now ready to receive the molten metal.

Investment (lost wax process) casting.   In this process a master pattern is used to make a die. This die cavity is then filled with wax and permitted to solidify. A "tree" is formed by fastening many wax patterns to wax runners and a sprue by using a heated spatula. Complicated shapes may be built up at this time. The tree is then dipped repeatedly into a fine refractory material to build up a thin-layered shell. Following this, the tree is placed into a refractory material such as plaster. This coarser refractory (investment) is vibrated to remove all air bubbles. The entire container is placed in a furnace. In the furnace the mold hardens and the wax melts and runs off, leaving the desired cavity in the plaster. This mold is heated and the molten material is poured into the cavity and allowed to solidify. Accurate, smooth castings may be made using this process.

Variations of this process use materials such as plastics, frozen mercury, or other low-melting-point materials which either melt and run off or vaporize. Recently, shells have been made by dipping the wax trees into a ceramic slurry. This slurry forms the mold cavity.

Plaster molds.   Molds made from plaster of paris and gypsum are poured over a pattern and permitted to set. The molds are made in two halves. The surface finish and dimensional accuracy of castings made using this process are very good. Since thermal conductivity of the mold is low, the liquid metal remains fluid for a longer period of time than for other methods. However, plaster molds are fragile and should be used for nonferrous materials. The *Antioch process* is a plaster process that develops porosity in the plaster, which in turn allows trapped gases to escape when the molten metal is poured into the mold.

Centrifugal castings.   In this process a mold, or several molds, are fitted to a rotating disk. If the center of an array of molds is the sprue, the molten metal will flow down this sprue and distribute itself to the farthest points in each of the molds. The density of castings made by this method is high. They also have a high degree of detail. Since the mass of the impurities is less than that of the molten metal, it will collect at the inner walls of the casting. There are several centrifugal casting methods, such as true centrifugal and semicentrifugal castings, the Watertown and DeLavand methods, and the sand-spun method.

Permanent-mold casting.   These methods use a mold cavity to form the workpiece. Molten metal is poured into the mold or allowed to flow in as a result of gravity. The finished part is removed from the die after it has solidified. This process is used to mold steel or cast iron castings. It is also used to make castings from materials that have a copper, aluminum, or magnesium base.

The *slush permanent weld* process is a variation of the process just described. In this process molten metal is poured into a cavity. At a given time the mold is inverted and the molten metal remaining is permitted to run off. The resulting solidified shell is the workpiece.

The *Corthias process* is another variation of the permanent mold process. In this process a precalculated amount of molten metal is poured into the mold. A plunger is inserted to force the metal into all corners of the mold cavity. As the metal sets, the plunger is withdrawn. A hollow casting results.

Die casting.   This process is another variation of permanent mold casting where the metal is forced into the mold under high pressure. There are essentially two basic die cast procedures.

The *hot-chamber die casting process,* Fig. 2.1a, uses a gooseneck container which is submerged in the melt. The small end of the gooseneck is attached to the mold. A plunger forces and holds the metal until it solidifies. Once it has solidified, the die is opened and the casting removed. Lead, tin, and zinc may be molded this way.

The *cold-chamber die casting process,* Fig. 2.1b, uses a ladle to pour the molten metal into the pouring basin. A plunger forces the melt into the die chamber. Once the metal solidifies, the core retracts, the mold opens, and the ejector pins free the casting.

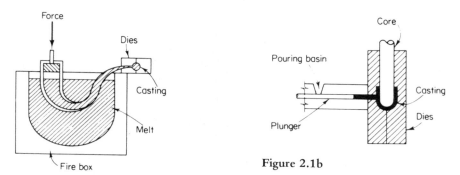

Figure 2.1a

Figure 2.1b

Continuous casting.   In this process the mold is kept full of molten metal at all times. The metal at the lower end of the mold is cooled, pulled out, and cut to desired lengths. Brass, zinc, copper and its alloys, aluminum and its alloys, and carbon and alloy steels may be cast using this process.

Electroforming.   A negative mold of the part is produced from plaster or machined from solid stock. A fusible low-melting alloy, wax, rubber, or other chemically soluble substance is poured into this negative mold. This results in a pattern which is then coated with an electrical-conducting material such as graphite, or silver. Once coated, the pattern is placed into an electrolytic solution and a $\frac{1}{2}$-in. shell is plated on its surface. When the pattern is removed (by melting or chemically dissolving) the shell resulting is the finished part. Laminating of different types of metal may be accomplished using this process.

Powder metallurgy.   Powder metallurgy was discussed to some degree in Chap. 1. To repeat, powders are compacted under high pressure into the desired shape. The compact is then heated at high temperatures in a controlled atmosphere to form the part. The temperature used should be high enough to melt the binder but should not affect the principal powder in the compact. In a true sense, this process is not a casting process.

2.3 HOT AND COLD FORMING OF METALS

Elastic deformation takes place when atomic forces cause atoms to return to their original positions in a structure once the deforming forces have been removed. *Plastic deformation* is said to take place if the atoms fail to return to their original positions once the deforming forces have been removed. Of course, when molten metal is used, the atoms are in total disorder. As the melt solidifies, the atoms take on an array characteristic of the metal being used.

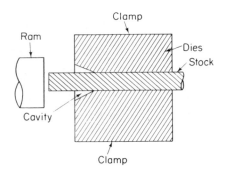

Figure 2.2

Since so many different types of operations are involved, in this book we merely make a statement about each so that readers will have enough knowledge of a particular operation to be able to refer to a book on those subjects that prove to be of interest.

Hot forging.   *Smith forging* is the process by which metal is hammered into a desired shape. The basic process uses the anvil and hammer. The more modern process is to raise a large hammer by steam, electric, or roller power, and drop it on the work. Gravity produces the needed kinetic energy. The operator rotates the work as it is being shaped.

The *drop forging* process employs two halves of a die, one carried by the ram and the other fastened to the table. Precalculated volumes of metal are heated, placed between the dies, and hammered into the shape of the die cavity when the die is closed. A variation of this process may be done using a horizontal machine. This machine has two horizontal rams which retract and strike the work.

Hot press forging causes a steady pressure to be applied to the workpiece. The hot metal is forced into the die cavity by steady pressure instead of an impact pressure.

Upset forging forces the end of a heated bar into a desired shape. The bar is heated, clamped, and upset into the die opening, Fig. 2.2.

Swaging uses two halves of a rotating die. The die opens and closes rapidly while striking the end of a heated tube or shell. This operation may reduce the diameter of the tube or taper its end.

Roll forging is a method that reduces the diameter of a bar and in the process makes the bar longer. Semicircular rollers receive the heated bar and reduce its diameter. Many passes of the bar through the rollers may be required before the finished diameter is achieved.

Extrusion.   *Direct pressure extrusion* is the process in which the ram forces the metal through a die. Figure 2.3 shows three different types of structural shapes. In all cases the ram moves forward and forces the metal through a die opening. Mandrels may be used to create a hole in the center of the extruded metal.

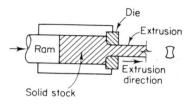

Figure 2.3

Figure 2.4 shows the *indirect pressure extrusion* process. In this process the ram moves forward and forces the metal back through itself.

Impact extrusion is another hot extrusion process. In this process the metal is forced to "splash" up the sides of the punch as it strikes a precalculated volume of material. Figure 2.5 shows a variation of impact extrusion called the *Hooker process*. In this process some of the metal is caused to flow into the small diameter at the bottom of the die opening.

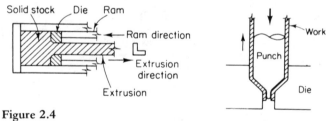

Figure 2.4

Figure 2.5

Hot rolling.   *Hot drawing,* Fig. 2.6, is a process of hot working whereby a heated blank is forced through a die opening by a punch under constant pressure. Cups or shells may be drawn using this process.

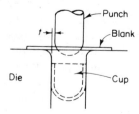

Figure 2.6

Hot spinning, Fig. 2.7, is a process that uses pressure against a metal disk while the disk is revolving. The pressure is applied by a blunt tool or roller which forces the metal to take the shape of a rotating form.

In *hot rolling,* billets, blooms, or slabs may be rolled into desired shapes or sizes by passing them through a series of rollers. Four such processes are shown in Fig. 2.8.

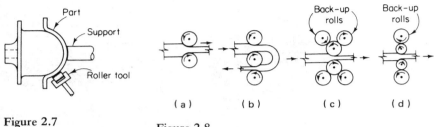

Figure 2.7

Figure 2.8

Piercing operations, Fig. 2.9a, shows a process for making seamless tubing; Fig. 2.9b and c show the material being passed through a bell die. This leaves a seam which is resistance welded.

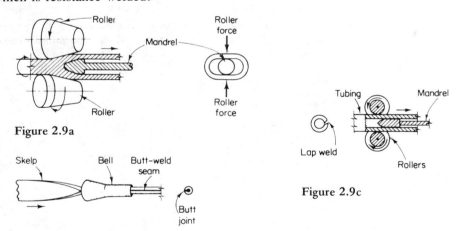

Figure 2.9a

Figure 2.9c

Figure 2.9b

Cold forging.   *Swaging,* Fig. 2.10, is a cold impact process that uses rollers to cause dies to shape the surfaces of work into points, tapers, or to reduce their diameters. The dies are rotated so that the work is struck many blows in a short period. In the *interform system* pressure is exerted on the material over a formed mandrel so that the outside of the work may be round and the inside may be formed.

Cold heading, Fig. 2.11, requires that the metal be confined in all directions. The metal under impact flows in all directions. Bolts, screws, rivets, and nails may be formed using this process.

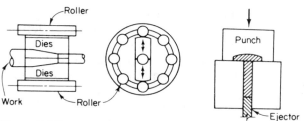

Figure 2.10

Figure 2.11

Riveting, Fig. 2.12, is a process of cold heading both ends of a slug and is used to fasten two pieces together.

Staking, Fig. 2.13, is another process that is used to fasten pieces together. The metal is caused to flow into grooves or against the face of one of the pieces.

Hobbing (hubbing), Fig. 2.14, is the process of forcing a pattern in the face of the punch into the surface of a workpiece. Plastic mold cavities are made this way.

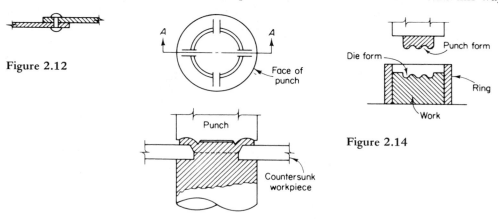

Figure 2.12

Figure 2.14

Figure 2.13

Coining, Fig. 2.15, requires that the work be completely confined. Both surfaces of the work receive impressions. The process is one of lateral material flow. Coins, medals, and jewelry are made this way.

Embossing, Fig. 2.16, uses a matched punch and die. The operation is a

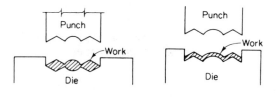

Figure 2.15   Figure 2.16

shallow drawing operation. Whereas the thickness of the material in coining is un-even, the thickness of the workpiece in embossing is uniform. The material flows in the direction of the applied forces.

Shot peening is a process of cold working that uses a large number of steel shots. These shots are fired at the surface of the work, giving a better finish and work hardening the surface.

Cold extrusion uses the same processes and procedures as those used for hot extrusion. The difference is that the material extruded must not work harden and must remain ductile without the application of heat.

Cold drawing.   *Cold drawn bars* are made from hot-rolled bars that have been pickled to remove the scale and drawn through dies. The process sizes the diameter. It produces a bar that has a smooth, but work-hardened surface.

Wire drawing, Fig. 2.17a, is accomplished by pulling a wire through a hard-ened die, usually carbide. Small-diameter wires are drawn through diamond dies. Figure 2.17b shows the reduction of the diameter when several passes are made through successive dies. Cold-drawn tubing may be reduced by pulling the tubing over a mandrel and through a die, Fig. 2.17c.

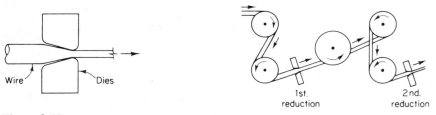

Figure 2.17a

Figure 2.17b

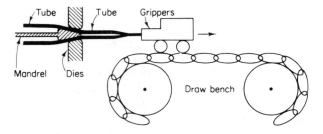

Figure 2.17c

Stretch forming, Fig. 2.18, is accomplished by stretching and wrapping a metal sheet over and around a form. The metal is stretched above its elastic limit as it is being wrapped.

High-energy forming, Fig. 2.19, uses a high explosive charge, or capacitor discharge across a gap, to generate shock waves which fan out in all directions and force the metal into a preformed die cavity.

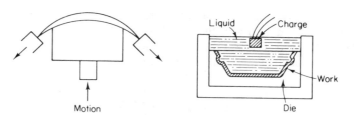

Figure 2.18   Figure 2.19

Cold rolling. *Form rolling* is a process whereby a number of strategically placed rollers form sheet metal into a desired shape. The sheet metal is passed through the rollers, which subject the metal to a series of bends. Figure 2.20a shows a drain gutter which may be formed by passing stock through three rollers, Fig. 2.20b.

   Seaming, Fig. 2.21a–b, may be done to produce single or multiple laps. This is accomplished with rollers so that the bends lap the material back on itself to close the form.

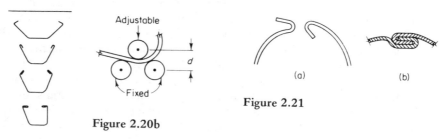

Figure 2.20a

Figure 2.20b

Figure 2.21

   Thread rolling, Fig. 2.22, uses two- or three-thread die rollers. The cylindrical work stock should have an outside diameter equal to the pitch diameter of the original thread. The metal between the outside diameter and the root diameter is forced up to form the addendum.

   Spinning metal, Fig. 2.23a, cold is the same process as spinning metal hot,

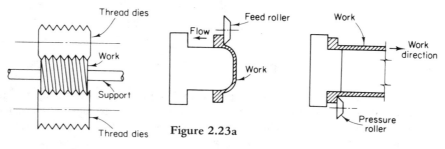

Figure 2.22

Figure 2.23a

Figure 2.23b

which was discussed earlier. The difference between the two processes arises out of the differences in ductility of the metal.

Shear forming, Fig. 2.23b, is the same process as direct spinning except that the metal is forced back in the opposite direction from the pressure exerted by the rollers.

2.4 SHEARING, BENDING, OR DRAWING PRESSWORK

Punch and die processes are probably the most widely used cold- or hot-working processes. Almost any quantity or quality of material may be ordered from the steel mill. Materials may be cut to special sizes, have special percentages of carbon, have special alloys, be specially heat treated, or may be straightened to other than standard straightness variations at additional cost. The standard straightness is $\frac{1}{4}$-in. variation from true flatness over 8 ft of length. Material thicknesses are listed in Table 2.1.

The manufacturers' standard gage sizes are used for carbon and alloy sheet steel. The table has been standardized at 41.82 lb/ft^2 for 1 in. of thickness. The U.S. standard gage, although still used for cold-rolled strip, nickel alloy steel, and stainless steel tubing, has been superseded by the manufacturers' standard gage size. Birmingham wire gage sizes are used for steel tubing and hot-rolled carbon alloy strips. The American wire gage or Brown and Sharpe wire gage sizes are used for copper, brass and bronze strip, and brass, bronze, aluminum, and magnesium sheet. The American National Standards Institute (ANSI) standards are used for stainless steel.

Shearing is the process that uses a punch and a die to separate metal. The process of shearing starts with a punch sinking into the metal and stressing that metal to its elastic limit. As the punch pushes farther into the metal, the elastic limit of the material is exceeded and rupture occurs, Fig. 2.24, at the cutting edge of the punch. This will occur if the space between the punch and the die is equal to a small amount of clearance. The theory of metal shear is discussed at some length in later chapters.

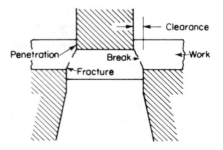

Figure 2.24

If the space between the punch and the die is equal to the thickness of the material, and the corners of the punch and the die are rounded, this material is *bent*, as shown in Fig. 2.25. The material at the outer radius is placed in tension, and the material at the inner radius is compressed. No cutting takes place. Since the elastic limit is exceeded, the material takes on a permanent set.

TABLE 2.1 WIRE AND SHEET METAL GAGES

Gage No.	U.S. Std.	Steel Wire Gage	Mfg. Std. Gage for Steel	Amer. Wire or B & S	Piano Wire Gage	Stub's Iron Wire	Stub's Steel Wire	Gage No.	Stub's Steel Wire
7/0	0.5000	0.4900						51	0.066
6/0	0.4687	0.4615		0.5800	0.004			52	0.063
5/0	0.4375	0.4305		0.5165	0.005	0.500		53	0.058
4/0	0.4062	0.3938		0.4600	0.006	0.454		54	0.055
3/0	0.3750	0.3625		0.4096	0.007	0.425		55	0.050
2/0	0.3437	0.3310		0.3648	0.008	0.380		56	0.045
1/0	0.3125	0.3065		0.3249	0.009	0.340		57	0.042
1	0.2812	0.2830		0.2891	0.010	0.300	0.227	58	0.041
2	0.2656	0.2625		0.2576	0.011	0.284	0.219	59	0.040
3	0.2500	0.2437	0.2391	0.2294	0.012	0.259	0.212	60	0.039
4	0.2344	0.2253	0.2242	0.2043	0.013	0.238	0.207	61	0.038
5	0.2187	0.2070	0.2092	0.1819	0.014	0.220	0.204	62	0.037
6	0.2031	0.1920	0.1943	0.1620	0.016	0.203	0.201	63	0.036
7	0.1875	0.1770	0.1793	0.1443	0.018	0.180	0.199	64	0.035
8	0.1719	0.1620	0.1644	0.1285	0.020	0.165	0.197	65	0.033
9	0.1562	0.1483	0.1495	0.1144	0.022	0.148	0.194	66	0.032
10	0.1406	0.1350	0.1345	0.1019	0.024	0.134	0.191	67	0.031
11	0.1250	0.1205	0.1196	0.0907	0.026	0.120	0.188	68	0.030
12	0.1094	0.1055	0.1046	0.0808	0.029	0.109	0.185	69	0.029
13	0.0937	0.0915	0.0897	0.0720	0.031	0.095	0.182	70	0.027
14	0.0781	0.0800	0.0747	0.0641	0.033	0.083	0.180	71	0.026
15	0.0703	0.0720	0.0673	0.0571	0.035	0.072	0.178	72	0.024
16	0.0625	0.0625	0.0598	0.0508	0.037	0.065	0.175	73	0.023
17	0.0562	0.0540	0.0538	0.0453	0.039	0.058	0.172	74	0.022
18	0.0500	0.0475	0.0478	0.0403	0.041	0.049	0.168	75	0.020
19	0.0438	0.0410	0.0418	0.0359	0.043	0.042	0.164	76	0.018
20	0.0375	0.0348	0.0359	0.0320	0.045	0.035	0.161	77	0.016
21	0.0344	0.0317	0.0329	0.0285	0.047	0.032	0.157	78	0.015
22	0.0312	0.0286	0.0299	0.0253	0.049	0.028	0.155	79	0.014
23	0.0281	0.0258	0.0269	0.0226	0.051	0.025	0.153	80	0.013
24	0.0250	0.0230	0.0239	0.0201	0.055	0.022	0.151		
25	0.0219	0.0204	0.0209	0.0179	0.059	0.020	0.148		
26	0.0188	0.0181	0.0179	0.0159	0.063	0.018	0.146		
27	0.0172	0.0173	0.0164	0.0142	0.067	0.016	0.143		
28	0.0156	0.0162	0.0149	0.0126	0.071	0.014	0.139		
29	0.0141	0.0150	0.0135	0.0113	0.075	0.013	0.134		
30	0.0125	0.0140	0.0120	0.0100	0.080	0.012	0.127		
31	0.0109	0.0132	0.0105	0.0089	0.085	0.010	0.120		
32	0.0102	0.0128	0.0097	0.00795	0.090	0.009	0.115		
33	0.00937	0.0118	0.0090	0.00708	0.095	0.008	0.112		
34	0.00859	0.0104	0.0082	0.00630	0.100	0.007	0.110		
35	0.00781	0.0095	0.0075	0.00561	0.106	0.005	0.108		
36	0.00703	0.0090	0.0067	0.00500	0.112	0.004	0.106		
37	0.00664	0.0085	0.0064	0.00445	0.118		0.103		
38	0.00625	0.0080	0.0060	0.00396	0.124		0.101		
39		0.0075		0.00353	0.130		0.099		
40		0.0070		0.00314	0.138		0.097		
41		0.0066		0.00280	0.146		0.095		
42		0.0062		0.00249	0.154		0.092		
43		0.0060		0.00222	0.162		0.088		
44		0.0058		0.00198	0.170		0.085		
45		0.0055		0.00176	0.180		0.081		
46		0.0052		0.00157			0.079		
47		0.0050		0.00140			0.077		
48		0.0048		0.00124			0.075		
49		0.0046		0.00111			0.072		
50		0.0044		0.00099			0.069		

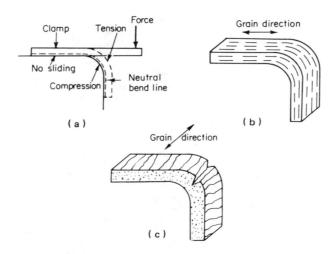

Figure 2.25

During the *bending operation,* material displacement takes place only at the bend. However, during the *drawing operation* a radical displacement of material occurs throughout its entire structure. The clearance between the punch and the die is equal to the thickness of the material, and the punch and die have radius corners to prevent cutting. The material is pulled over these rounded corners so that the material slides over the die radius. Every molecule is displaced during the operation. The displacement is analogous to water flowing over a dam. The operation is shown in Fig. 2.26. In subsequent chapters we deal more fully with the design of drawing dies.

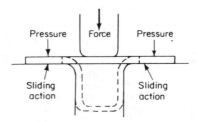

Figure 2.26

2.5 WELDING

Welding may be classified into four categories: (1) processes requiring heat and pressure, (2) processes requiring heat and no pressure, (3) low-temperature alloying processes, and (4) processes requiring chiefly adhesion. In the first two categories, and to some extent in the third category, alloying takes place between the filler rod and the base metal. The first two processes require that both the base metal and the filler material become liquids.

Pressure welding.   Processes that require heat and pressure are hammer, spot, seam, projection, cross-wire, flash and upset butt, and percussion welding.

Hammer welding is the most primitive of the welding processes. Two pieces to be joined may be heated and hand hammered together. When a large machine is used, a gravity drop hammer furnishes the energy.

Spot welding, Fig. 2.27a, consists of two electrodes, one of which is spring loaded, applying a force against the two overlapping workpieces to be joined. After the force has been applied, a high current is sent through the electrodes. The resistance between the workpieces is high, which causes high localized heat to be generated. The material becomes plastic at the point of contact. The force applied to the electrodes causes fusion of the material.

Seam welding, Fig. 2.27b, is a continuous spot welding process. The pressure is applied by two electrode rollers. The current is intermittent and regulated by a timer. The timer may be set to produce watertight overlapping welds, or equally spaced spot welds.

Projection welding, Fig. 2.27c, uses multiple projections to concentrate the current. The localized current and the multiple projections create a series of spot welds.

Cross-wire welding is a variation of projection welding. Touching wires at their point of crossing act like projections. Large cross-sectional flat-faced electrodes apply the pressure.

In *flash and upset butt welding,* Fig. 2.27d, the ends of two pieces are brought together and placed under pressure. The air gap between the ends of the two pieces provides the resistance, which, together with the current, provides the heat that makes

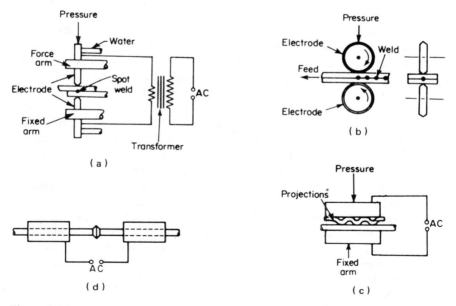

Figure 2.27

the materials plastic. Under pressure the weld is completed. *Flash butt welding* requires that the current be turned on first, then as arcing takes place the pressure is applied. *Upset welding* requires first that pressure is applied, then the current is turned on.

In the *percussion welding* process two pieces are clamped a short distance apart. In one variation, electrical energy is stored in a capacitor. In another, the energy results from the rapid breakdown of a magnetic field. The pieces are caused to move toward each other. At an exact instant the charge is released, which causes arcing. The impact completes the weld.

Fusion welding.   These are the processes that require no pressure as part of the operation of welding: carbon arc, electric, acetylene, inert gas, atomic hydrogen, submerged arc, thermite, electroslag, electron beam, friction, ultrasonic welding, and gas metal-arc welding.

Carbon arc welding, Fig. 2.28a, relies on a carbon arc electrode to provide the path for the dc electric current. The air in the gap between the end of the rod and the work creates the resistance which generates the necessary heat for melting the metal and the filler rod.

Arc welding, Fig. 2.28b, relies on an electric arc that is generated between the filler rod and the workpiece. The filler rod acts as a source of filler material and as an electrode. It may or may not be coated with a flux. The current may be ac or dc. When dc welding, it should be noted that if the electrode holder is connected to the negative terminal of the welding machine, and the work to the positive terminal, the operation is said to be *straight polarity welding.* If the terminals are reversed, the operation is said to be *reversed polarity welding.* The polarity is not important in ac welding.

Gas welding processes use acetylene, hydrogen, and natural or producer gas in combination with oxygen. The oxyacetylene process, Fig. 2.28c, which uses oxygen and acetylene gases, is by far the most widely used combination. The gases are mixed in the desired proportion and when ignited produce temperatures well above that required to melt both the base metal and the filler rod. It should be noted that when using a special torch, oxyacetylene cutting may also be done.

In *inert-gas arc welding,* Fig. 2.28d, the gas used—helium (heliarc), argon, or carbon dioxide—is introduced around a tungsten or carbon electrode and forms a protective cloud about the arc when aluminum or magnesium is being welded. The arc is generated between the electrode and the work.

In *atomic hydrogen welding,* two tungsten electrodes are housed in a multiple chamber which transports molecular hydrogen past them. An ac current between the two electrodes breaks down the molecules of hydrogen, absorbing large amounts of heat in the process. These atoms release this heat energy on contact with metal. A filler rod may be used if desired. The gas acts as a shield during the operation.

Submerged arc welding, Fig. 2.28e, uses a consumable bare rod in combination with a flux feeder tube. The arc is generated under a blanket of flux. Both tubes are fed automatically. The flux is melted as the base metal is melted. The weld is therefore completed under a blanket of flux. This protects the weld.

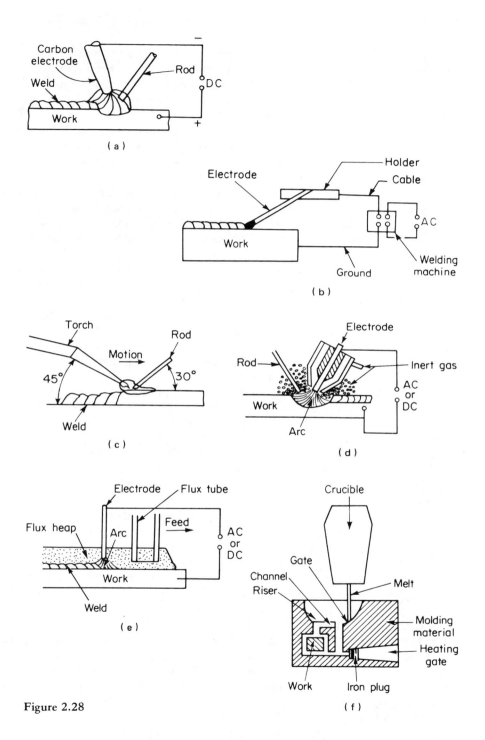

Figure 2.28

Thermite welding, Fig. 2.28f, is a process that uses a mixture of iron oxide and granular aluminum. A mold is prepared around the section of the pieces to be welded. The powdered mixture is poured into the mold and ignited with a magnesium strip or some special ignition power. The aluminum transforms into aluminum oxide and the iron oxide into iron which consummates the weld.

The *electroslag process* is used to join large plates. A water jacket moves with a molten pool of metal along the joint to be welded. The molten metal is caused to solidify as it moves. The pod-metal is replenished by feeding a welding wire into the pool of molten metal.

The *electron-beam process* makes use of a highly evacuated chamber in which an electron beam is started. This beam is directed at the joint to be welded.

Friction welding is accomplished by rapidly rotating one of the two members to be welded in contact with the other. The friction generates the needed heat to achieve the weld.

Ultrasonic welding employs high-frequency vibrations parallel to the surface to be welded. Two pieces that are clamped together will join when brought near the vibrating machine.

The *micro-wire gas metal-arc process* uses a small-diameter filler rod and carbon dioxide to weld very thin materials.

The *self-shielding electrode wire process* uses a tubular electrode wire which produces a core of deoxidized flux and a gas to weld thin materials.

Brazing. The brazing of two materials requires that the temperature of the pieces to be joined remain below their melting point but above the melting point of the filler material. The brazing material must wet the surface to be joined and by capillary action draw the melt into the space between the two parts. Capillary action takes place when the attracting forces between the braze material (or solder material) and the base material is greater than the attraction between the molecules of the braze material itself. This causes the braze material to spread over the work pieces. It is called *wetting.*

Brazing may be accomplished using the following materials: silver (silver soldering), copper, Tobin-bronze, phos-copper, aluminum, nickel-chrome (heat resisting), and silver-manganese (heat resisting).

The fluxes that may be used are boric acid for joining brass, copper, bronze and low-alloy steels; boric acid or a fluoride plus a wetting agent for joining cast iron; a water mixture of fluoride and chloride to form a paste for joining aluminum and magnesium. Fluxes may be in a solid, powder, paste, or liquid form.

Torch brazing uses a neutral flame of oxyacetylene, oxyhydrogen, or natural gas to heat the parts. These are the most commonly used types of brazing operations.

Resistance brazing requires that an electric current be passed through the parts being brazed. This resistance to the passage of current causes the braze material and the flux to melt while consummating the weld. The weld time is controlled accurately by a timing device.

Induction furnaces, Fig. 2.29, makes use of high-frequency alternating current which generates a rapidly oscillating magnetic field. The parts to be brazed are

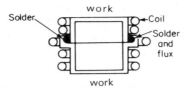

Figure 2.29

surrounded by specially designed water-cooled coils. The current is turned on for a preestablished length of time, and the braze is consummated. Heating is very localized when this method is used.

Furnace brazing uses a controlled-atmosphere furnace. Where oxide formation is needed, a controlled atmosphere is used. The entire assembly fitted with the brazing material and the flux is heated in a furnace.

Dip brazing is accomplished by dipping the assembled parts into a molten bath of the filler material. This is called *metal bath brazing*.

In chemical bath brazing the assembly is fitted with the filler material and the entire assembly is dipped into a neutral salt solution.

2.6. MACHINING

The basic machine tools used for the machining of materials are shown in the block diagram in Fig. 2.30. A thorough discussion of machine tools can be found in any good manufacturing textbook. It is the intention of this discussion to provide the tool designer with an overview of the various methods used to fabricate materials.

Although Fig. 2.30 does not arrange the basic machine tools according to any plan, a good way to classify machine tools is as rotary or reciprocating. Thus a lathe, a milling machine, and a drill press are all classified as rotary machines, while shapers and planers are classified as reciprocating. Another way of saying the same

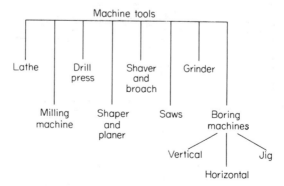

Figure 2.30

thing is to distinguish between continuous motion and interrupted motion cutting. Thus a lathe is continuous motion, whereas a shaper cuts on only one part of the stroke. A hacksaw would have interrupted motion, whereas a band saw would have continuous motion cutting.

Lathe.   There are many types of lathes available: toolroom, engine, bench, jewelers, production, turret, automatic screw, and special lathes. The distinction in these machine tools is one of precision, degree of automation, and adaptability to production. In all cases the cutting is single-point cutting. It is true that several tools may be cutting at the same time, but not in the same manner as a milling cutter cuts.

The basic lathe is shown in Fig. 2.31a and the corresponding powered components and manual components are shown in Fig. 2.31b. The turret lathe substitutes a hexagon turret for the tailstock and a square turret for the compound rest. A tracer lathe substitutes a power supply operated by a tracer mechanism and template which causes the tool to follow a predetermined path that has had the form machined into a template. In all cases the work revolves and the tools feed while in contact with the work.

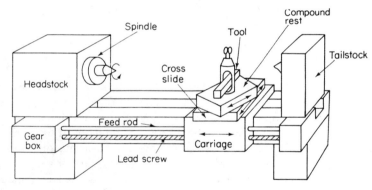

Figure 2.31(a)

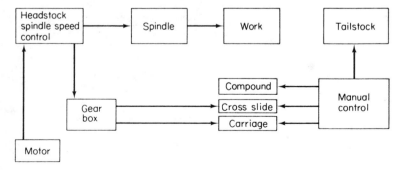

Figure 2.31(b)

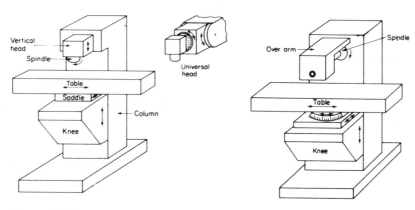

Figure 2.32(a)   Figure 2.32(b)

Milling machine.   Figure 2.32a shows a block diagram of a plain vertical milling machine. Figure 2.32b shows a universal horizontal milling machine. There are many types of milling machines available: plain and universal vertical, plain and universal horizontal, bed type, planer type, thread, and duplicating milling machines.

In all cases the basic cutting tool is the milling cutter, which in most cases operates perpendicular to the line of action of the work. These machines may cause the work, when fastened to the table, to move to the left or right, in or out, up or down. Universal machines are equipped with a swivel mechanism of the table, which causes the work to move past the cutter at an angle. The universal vertical miller may permit swiveling of the vertical head in two planes, Fig. 2.32a, as well as the table. Bed (planer) type millers are production-type machines and provide for table movement in front of one column or between two columns. Thread and duplicating-type machines are special purpose, as their names indicate.

Drill press.   These machines provide a method for holding the cutting tool, which in this case is the drill bit. This type of machine tool allows either the work or the drill to feed while the drill bit is revolving. The types of drill presses are: sensitive, power-fed standard, radial, production, gang, multiple spindle, or horizontal drill presses.

A sensitive drill press is responsive to feel through the feed wheel. A power-fed drill press may be fed by hand or automatically through the power feed mechanism. Such a drill press is shown in Fig. 2.33a. When the power feed is used, the operator is not able to "sense" the activity at the drill point.

When heavy workpieces need many holes drilled into them, moving and positioning the workpiece becomes very difficult. Radial drill presses, Fig. 2.33b, provide for moving the head into the desired position. The over-arm may be moved up or down, or swiveled. The head moves in or out from the column. The spindle, which mounts the drill bit and rotates, can be manually or power fed into the work.

Gang and multiple-spindle drill presses are used in production. Gang drill

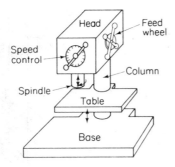

Figure 2.33(a)

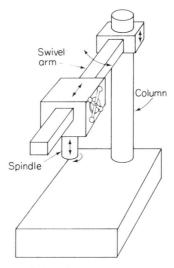

Figure 2.33(b)

presses have several drill spindles, each powered separately. Multiple-spindle drill presses have several spindles driven from a central spindle. Horizontal drill presses are used to drill long holes. This has the advantage over vertical drill presses of permitting the clearance of chips more easily.

Shaper and planer.   These machines are reciprocating-type machines. Either the work oscillates and the tool feeds; or the tool oscillates and the work feeds. These machines are the horizontal, vertical, or special shapers; and the open side or double-housing planers. Gear shapers and generators are also available.

The standard shaper is shown in Fig. 2.34a. In this machine the tool is caused to oscillate. With each stroke of the ram the table feeds a fixed amount, positioning the work for the next cut. Surfaces machined are essentially flat, although contouring is possible. It should be noted that draw shapers may also be purchased. In this type of machine the cutting takes place on the "pull" part of the stroke.

The planer is constructed so that after each cutting stroke, the work oscillates under the tool bit as it feeds. Figure 2.34b shows a double-housing planer. The toolheads may be set to a particular depth. They will then hitch feed after each cutting stroke is completed. Open side–single housing plans are also common.

Shaver and broach.   These tools also operate by oscillating the cutting tool. The shaver takes very light cuts and is used to achieve high finish and dimensional accuracy. The broaching machines generally operate on the one-stroke principle. That is, a long bar has many teeth cut into its surface, each designed to take a small cut as the broaching tool is either pushed or pulled through the work. Figure 2.35 shows a push-type broaching machine.

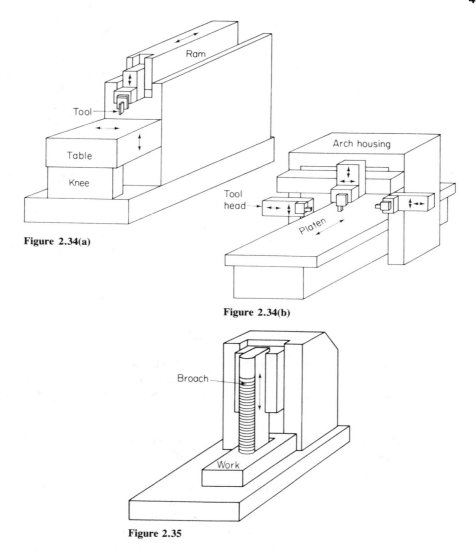

Figure 2.34(a)

Figure 2.34(b)

Figure 2.35

Saws. These machines are used extensively for cutting materials to a desired length. Sometimes saws are used to cut contour shapes. The types of power saws are cutoff hack and continuous band, and contour continuous blade band saws. Other types of cutoff machines are those which use abrasive banded disks and those which use large multiple-tooth circular saws which have milling-type cutter teeth.

Hacksaws have a flat blade fastened to a U-frame as shown in Fig. 2.36a. The U-frame is pulled back and forth by a crank mechanism. These machines generally cut on the pull stroke. A cam mechanism, or some other type of mechanism, is provided to lift the frame on the noncutting part of the stroke to preserve the teeth on the blade.

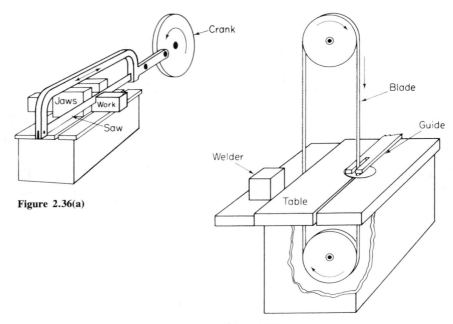

Figure 2.36(a)

Figure 2.36(b)

Band saws may be horizontal or vertical. The horizontal continuous band saw is used as a cutoff machine. The vertical continuous band saw, Fig. 2.36b, is generally used for contour cutting. For this reason it is generally equipped with a combination butt-welding and grinding unit.

Grinding machines.   Grinding machines are those which use an abrasive grit as the cutting tool. They may be classified as cylindrical, surface, and special grinding machines.

Cylindrical grinders may be of the external, internal, or centerless type. Figure 2.37a shows an external cylindrical grinder. With the work rotating between centers, the wheel is brought into contact with the work, thus removing material. The internal cylindrical grinder, Fig. 2.37b, uses a chuck to hold the work. With the grinding wheel revolving the table moves from right to left, causing the wheel to enter the work and grind its internal surface.

A centerless grinder uses a drive wheel, a grinding wheel, and a guide support blade, as shown in Fig. 2.37c. With this method workpieces need not be held between centers. Through-grinding will produce ground cylinders. Tapers, shoulders, threads, and contours may be ground using this method.

Figure 2.37d shows a surface grinder which uses a magnetic chuck to hold the workpiece while it is being ground. The machine shown has a horizontal spindle in combination with a flat magnetic chuck. Others have horizontal spindles in combinations with round chucks, and vertical spindles in combination with flat and round magnetic chucks.

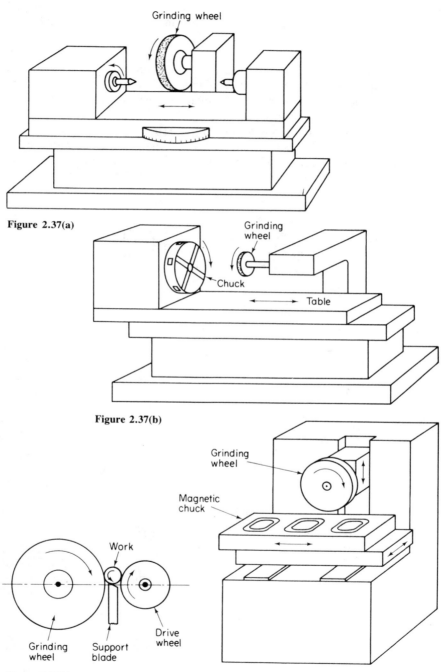

Figure 2.37(a)

Figure 2.37(b)

Figure 2.37(c)

Figure 2.37(d)

Tool and cutter grinders are used for grinding cutters of all types. Abrasive belts are used to grind certain materials. Hones and laps are used to produce fine finishes to accurate dimensions.

Boring machines.   The most common types of boring machines are the vertical, horizontal, and jig borers. Figure 2.38a shows a block diagram of a vertical boring machine. The workpiece is rotated on the table as the toolheads, which mount the cutters, machine the work.

The horizontal boring mill in Fig. 2.38b is probably the most versatile of all the machine tools. It may be used as a boring machine, milling machine, or jig borer. Cutters, boring heads, drills, reamers, and so on, may be mounted directly in the spindle. In addition, long boring bars, supported at the end and mounting boring

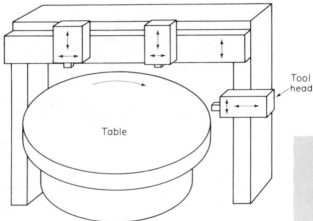

Figure 2.38(a)

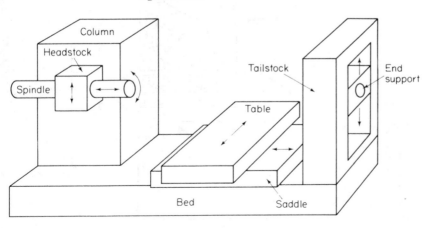

Figure 2.38(b)

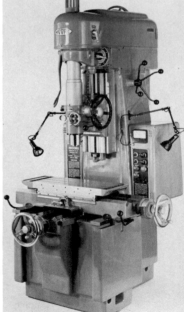

Figure 2.38(c)

blocks, may be used to line bore holes in large castings. Some of these machines are equipped with a rotary table instead of the square table shown.

Jig borers, Fig. 2.38c, are the toolmakers' machines. They are used to position work accurately under the spindle. Methods for boring hole diameters to very close tolerances are then used to machine holes in jigs and fixtures or in other workpieces which require these accuracies.

2.7. AUTOMATED MACHINES

Numerical control (NC).   These machines are generally controlled either by manual input or by an eight-channel tape, Fig. 2.39. They are not controlled by a computer.

If we assume a two-axis machine milling machine, the table will move according to the x and y data that have been programmed. Thus, if the cutter is to move away from the operator, the direction is programmed plus (+). If it is to move toward the operator, it is programmed minus (−). A left-to-right movement of the table is plus (+). A right-to-left movement is minus (−)* (see Fig. 2.40).

Figure 2.39   Courtesy of Ex-Cell-O Corporation, Remex Division.

Cutter Movement

Plus (+) direction	Minus (−) direction
X to right	X to left
Y toward rear	Y toward front
Z down	Z up
W clockwise	W counterclockwise

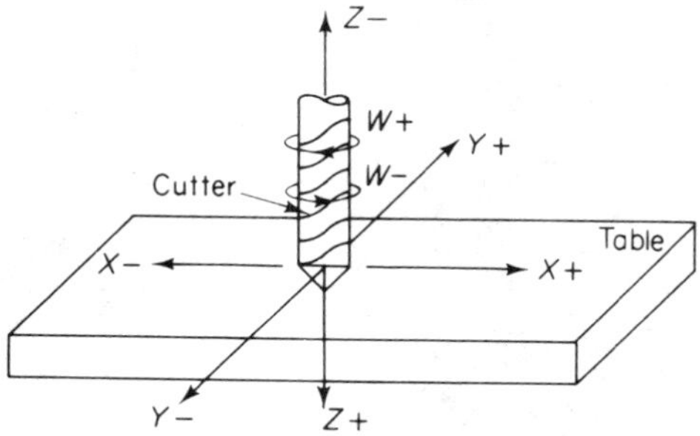

Figure 2.40

Computer numerical control (CNC).   These machines may be programmed directly with a teletype, by tape, or through a computer. Unlike the NC

*There are systems that reverse the +/− z-movement of the spindle.

Figure 2.41   (a) CNC Machining Center. Courtesy of Kearney & Trecker Corporation; (b) CNC Lathe. Courtesy of The Pratt & Whitney Company, Inc. P.O. Box 106221, Charter Oak Blvd. West, Hartford, Ct. 06110-6221.

Figure 2.41(b)

machine, the signal is fed into a computer, which operates the machine. Figure 2.41a shows a CNC milling machine, and Fig. 2.41b shows a CNC lathe.

PROBLEMS

2.1. List the processes that could be included under the general heading of casting.

2.2. (a) Discuss the roles played by a pattern and mold when making a sand casting.
(b) Assume that you are to sand cast a simple cylinder that has a hole parallel to its central axis. Describe the process for making this pattern, mold, and casting.*
(c) Which allowance must you consider?

2.3. Describe the process called shell molding.

2.4. Describe the process for making casting called investment casting.

2.5. (a) How are plaster molds made?
(b) What is the advantage of the Antioch process over the plaster of paris method?

2.6. (a) Describe centrifugal casting.
(b) What are some of the desirable features of these castings?
(c) List other centrifugal methods available.

2.7. (a) Describe permanent mold casting,
(b) slush permanent molding, and
(c) the Corthias process.

2.8. Describe the hot and cold chamber permanent mold process.

2.9. Describe the continuous process of casting.*

2.10. Describe the electroforming process of laminating castings.

2.11. Describe
(a) elastic deformation and
(b) plastic deformation.

2.12. (a) Describe the process of hot forging of steel when smith forging is used.
(b) What is drop forging?
(c) How does it differ from smith forging?

2.13. How does drop forging differ
(a) from hot press forging and
(b) from upset forging?

2.14. Describe the hot forging of steel when swaging is used.

2.15. How does roll forging reduce the diameter of bar stock?

2.16. (a) Describe direct and indirect extrusion of metal.
(b) What are the basic differences between hot and cold extrusion?

2.17. What is impact extrusion? How does it operate?

2.18. Describe the three processes for hot rolling steel.

2.19. (a) Discuss cold swaging.
(b) What is the interform system?

2.20. Describe cold heading, riveting, and staking. How do they differ?

2.21. Describe the processes of hobbing, coining, and embossing. How do they differ?

2.22. Discuss the cold drawing of bars, wire, and tubing.

2.23. Describe stretch forming.

2.24. How does high-energy forming work?

2.25. (a) Describe form rolling of sheet metal.
  (b) What is seaming?

2.26. Discuss thread rolling. In your own words, explain the statement that ''the bar diameter will become the pitch diameter after rolling is finished.''

2.27. (a) Discuss the operations of spinning sheet metal hot and cold.
  (b) How do they differ?
  (c) What is shear forming?

2.28. What is the standard straightness of material when ordered from the steel mills?

2.29. Which of the standardized columns in Table 2.1 is used for
  (a) cold-rolled strip,
  (b) stainless tubing,
  (c) carbon sheet steel,
  (d) hot-rolled alloy strip,
  (e) cold-rolled alloy strip,
  (f) hot-rolled carbon strip,
  (g) cold-rolled carbon strip,
  (h) steel tubing,
  (i) copper or brass strip,
  (j) stainless steel sheet, and
  (k) aluminum sheet.

2.30. (a) Discuss the mechanics of shearing a piece of metal.
  (b) How does this differ from the drawing operation?

2.31. How does drawing a metal blank into a cup differ from bending a strip of that metal into a U shape?

2.32. Discuss the four categories of welding.

2.33. (a) List the welding processes that require heat and pressure to join metal.
  (b) Make a statement about each which is characteristic of that process.

2.34. Repeat Prob. 2.33 for fusion welding.

2.35. In dc welding, how should the terminals be connected for reversed-polarity welding?

2.36. (a) Research the oxyacetylene process of gas welding and describe it.*
  (b) Discuss the type of welded joints that may be made.
  (c) List some of the causes and cures of poor results when using this process.

2.37. Discuss the requirements of temperature, wetting, and capillary action necessary to good brazing practice.

2.38. (a) What are some commonly used brazing materials?
  (b) List the appropriate flux to be used when brazing cast iron, aluminum, copper, brass, and alloy steels.

2.39. Discuss the various brazing processes listed in this chapter.

2.40. List at least five
  (a) rotary-type machine tools and
  (b) reciprocating machine tools.

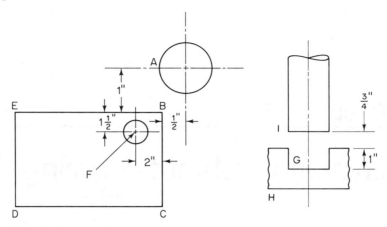

Figure 2.42

2.41. (a) List the various types of lathes.
(b) What is the fundamental principle common to all lathes?

2.42. (a) List the various types of milling machines available.
(b) What is a unique characteristic of each?

2.43. (a) List the various types of drill presses available.
(b) Discuss a unique characteristic for each.

2.44. (a) Describe the operation of the shaper.
(b) How does its operation differ from that of the planer?

2.45. Describe the operation of
(a) a shaving machine and
(b) a broach.

2.46. (a) Describe the uses of hack and band cutoff saws.
(b) What is a power contour saw?

2.47. (a) How many types of cylindrical grinders are listed in this chapter?
(b) How do they differ?

2.48. List several types of surface grinders.

2.49. Describe the operation of a vertical boring machine.

2.50. Describe the operation of a horizontal boring mill.

2.51. Describe the operation of a jig borer.*

2.52. Describe the difference in the way the program is used by an NC and a CNC machine.

2.53. Program the directional movement in Fig. 2.42.

*You may have to refer to a book on manufacturing processes to answer this question.

Chapter 3

Systems Manufacturing

3.1 UNIT VERSUS MASS PRODUCTION

The process by which a single or small quantity of work is manufactured is called *unit production*. The first step of this process is to lay out on the workpiece the dimensions that appear on the blueprint. The proper tools and machines are then selected and the workpiece is machined, operation by operation. It may be that several pieces of work have to be made and then assembled into a mechanism. In the purest form of unit production, one part is machined as close to the print dimensions as possible. The next part is then machined to fit this finished part. This is repeated until the assembly has been completed.

Note that if all the part prints are assigned tolerances, and if the machining is done according to these dimensions and to the stated tolerances, but machined one or several pieces at a time, the process would still be classified as unit production. Probably a good measure of determining whether or not unit production is being used to produce a part, or parts, is to examine the machines, tools, gages, and so on, that are being used. More important than that criterion is to seek answers to the question: How much of the machining depends on the skill of the operator? Jigs, fixtures, and gages are only a means for transferring the operator's skills to the machine. Since the machine lacks thinking processes, the prior operations should be easy to repeat. Unit production places the emphasis of decision making on the operator.

Mass production means the making of assemblies or parts in large quantities. This implies that the principles of interchangeability must apply. The principle, stated simply, means that any finished part, picked at random from a quantity of parts, should assemble with its mating part picked at random from a quantity of the latter

58

parts. Therefore, if a part of an automobile breaks, it should be possible to interchange a part having the same part number with the broken part.

This principle of interchangeability requires the parts be machined to strict standards which are set before the part is first manufactured. Once these standards have been established and if a quantity, large or small, is manufactured, it becomes possible to machine all parts to completion before starting the second part, or third part, and so on. It also makes possible the simultaneous machining of all operations of a particular part, and the simultaneous machining of all parts that are to go into an assembly.

With the advent of the interchangeability principle came the development of the concepts of allowance and tolerance. If two parts must be assembled in a certain way and meet certain performance standards, they must be sized so as to produce a certain clearance between the parts. The intentional dimensional difference between two mating parts is called the *allowance*. This produces a clearance, or interference, which has upper and lower limits. Thus each part must be manufactured to limits in order to preserve the maximum and minimum clearances. These limits may be written as tolerances for a dimension. Therefore, the *tolerance* is the allowable variation in the size of a part. Figure 3.1 illustrates these terms.

Once it becomes necessary to produce parts to a preestablished dimension and within certain limits, it becomes necessary to plan the procedure to be used when building fixtures and gages. Planning leads to the investigation of the speed of production, the control of quality, the development of machine tools for greater precision, and the analysis of the entire area of systems manufacturing. The objective now becomes *to produce the greatest number of functional parts per unit time at the lowest cost*.

The obvious next step is to automate the production of the parts, their assembly, and their testing. Automation is to make something operate by itself from start to finish. The development of automation started with individual machines, then several machines were linked into an automatic line, and now groups of lines are controlled by tape or punched cards.

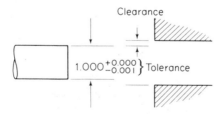

Figure 3.1

3.2 PLANNING A PRODUCT

Figure 3.2 is a flowchart that shows the route taken by a part from the time it is an idea to its completion as a finished product. The first step that must be taken before the parameters of a design can be stated is a decision by management to produce a

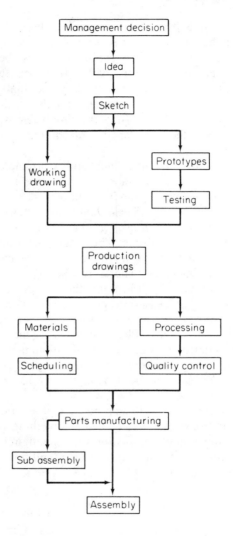

Figure 3.2

particular project. Once this decision is made, the cost, purposes, and parameters of the design are established. The various methods for achieving these decisions are then developed. Sketches are made and the advantages of one design over another are discussed. Management will usually make a decision to produce one of the designs, although it is not uncommon for companies to make the decision that two designs will be investigated simultaneously, prototypes built, and a decision made after the prototypes have been tested and evaluated.

At this point working drawings are made; at the same time, prototypes of subassemblies may be built and tested to investigate particular principles. Changes are made on the working drawings as principles are supported or rejected. Once all testing has been completed, production drawings are made from the working drawings. At this point management is committed to a particular design.

One set of production drawings may be sent to the processing department for an analysis of the sequence of operations, machines, jigs, fixtures, gages, and quality control to be used. In most instances the jigs, fixtures, and gages will be designed by the tool engineering department in accordance with the management-accepted process sequence. An additional set of prints goes to the production department, which may order the materials and prepare the factory schedules.

When all of these details have been achieved, manufacturing is ready to proceed. Parts are manufactured and routed to arrive at the assembly department when needed. These parts may be put together as sub-assemblies and then assembled into the final product. In other instances the parts go directly into the final assembly. Quality control steps are placed strategically throughout this procedure. Of course, the procedure just described has been generalized. Every company sets up its management structure to suit its own particular needs.

The tool engineer's position in this process is to plan and design. In large companies the planning is done by product engineers and the design by tool engineers. In small plants the tool engineer may be responsible for both. The tool engineer's role in all of this is discussed further in Sec. 3.7.

3.3 AUTOMATION OF SIMPLE MACHINES

The factors that control the type of and degree to which automation is to be used are quantity and quality. The number of parts per lot, that is, the length of a particular "run," is an important factor in determining the type of automation implemented. Let us assume that to support a particular assembly department, a lathe department needs to produce 200 parts a month and that the 200 parts can be produced in 1 week. A decision is made to run this part 1 week out of each month. To produce more would mean storing the excess parts. Storing these parts means that the cost of storage must be considered. Moreover, the cost of paying labor, and all other production costs, are capital outlays which will not be recouped until the parts are used. Money invested in stored parts should be earning income for the company in other ways. In the storehouse it ties up capital.

Thus a decision is made to run 1 week each month. If a special machine is made, or purchased, which is capable of running only this part, the machine will be idle 3 weeks each month. The better decision would be to use one of the general-purpose machines (lathes), and to develop a setup and tooling which make it possible to set up and tear down rapidly. Thus the machine can be used to run the 200 parts 1 week out of each month, leaving it free for other uses the rest of the month.

At some point the production schedule may tie up the general-purpose machines to the point where they are no longer being used as general-purpose machines. At that point a study should be made to determine whether or not it would be better to have a special machine for that job, and how long the job is to run.

If a special-purpose machine can be amortized over the length of the job, it may be wise to design or purchase such a machine. The quantity and quality of the work output are usually improved when special-purpose machines are used instead of general-purpose machines. If, however, expensive special-purpose machines are purchased and used for the run of the job, when the job is over, often no further use can be found for the machine.

The quality of the work produced is also a factor. As indicated, special-purpose machines generally increase the quality of the output along with the quantity produced. However, quality costs money. Although quality in a product is important, that quality should not exceed the functional requirements agreed upon by management. It might be that some part may have only one surface which needs to be machined. It would be foolhardy to machine all surfaces. It may be that a casting or a weld structure may do the job instead of a part machined from solid stock. If a cylinder may be turned on a lathe and used without grinding, it should not be ground.

In any event once the decision is made to automate a general-purpose machine and restrict its use to the production of that part, the machine may have fixtures mounted to hold the raw stock, fixtures for the tooling, and fixtures for the gaging. It may be equipped with automatic feeds, loading, and unloading devices. In some instances the part may be clamped, machined, and unclamped automatically. If the loading of the part into the holding device is by hand, the machine cycle is classified as semiautomatic.

3.4 AUTOMATION WITH STANDARD MACHINE TOOLS

Automation with standard machine tools leads to the two principles of mass production. The first principle is that of *simultation,* which applies to setups where two or more machine operations are done at the *same* time. The second is the principle of *integration,* which applies to the taking of several cuts in *sequence.* A turret lathe setup is a good example of both principles. The turret may have several toolbits mounted in a block, cutting several diameters on the workpiece in one pass. This is an illustration of simultation. The eight stations, each being used separately, would be an example of integration of cuts, as shown in Fig. 3.3.

In the principle of integration, several standard machines are lined up and set up to perform a series of related sequential operations on a workpiece. In its simplest form this may be accomplished by the addition of material-handling equipment between two or more standard machine tools. The simplest example of the adaptability of such a machine to this type of automation is the series of drill presses mounted on one long table shown in Fig. 3.4a. A block diagram of a series of machine tools connected to produce parts is shown in Fig. 3.4b. The sequence of operations may call for centering, rough turning, finish turning, and grinding. To cut down on the scrap produced, it would be necessary to install an automatic gaging system of the in-process parts. This could be included as part of the transfer mechanism between machines, or could be on the machine and working while the operation is being performed.

Figure 3.4a (Courtesy of Rockwell Manufacturing Company, Power Tool Division.)

Figure 3.3

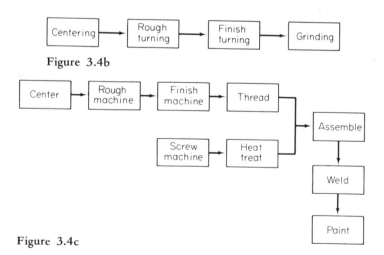

Figure 3.4b

Figure 3.4c

Another possibility is the production of two workpieces that are to be assembled. Figure 3.4c is a block diagram of two parts being manufactured. They are assembled at the end of the sequence. After assembly the two parts are welded and the entire assembly is dipped in paint.

It should be noted that the integration of several machines into a production line makes possible many types of arrangements. Figure 3.4b shows what is sometimes called a *series integration*. In series integration all machines are connected so that one machine feeds the next until the production objective is met.

Figure 3.4c shows an example of *parallel integration,* where two series arrangements are operational at the same time. The parts then come together into a series arrangement to be assembled, welded, and painted.

Because of the differences in production rates of the various operations, and because of the need to stop a particular machine to change or sharpen its tools, provision must be made for in-process storage of parts. If a machine in the line breaks down or is shut down to change tools, provision must be made to insert these parts back into the line.

The integration of standard machine tools into series, parallel, or series-parallel automated production lines requires an in-depth study of the types of machines needed, a time study of each operation, a study of the automatic transfer system needed between machines, a study of the automatic loading and unloading devices, a study of the possible automatic storage of parts between machines, a study of in-process quality control (inspection) of work, and a study of the possible use of a feedback system which relies on the gaging methods used to supply the data of in-process quality.

The introduction of robotics, computer numerical control machine tools, made possible such flexible manufacturing systems as computer-aided manufacturing (CAM), computer-aided design and manufacturing (CAD/CAM), integrated computer-aided manufacturing (ICAM), and many others.

Through the use of the computer, the strategy is to link various components into a continuous production/inspection system so that short- or long-run workpieces can be processed. The linkages that need to be considered in integrated manufacturing systems are the machines, tools, fixtures, parts, transfer units, robots, pallets, and storage. The consideration of these components can be called computer-integrated flexible manufacturing (CIFM).

3.5 AUTOMATION WITH SPECIAL MACHINE TOOLS

It should be noted that the discussion in Sec. 3.4 which relates to the design of automatic production lines using standard machines also applies to the design of automated production lines using component machine parts (Fig. 3.6a) or machines designed to do a particular job in a production line.

The flowchart of special machine tools, shown in Fig. 3.5, is graded from those which have the least complex degree of automation to those which are the most complicated. Thus the least complex of the automated special machines is one that holds the workpiece and has tools fed into it to accomplish the machining operations. The most complex of the machines moves the workpiece past the tools as they feed from all directions to machine the workpiece.

Figure 3.6a shows the use of three special air-hydraulic drill units. They are

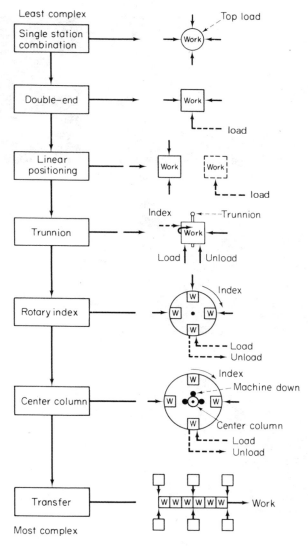

Figure 3.5

mounted in a vertical plane for drilling three equally spaced holes. This class of special machines does *not* transfer the workpiece from one machine to another, but rather, does the operations at one station without revolving the workpiece. Operations such as drilling, tapping, countersinking, spot facing, milling and reaming may be done using this method. This type of automation may be called a *single-station combination-operation machine*.

The *double-end machine* used in automation is shown in Fig. 3.6b. Two machines have been tied together, end to end, to perform operations on both ends of

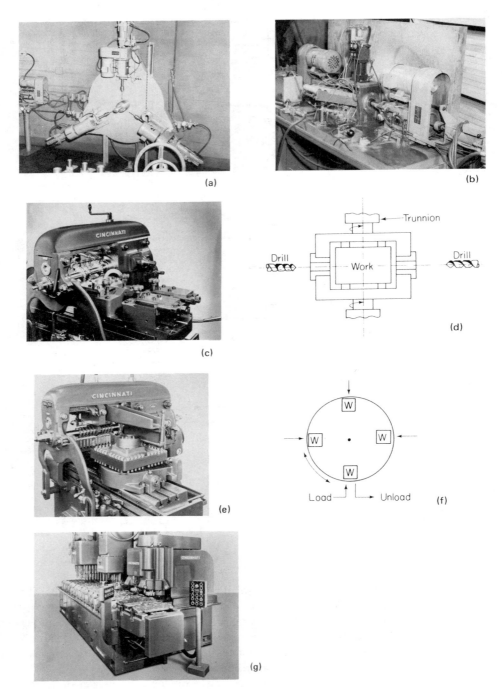

Figure 3.6

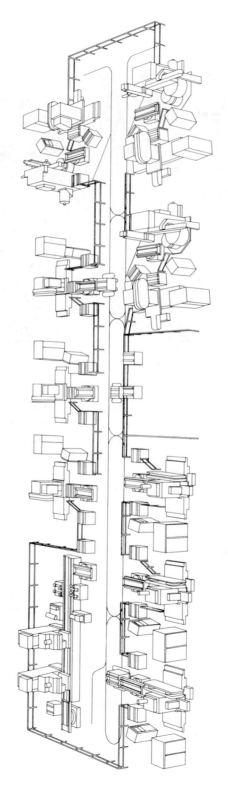

Figure 3.6(h) (*continued*) (Courtesy of Giddings & Lewis, Mfg. Systems, a unit of AMCA International Wi.)

Figure 3.6(i) (*continued*) (Courtesy of Giddings & Lewis, Mfg. Systems, a unit of AMCA International Wi.)

pieces of tubing. Operations such as chamfering, facing, centering, drilling, milling, boring, spinning, and so on, may be done with double-end machines. All of the loading, unloading, and inspection devices may be used with these machines.

Linear positioning special automated systems permit mounting the workpiece on a table, which then moves in between single or double rows of machining heads. If the table is long enough, one fixture may be unloaded and reloaded as the workpiece in the fixture is being machined. Such a setup is shown in Fig. 3.6c.

A *trunnion-type* machine Fig. 3.6d holds the workpiece in a fixture which is mounted in a trunnion, or on a table equipped with a trunnion. The fixture is indexed so that the faces of the work may be machined by the tools mounted in the machining heads. A simple setup of this type is shown in Fig. 3.6d.

Another type of automated special machine setup is the *rotary-index type* Fig. 3.6(e). This setup utilizes a rotary index table with several fixtures mounted on the table so that each fixture always has a partially completed workpiece. The first station is the load-unload station. The table is indexed so that the workpieces rotate into position in each station in front of horizontal, or vertical, machining units. With each index, a finished piece is unloaded, and a finished piece loaded in its place. A simplified version is shown in Fig. 3.6f.

The most complex of the special machines are those with a *center column* on which machining heads are mounted. Thus not only may tools be mounted above the rotary index table but also around its periphery. Again, several fixtures are mounted, as shown in Fig. 3.6g. Using this system, machining may also be done from above the table by the machining heads mounted on the center column. All machining is done simultaneously so that with each index motion, a finished part is removed and replaced by an unfinished part.

All of the machines discussed to this point in the section have operated according to the principle of simultation. The last of the special categories is that which deals with special *transfer machines*. The procedure of passing the workpieces between various machine units provides the greatest amount of flexibility for automation. Virtually any type of operation may be included in the transfer line. The indexing is linear and provides a finished workpiece with each index. Such transfer lines are shown in Figs. 3.6h and 3.6i.

3.6 FEEDBACK AND CONTROL

Automated systems are controlled by either one of two systems: open-loop or closed-loop. The *open-loop system* of automation is actuated by input signals, or commands. The signals, whether manual, electrical, mechanical, hydraulic, or pneumatic, feed the commands in and rely on preset tools or stops to produce the proper results on the workpiece being machined.

A *closed-loop system* may be thought of as an open-loop system with feedback. Thus a sensing device senses what is happening at a particular point in the system and feeds the signal back to the control center, where it is compared to a desirable set of data. If the feedback signal matches the desired data, the process is allowed to continue. If it does not match the data, the control system may either make a correction, flash a warning, or shut the system down.

There are six elements of control* which should be present for a closed-loop system.

1. *Variables.* These are picked up, amplified, and used to actuate a controller.
2. *Sensors.* A device that picks up the variables as discrete signals which are fed into the controller.
3. *Memory.* A device that fixes the limits within which the system operates. Such a system may be a stop, punch tape, and so on.
4. *Collators.* This device compares the signal from the sensor with the information in the memory.
5. *Actuators.* This device uses the information from the collator and sends out the appropriate command signal to make any desired corrections.
6. *Controllers.* A device that applies the controls to the system. Such controls may be pumps, dampers, or valves.

When these elements of control are actuated, the feedback system must sense the effect of the controllers on the system, feed it back, and make another comparison to ensure that the appropriate action was taken.

---

*From "Control Aspects of Automation," in ASTME, *Tool Engineers Handbook*, 2nd ed. (New York: McGraw-Hill, 1959), pp. 10–21.

An automatic turret lathe is an example of an open-loop automated system. The machine will run through its entire cycle without reference to the quality of its output. Indeed, the raw stock may not even be in the machine while the cycle is being run. When the start button is pushed, the machine will go through its entire cycle and shut off.

A closed-loop system is capable of immediately sensing that there was something wrong. Thus gun drills have pressure-sensing devices that permit the drilling to continue as long as the pressure at the drill point is below a desired level. As soon as the drill gets dull, or if chips clog the hole being drilled, the forces build up and a signal shuts off the electric power.

A temperature control thermostat, attached to a heating furnace equipped with an air conditioner, will sense the temperature in a room and send an appropriate signal to the furnace controls. If the room temperature is below a preestablished thermometer reading, it will direct the signal to the heating unit. The heat will stay on until the desired temperature is reached, at which time it shuts off. If the room temperature is above the preestablished thermometer reading, the feedback signal will be directed to the air conditioner, which will operate until the temperature drops to within the limits of the thermostat setting. This is a closed-loop system.

It should be understood that the feedback of commands into the control system is essential to automation. The input of signals whcih control the operation of a process is one part of an automated system which will produce workpieces. Feedback and control is the other part of an automated system which produces not only workpieces, but produces *usable* workpieces. The input commands produce the manufacturing cycles, whereas the feedback commands control the quality.

3.7 THE TOOL ENGINEER

From the previous discussions it is evident that the production of a part or an assembly can be a very complicated endeavor. The tool engineer is a professional whose job it is to assist the product engineer in the design and refinement of the product, in designing or purchasing the machinery, tools, fixtures, dies, and gages to be used in producing the parts and in their assembly into the final product. He is responsible for maintaining the quality consistent with minimum cost of the product.

He must be thoroughly familiar with the machines available in the plant and those that may be purchased, including their capabilities and their limitations. He must then be thoroughly familiar with the product to be manufactured, the precision desired, quantities to be produced, job lot sizes, lead times, and cost limitations. All these will determine how the product will be manufactured, and this, in turn, will determine the kind of tooling needed and how much can be spent on this tooling.

It should be noted that the tool engineer must be familiar with all available standard parts; what is available as standard but can be altered to meet job needs; what can be cast, welded, or stamped; and what needs to be machined. Unless parts are needed in very large quantities, in most instances it is cheaper to buy standard parts than it is to make them. Thus the cost of a part increases as one reads down the list of choices at the beginning of this paragraph; that is, purchasing standard parts costs less than buying a standard part and altering it.

When it comes down to the actual design of the tools, jigs, fixtures, and machines, the tool engineer must follow a plan of action or design procedures. Thus he will first identify what it is that he wants to do and prepare a statement of the problem that is to be solved.

The next step is to research the design and to gather information and data. This requires a close examination of the blueprints to determine the importance of the dimensions, the type of material to be used, how the part is to be held, a preliminary sequence of operations, and which machines are capable of doing the desired job. This could lead to library research, contacting suppliers or experts on particular problems, or making models and running tests on sections of the problem. All sketches, calculations, and test data are saved.

The sketches, calculations, and data are then evaluated and a preliminary design of the tool, jig, or fixture is prepared. At this point, creativity is an invaluable asset to a tool engineer. It may be desirable to build a prototype of the tool, run tests, and evaluate the results.

Once a design is decided upon, the final drawings are made. Although changes may still be made, at this point the design is considered ready for the toolmaker.

PROBLEMS

3.1. (a) Define the concept of unit production.
(b) Give several examples and explain how they meet your test.

3.2. (a) State the principle of interchangeability.
(b) How does it affect the machining of parts?
(c) Support your answer with an example, using a simple assembly with which you are familiar.

3.3. (a) Define mass production.
(b) Give several examples.
(c) State how the examples meet your definition.

3.4. Define
(a) allowance,
(b) tolerance, and
(c) limits. Explain the function of each.

3.5. What are the implications of the principle of interchangeability for mass production?

3.6. (a) Define automation.
(b) Give examples from your experience.
(c) Explain how they fit your definition.

3.7. Explain each of the steps in Fig. 3.2.

3.8. Describe what you understand to be the purpose of building a model or prototype. Why is it valuable to the design or research engineer?

3.9. (a) Describe a working drawing.

      (b) What is its purpose?

      (c) How does it differ from production drawings?

3.10. What role in the development of a product is played by

      (a) the process engineer and

      (b) the production engineer?

3.11. (a) Can you state the importance of having finished parts arrive at the assembly department when they are needed?

      (b) What are the implications of this statement for the people who schedule and route materials?

3.12. What factors determine the type of automation that is to be used to produce a particular part? Explain.

3.13. What factors determine the degree to which automation is to be used to produce a part? Explain.

3.14. Why is it unprofitable for a company to overproduce parts that must be stored?

3.15. Discuss the advantages and disadvantages of using a special-purpose machine versus a general-purpose machine.

3.16. Explain the statement: ''. . . quality should not exceed the functional requirements . . . .''

3.17. What are some of the steps that can be taken when it becomes necessary to automate a general-purpose machine? Discuss your answer.

3.18. Define the principle of simulation. Give at least two examples and explain.

3.19. Repeat Prob. 3.18 for the principle of integration of cuts.

3.20. Lay out an integrated production line (block diagram) for producing a part, or small assembly, with which you are familiar. Include the gaging points in the line. Justify each block in the diagram.

3.21. Explain

      (a) series integration and

      (b) parallel integration as related to production lines, and the

      (c) series-parallel production line.

3.22 **Check the technical magazines in your library. Find examples of problem 3.21 a, b, and c.**

3.23. Discuss CIFM. (Check your library).

3.24. Which components need to be controlled to develop a flexible production system?

3.25. Describe

      (a) CAD/CAM and

      (b) CIM/ICAM.

3.26. Devise a series-parallel production line for the manufacture of a common small assembly with which you are familiar.

3.27. ''Sharpening of tools, or machine breakdown, are two justifications for storing parts which have been stocked up while the setup is being corrected.'' Explain this statement.

3.28. A paragraph in Sec. 3.4 reads: ''The integration of standard . . . quality.'' Explain each of the in-depth studies recommended and give a reason for their need.

3.29. Explain the flowchart of Fig. 3.5.

3.30. Describe the single-station combination-operation type of automated machine. See if you can find a machine tool setup of this nature in a professional journal.

3.31. (a) Explain double-end machine-type automation.

(b) Devise an illustration, or find a similar machine tool setup in a professional journal.

3.32. Repeat Prob. 3.31 for the linear-positioning special automated machine class of machine tools.

3.33. Repeat Prob. 3.31 for the trunnion-type automated class of machine tools.

3.34. Repeat Prob. 3.31 for the rotary index automated class of machine tools.

3.35. Repeat Prob. 3.31 for the center-column automated class of machine tools.

3.36. (a) Repeat Prob. 3.31 for the transfer automated class of production lines.

(b) How does this class differ from all the classes in Probs. 3.30 through 3.35?

3.37. (a) Define and discuss the open-loop system of automation.

(b) Give at least two examples of open-loop systems and show how they match your definition.

3.38. Repeat Prob. 3.37 for a closed-loop system of automation.

3.39. List and discuss the six elements needed for the control of a closed-loop system.

3.40. Describe the duties and responsibilities of the tool engineer.

3.41. What must the tool engineer be familiar with before he can logically start to build the fixtures for a job?

3.42. List the order of priorities, starting with standard parts, in terms of the cost of production.

3.43. List and describe a tool engineer's plan of action or design procedure when he is preparing to embark on a design project.

Chapter 4

Toolroom Drafting

4.1 TOOLROOM DRAWING PROCEDURES

It is taken for granted that the reader is familiar with standard drafting procedures. However, toolroom drafting differs in some respects from standard practice. For example, the toolmaker is very highly skilled. Every refinement does not need to be spelled out for him. Furthermore, a tool drawing is generally used once and not referred to again unless the tool is redesigned for better performance, or altered for use on another job. In any event, much of the information found on a working drawing need not be included on a tool drawing.

In very large companies a set of toolroom drawings include an assembly drawing and all the details. It is possible for several people to work on the tool, each making his own part, or parts, which will eventually be assembled into the desired tool.

In companies where one tool and die maker is responsible for the entire tool, a set of drawings consists of an assembly drawing and details only if they are necessary. In this process the company relies on the skill and knowledge of the toolmaker to produce the tool or die.

There are some rules that will result in a saving of design time and will aid in making the drawing clear.

1. Use only those views that are absolutely necessary to convey all the necessary information. Descriptive notes on a drawing in many instances can replace an entire set of lines, dimensions, and sometimes entire views. Avoid repeating details in various views. They can be confusing.

2. If the shape of the part can be described, it may be unnecessary to draw the part.

74

Standard parts should be outlined on a tool drawing to show their location and possibly, a location dimension included. Thus standard components, such as screws, washers, nuts, dowels, locator pins, die sets, and so on, need not be dimensioned, particularly if they are listed and dimensioned in the bill of materials.

3. Standard parts that need to be altered should show only the dimensions of the alteration and one tie-in dimension. The drawing should be a simple as possible.

4. Dotted or dashed lines should be avoided if they do not contribute to the clarity of the drawing. Avoid making section views, except when drawing dies.

5. If an object is symmetrical, a partial view will usually serve the purpose. The same practice should be followed when it is necessary to draw a thread or a gear configuration, Fig. 4.1. Drawing several threads or gear teeth is a great deal easier than making a complete drawing; it saves time and is clearer to the person reading the print.

6. Avoid repeating dimensions in different views. Errors can be made so that the dimensions are not consistent. If a dimensional change must be made, it will most assuredly be made in one view and not in the other.

7. Tool drawings should carry critical size dimensions and location dimensions. It is vital that the toolmaker not be required to make any calculations in order to determine critical size or location dimension. The location of screws, locator pins, and dowels should be indicated but not dimensioned.

8. In some instances, parts may vary only slightly. If there are several parts that are of this nature, a descriptive box may be made and the variations listed in the box. The drawing should carry the dimensions which are the same for all parts. If two parts are very similar, one drawing should be made and a note explaining the differences should be added. It could read: ''Same as except. . . .''

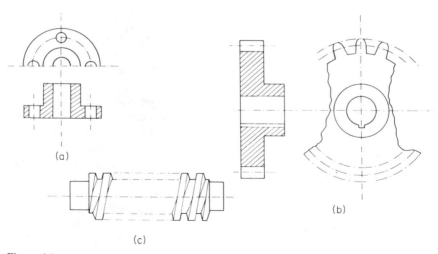

(a)

(b)

(c)

Figure 4.1

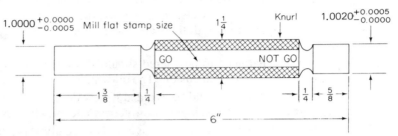

Figure 4.2

9. If balloon leader lines are used, the leader lines should be curved or at an angle with the dimension lines so that they are not confused with dimension lines. Balloons should be ordered between or next to views so that they can easily be found.

10. The designer should outline the workpiece in red for which the tool, jig, fixture, or die is to be made. This not only helps the designer but in most instances helps the toolmaker and anyone else reading the print.

11. Another timesaver when designing tools is the use of standard part templates whenever possible.

Figures 4.2 and 4.3 show detailed drawings that embody many of the rules just discussed.

There are many types of title blocks. The title block shown in Fig. 4.3 is probably as simple as it is possible to get without leaving out important information. It has a place for the following items:

1. The company name
2. The tool name and number
3. Part name and number
4. Machine number

5. Drawn by and date
6. Checked by and date
7. Scale of the drawing
8. Tolerance box

In some companies a bill of materials is started at the top of the title block and carried up the right side of the sheet. The part number should be close to the title block as shown in Fig. 4.3. The numbers that have been assigned increase in increments of 5. This permits changes or additions to be made without the need of inserting a number such as 5A, and so on. Writing the stock list from the bottom up also leaves room at the top for adding additional numbers at a later time.

Because other personnel in the company will be using the stock list, it is a good idea to group numbers and therefore parts which will be handled by one person. Thus if all purchased parts are grouped, it makes it easier to order and check the order when it arrives. The sequence of numbering used by one company is as follows:

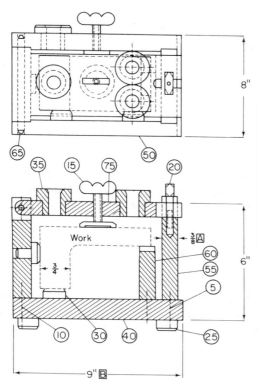

Revisions

Symbol	Description	Date	Made by	Date	Check by
A	$\frac{3}{8}$ Thick was $\frac{1}{2}$"	–	–	–	–
B	9 Lg. was $8\frac{1}{2}$	–	–	–	–

65	1	$\frac{1}{4}$ Dia x 8 Pin		Dr. Rod
60	1	$\frac{1}{2}$ x 4 x $7\frac{1}{2}$ Support Block		1020 CRS
55	2	$\frac{3}{8}$ x $5\frac{1}{2}$ x $7\frac{1}{2}$ End Plate		1020 CRS
50	2	$\frac{3}{8}$ x $5\frac{3}{4}$ x 9 Side Plate		1020 CRS
45	1	$\frac{3}{8}$ x $7\frac{1}{2}$ x $9\frac{1}{2}$ Leaf		1020 CRS
40	1	$\frac{3}{8}$ x 8 x 9 Base Plate		1020 CRS
35	4	$\frac{1}{2}$ in. Dr. Brush		Std
30	4	Support Pin		Std
25	3	Legs		Std
20	1	Lockscrew $\frac{1}{2}$ – 13		Std
15	2	$\frac{3}{8}$ x 16 Wing nut		Std
10	8	$\frac{1}{4}$ Dia. x 1" Dowel		Std
5	8	$\frac{1}{4}$ – 20 x $\frac{3}{4}$		Std
Det. No.	No. Req.	Description		Mat.
Company Name				
Tool Name		Tool No.		
Part Name		Part No.		
Machine No.		Scale		
Drawn by		Date		
Check by		Date		
Tolerance				

Figure 4.3

Purchased Parts	Lowest Numbers
Purchased but altered	*200 series*
Casting, forgings, etc.	*300 series*
Stamping, drawn shells, etc.	*400 series*
Parts machined from raw stock	*500 and over*

The bill of materials should have at least four columns. The columns should show the balloon numbers, a description of the parts, the quantity required for *each* assembly, and comments. It should be noted that when the size of the material is given in the column headed "description," that size should be the *rough size*.

When changes are to be made on a drawing after its release for production, these changes should be entered in a *change block* at the upper right of the drawing and it should read from the top down. It should include the following information:

1. A capital letter symbol
2. A description of the old information and the change
3. The date the change was made
4. Who made the change
5. The date the change was approved
6. Who approved the change

It should be noted that if at all possible, the old dimension, balloon, and so on, should have a line drawn through it and a reference to the change symbol. The old information should not be erased. If it becomes necessary to erase a great deal of information, the balloon number should be struck and the change symbol added, the print removed and stored, and a new drawing made which shows a new balloon number and a reference to the withdrawn number.

4.2 TITLE BLOCKS AND BILLS OF MATERIALS

Figure 4.3 shows a tool drawing for drilling four holes into an angle plate. The drawing is ballooned so that the numbers in each balloon correspond to the numbers in the bill of materials. The workpiece, outlined in the drill jig, shows all the important dimensions and the functional parts of the drill jig.

4.3 THE TOOL DRAWING

Tool drawing, such as jigs, fixtures, gages, cutting tools, and so on, follows conventional practices. If possible, two views should be used. Three views should be used only if the design becomes complicated. Sections and enlarged views should be added to the assembly drawing before detail drawings are made.

Die drawing is arranged somewhat differently. A die set, in the conventional position, shows the punch holder as the upper part. It carries the punches. The lower part is the die shoe. It carries the die openings. This is shown in Fig. 4.4a. A sectional view of the die set is shown in the lower left of Fig. 4.4a. It is shown in *closed position* and *sectioned* to reveal as many of the working parts of the die set as possible.

At times punches and dies are shown half-open and half-closed. This is the case with the simple drawing die in Fig. 4.4b. The plan view of the die shoe is shown in the upper left of the drawing. It is shown looking down at the die openings *with the punch holder removed*. The punch holder is shown to the right of the die shoe but rotated 180° from the plan view of the die. It is as if there were a hinge between the two views (Fig. 4.4a) and the die set were opened like a sandwich. Front and side views may be used for clarity.

Abbreviations used on tool drawings are no different than those used on conventional drawings. Table 4.1 shows some of the standard abbreviations.

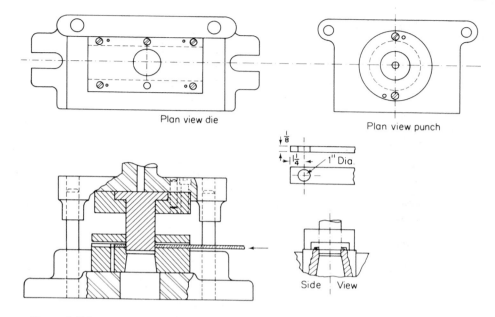

Plan view die

Plan view punch

Side   View

Figure 4.4(a)

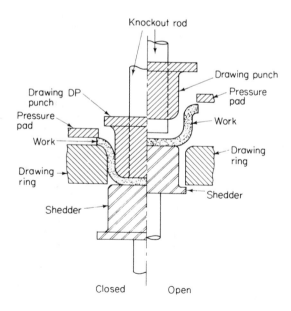

Knockout rod

Drawing punch

Pressure pad

Work

Drawing DP punch

Pressure pad

Work

Drawing ring

Drawing ring

Shedder

Shedder

Closed   Open

Figure 4.4(b)

TABLE 4.1 ABBREVIATIONS

Word	Abbreviation	Word	Abbreviation
Allowance	Allow	Castle nut	CAS NUT
American Wire Gage	AWG	Cast steel	CS
Approximate	Approx	Center line	CL
Bearing	BRG	Center of gravity	CG
Bevel	BEV	Center punch	CP
Bill of materials	B/M	Center to center	C to C
Birmingham Wire Gage	BWG	Chamfer	CHAM
Brass	BRS	Chromium plate	CRPL
Brinnell hardness	BH	Circle	CIR
British thermal units	BTU	Circular pitch	CP
Brown and Sharpe	B&S	Circumference	CIRC
Bushing	BUSH	Clamp	CLP
Cadmium plate	CD PL	Clearance	CL
Cap screw	CAP SCR	Cold drawn	CD
Carburize	CARB	Cold-drawn Steel	CDS
Case harden	CH	Cold finish	CF
Cast	C	Cold-rolled steel	CRS
Cast iron	CI	Concentric	CONC
Cast iron pipe	CIP	Counterbore	C'BORE
Countersink	CSK	Motor	MOT
Cubic feet per minute	CFM	Mounting	MTG
Detail	DET	Music wire gage	MWG
Diameter	DIA	New British standard	NBS
Diametral pitch	DP	(Imperial wire gage)	
Dowel	DWL	Nipple	NIP
Drill	DR	Nominal	NOM
Drill rod	DR	Not to scale	NTS
Drive fit	DF	Number	NO
Extrude	EXTR	Obsolete	OBS
Fillet	FIL	Octagon	OCT
Finish all over	FAO	On center	OC
Flat head	FH	Outside diameter	OD
Forge steel	FST	Overall	OA
Forging	FORG	Parallel	PAR
Gage	GA	Perpendicular	PERP
Ground	GND	Pierce	PRC
Half-hard	1/2 H	Pipe tap	PT
Half-round	1/2 RD	Pitch	P
Hard	H	Pitch circle	PC
Hard-drawn	HD	Pitch diamter	PD
Harden	HDN	Press fit	PF
Heat treat	HT TR	Quantity	QTY
Hexagon	HEX	Quarter-round	1/4 RD
High-speed steel	HSS	Radius	R
Hot-rolled steel	HRS	Ream	RM
Inside diameter	ID	Reference	REF
Internal	INT	Register	REG
Intersect	INT	Relief	REL
Iron-pipe size	IPS	Required	REQD

TABLE 4.1 (*Continued*)

Word	Abbreviation	Word	Abbreviation
Left hand	LH	Requisition	REQ
Machine steel	MS	Right hand	RH
Malleable iron	MI	Rivet	RIV
Material	MATL	Rockwell hardness	RH
Material list	ML	Root diameter	RD
Maximum	MAX	Root mean square	RMS
Minimum	MIN	Round	RD
Miscellaneous	MISC	Screw	SCR
Morse taper	MOR T	Setscrew	SS
Shaft	SFT	Washer	WASH
Sheet	SH	Width	W
Shoulder	SHLD	Woodruff	WDF
Specifications	SPEC	Wrought Iron	WI
Spotface	SF		
Spring	SPG	Threads	
Square	SQ		
Stainless steel	SST	American National	
Standard	STD	Coarse	NC
Steel	STL	Fine	NF
Stock	STK	Extra fine	NEF
Strip	STR	Special pitch	NS
Structural	STR	Pitch thread	8N, 12N,
Supersede	SUPSD		or 16N
Surface	SURF	Acme	NA
Symmetrical	SYM	Taper pipe	NPT
Tangent	TAN	Dryseal	NPTF
Taper	TPR	Failing fixture	NPTR
Teeth per inch	TPI	Straight pipe	
Tensile strength	TS	Coupling	NPSC
Thick	THK	Dry seal	NPSF
Thousand	M	Mechanical joint	NPSM
Thousand pound	KIP	Locknut	NPSL
Thread	THD	Hose couple	NPSH
Threads per inch	TPI	American Truncated Whitworth	
Through	THRU	Coarse	TWC
Tolerance	TOL	Fine	TWF
Tool steel	TS	Special	TWS
Total indicator reading	TIR	Unified National	
Tubing	TUB	Coarse	UNC
Typical	TYP	Fine	UNF
United States Standard	USS	Special	UNS
Vertical	VERT		

4.4 RULES FOR DIMENSIONING

Dimensioning provides the basis for interchangeable manufacturing, for the inspection of each workpiece and its assembly, and provides the desired functioning of that assembly. If prints are not dimensioned properly, the results are poor production, destruction of functional requirements, loss of interchangeability of parts, and failures at assembly, which could result in excessive and expensive rework, redesign of parts, fixtures, or gages, revamping of drawings, and so on. Any one of these items by itself adds cost to the project, loss of time, and loss of company prestige.

It is therefore very important that dimensions when placed on a print be accurate and clear. There are several rules that should be observed when dimensioning prints. These are discussed below.

Lead dimension.   The lead dimension should be given as that dimension to which the part should be machined if it were possible to make the dimension without variations. Dimensions cannot be made exactly, and that is the reason for tolerances. Table 4.2 shows the various ways of writing 4.000 in. to a tolerance of ± 0.002.

The manner in which the dimension is written depends on the ideal dimension desired. That is, 4.000 ± 0.002 means that the best functional conditions will exist if the toolmaker can produce a 4.000 dimension exactly. It also means that if he cannot, the part will be acceptable if his dimension is within the limits of 4.002 and 3.998 in.

The dimension $4.002 {}^{+0.000}_{-0.004}$ means that 4.002 in. is the ideal dimension and that anything larger will be unacceptable. The dimension $3.998 {}^{+0.004}_{-0.000}$ means that 3.998 is the ideal dimension and that anything smaller will be unacceptable.

TABLE 4.2

$4.002 {}^{+0.000}_{-0.004}$	$4.001 {}^{+0.001}_{-0.003}$	$4.000 {}^{+0.002}_{-0.002}$
$3.999 {}^{+0.003}_{-0.001}$	$3.998 {}^{+0.004}_{-0.000}$	

Limits.   Limits are a system of tolerances that indicate the maximum and minimum sizes only. This type of dimensioning is for the purpose of showing outer dimensions without worrying about preferred sizes. It is an ideal way of tolerancing GO/NO GO gages. This type of dimensioning is shown in Fig. 4.7b and c.

Chain dimensioning.   In chain dimensioning, one of the dimensions in the chain is superfluous if the chain is indicated with an overall dimension. This stems from the fact that only one dimension should be used between two points, lines, or surfaces. In Fig. 4.5a one of each of the 1.000 in., 0.500 in., and the 3.000 in. dimensions and the 90° angle dimension all shown crossed out should not be given.

In the case of the 3.000 dimension, the tolerance is ± 0.001. If the toolmaker

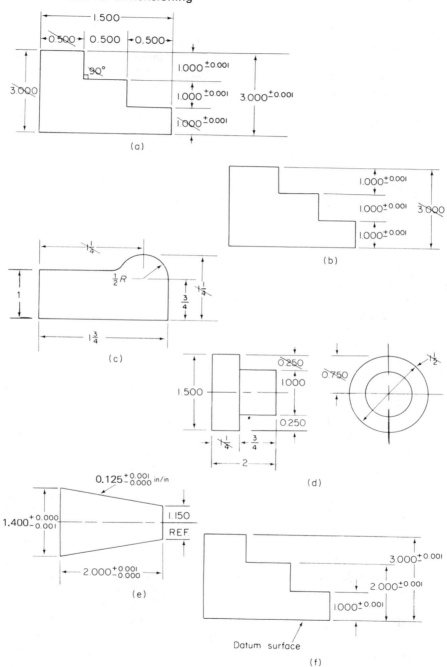

Figure 4.5

should machine every step to the plus tolerance, the sum of the three steps would accumulate to +0.003. Of course, this would not be acceptable since the overall tolerance that is permitted is ±0.001. The same reasoning may be applied to the negative tolerance. On the other hand, if it is not necessary that the 3.000 in. dimension be held to ±0.001, it would be better to dimension the drawing as shown in Fig. 4.5b. In this case the 3.000 in. dimension is superfluous. Figures 4.5c and d show other dimensions that are not needed because at least one dimension on the drawing has already defined the surfaces.

The four basic dimensions that define a taper are as follows:

1. The taper per inch of length, or the included angle
2. The diameter of the large end or of the small end
3. The length of the axis of the tapered portion
4. The reference note shown in Fig. 4.5e

Any 1 of the first three values will completely define a taper. The fourth dimension is usually included as a reference dimension. This is shown in Fig. 4.5e.

Related surfaces.   The accumulation of tolerances in chain dimensioning has already been discussed. One way to avoid the accumulation of tolerances is to select a point, line, or surface and using one of these as a reference, dimension any other point, line, or surface to it.

The reference point, line, or surface is then called the *datum* point, line, or surface. This is shown in Fig. 4.5f. Ideally, the datum surface should be a functional surface.

Repeating dimensions.   It is of the utmost importance that a dimension that appears in one view should *not* be repeated in another view. In complicated drawings it may take a little extra time to find a particular dimension. Yet the student should realize that if the dimension is repeated in another view, a designer will probably not take additional time to look for the second dimension if he is making a change. He will simply assume that it does not exist, make his change, and go on to the next job! If a dimension is to be repeated in another view, it should be marked REF.

Defining a part.   The student should understand that a drawing is a pictorial definition of a part and that dimensions define that part further. Therefore, to define a part fully, there should be present on the print all necessary views and these views should be dimensioned. The dimensions, as already stated, should be referenced to functional surfaces. To determine which surfaces are important, four types of surfaces should be defined. An analysis of a part and its function in an assembly will reveal to the designer the functional surfaces on the workpiece. The surfaces to be defined are indicated by datum, functional, clearance, and atmospheric surfaces, lines, or points.

A *datum surface* is one that may be used as a reference from which measurements may be taken. All objects occupy space in three planes, which mathematicians usually define as the *x, y, z* planes. These planes may be on the assembly, as shown in Fig. 4.6(a), or may be planes chosen off the assembly, as in the case of numerical control (NC) when a reference point is taken off the part.

Three datum planes are shown in Fig. 4.6(a). The vertical plane determines all right-to-left locations. The position and dimensioning of the bearing, retainer rings, gear, key, and so on, all depend on where the datum plane is located. Horizontal and vertical planes through the centerline serve the same purpose when chosen as the datum planes.

When considering the dimensioning of assembly drawings, the datum planes of the various subassemblies, because they are related, should be dimensioned to a datum plane on the assembly drawing. This plane may be a plane of the main housing, or one of the datum planes of a subassembly, or a completely new plane. These reference planes (or surfaces) tie the assembly together.

As noted above, the plane may be one that is arbitrarily chosen anywhere by the designer if it suits his purpose. Care must be exercised when choosing the plane (or points) off the drawing because it is not an easy matter for a toolmaker to take measurements from a point in space. Numerical control machines are programmed to start at some point and proceed progressively from point to point.

A *functional surface* is one that enters into the operation or location of other parts. Thus in Fig. 4.6(b) the small shaft diameter locates the bearing; the retainer ring grooves hold the bearing in position; the shoulder on the shaft locates the bearing in a right-to-left plane; the keyway is functional; and so on. These are all functional surfaces. It should be noted that the vertical datum plane is also a functional surface which lends itself to being chosen as a reference plane for dimensioning. The horizontal centerline is not a functional surface, although it is a very important line that establishes two datum planes on the drawing.

A *clearance surface* is one that has no contact with the surface of a mating part, but which supports a functional surface. These surfaces are generally in close proximity to other surfaces. Figure 4.6(b) shows a clearance surface to the left of the bearing. It clears the shaft entirely, yet it must be in existence; otherwise, the datum

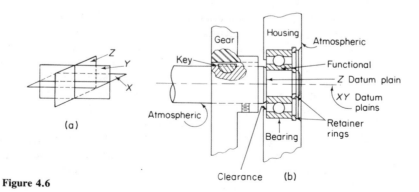

Figure 4.6

plane, which is a functional surface, would have no thickness and, in fact, could not exist as a functional surface. These surfaces may or may not require machining.

An *atmospheric surface* is one that bears no relationship in terms of position to any other part in the assembly. It is a surface that is completely exposed to the "atmosphere." Thus the rough surface of a casting is an atmospheric surface. Figure 4.6(b) shows the front of the bearing housing as an atmospheric surface. The exposed part of the shaft is also an atmospheric surface even though the same surface is classified as functional where it supports the gear. Except for aesthetic reasons or as machining locator surfaces, these surfaces would ordinarily not be machined.

4.5 FIT DEFINITIONS

At this time the discussion will center around the fact that it is impossible to machine a part to an exact size and the limits will be used to ensure proper functioning of the part. Another area of concern to the designer is the amount of clearance, or interference, which is permitted between two mating parts and the way this can be achieved. When selecting the tolerances and allowances, considerations are: the size of the part, the length of engagement, the type of material in both parts, the surface roughness permitted, the relative speeds of the mating parts in contact, the lubrication used, the loads, and temperature. Necessary definitions that should be discussed before any explanation of fits can take place are given below.

Nominal size is a size number used to identify the physical size of a part in general terms. An example would simply be $1\frac{1}{2}$ in.

Basic size is the exact size from which the limits are figured considering the allowances and tolerances. It is the ideal dimension that would be desired if it were possible to machine without tolerances. An example would be the nominal size of $1\frac{1}{2}$ in., written as 1.500 in. when applied to mating parts.

Design size is the size from which the limits are calculated by the use of tolerances. When there is no mating part, and therefore no allowance, the design and basic sizes are the same. An example would be the nominal size $1\frac{1}{2}$ in., written as 1.500 in. when applied to a single part.

Allowance is the intentional least difference *between two mating parts* or the *tightest fit permitted*. It is defined as the *least clearance* or the *greatest interference* that exists when two parts are assembled. It deals with positive allowance when there is minimum clearance, and with negative allowance when there is maximum interference between mating parts. Both conditions are shown in Fig. 4.7a. If the nominal size is $1\frac{1}{2}$ in., the basic size is 1.500 in., the interference allowance is 0.002 in., and the clearance allowance is 0.004 in.

Limits are the maximum and minimum permissible sizes. Examples are limits applied to a hole, Fig. 4.7b, and to a shaft, Fig. 4.7c. It should be noted that when writing limits for *internal* dimensions, the *smallest limit* is placed at the top; and for *external* dimensions, the *largest limit* is placed at the top.

Tolerance is the total allowable difference from a size. *Unilateral tolerances* are

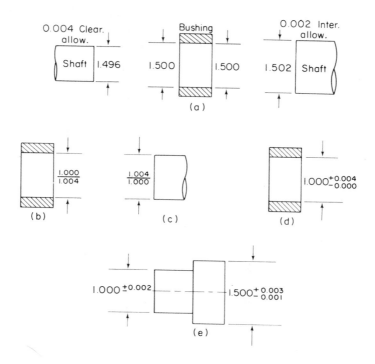

Figure 4.7

those which are applied in one direction as a variation from the basic size. This is shown in Fig. 4.7d. *Bilateral tolerances* are those which are applied in both directions as a variation from the basic size. This is shown in Fig. 4.7e. Note that the distribution of the total tolerances (0.004) need not be even for bilateral tolerances.

The *basic hole system* uses the *minimum hole size* as the dimension from which to calculate the allowance and to establish the tolerances that will produce the desired fit. Thus, in Fig. 4.8a, the basic dimension is 2.000 in. This dimension has been applied to the hole. A *clearance allowance* of 0.003 in. has been subtracted from the

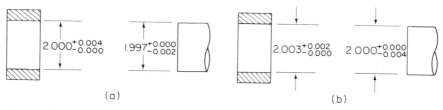

Figure 4.8

basic size to produce a shaft size of 1.997. Tolerances of +0.004 and −0.000 are applied to the hole size. Tolerances of +0.000 and −0.002 are applied to the shaft. An *analysis* of these tolerances shows the minimum and maximum clearance. A plus sign indicates clearance, a minus sign indicates interference.

Minimum clearance		*Maximum clearance*	
Smallest hole	2.000	Largest hole	2.004
Largest shaft	1.997	Smallest hole	1.995
Minimum clearance	+0.003	Maximum clearance	+0.009

Figure 4.8b shows the same allowance and tolerances applied in the *basic shaft system*. In this system the *maximum shaft dimension* is the basic size. If the allowance is 0.003 in., the minimum hole would be 2.003. Tolerances of −0.004 and +0.002 in. are applied to the shaft and the hole, respectively.

Minimum clearance		*Maximum clearance*	
Smallest hole	2.003	Largest hole	2.005
Largest shaft	2.000	Smallest shaft	1.996
Minimum clearance	+0.003	Maximum clearance	+0.009

Basic hole sizes are used more frequently because it is easier to machine the outside diameter of a shaft than the mating bore. Holes are generally machined with solid reamers, thus making it advantageous to hold the hole size constant. However, where several parts, all having the same basic size but requiring different fits, are mounted on one shaft, it is better to use the basic shaft size.

4.6 FITS

Fit is the degree of looseness, or tightness when mating parts are assembled, and it results from the application of allowances and tolerances. There are three general classes of fits: clearance, transition, and interference fits.

Figure 4.9

nominal size range inches	Class RC4		
	Clearance M	Standard tolerance limits	
over-to		Hole	Shaft
1.97–3.15	1.2	+1.8	− 1.2
	4.2	0	− 2.4

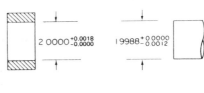

RC—running and sliding fits (Fig. 4.9).   RC1 through RC9 are all sliding or running fits, with the closest fit being RC1 and progressing to the more open running fit RC9. All fits in this class are intended to operate under running performance conditions when suitably lubricated.

Example 4.1

A 2-in. shaft and bushing are to operate with a class 4 running fit. Make a drawing of the bushing and shaft and do an analysis of the allowances.

Solution   Table 4.3 has been extracted from American Association publication USAS B4.1–1967 tables, which are reproduced in Table 4.4. The nominal size is 2 in. The basic hole size is 2.000 in. It should be noted that the tables read in thousandths of an inch. Therefore, numerals that appear in Table 4.4 are written as follows:

1.8   is written as   0.0018
0.7   is written as   0.0007

The use of these numbers from Table 4.3 in Fig. 4.7a is as follows if a 2-in. nominal size and RC_4 is the class of fit:

1. The basic size of 2.0000 in. is found in the *nominal size range* column under 1.97–3.15 in.
2. Move to the left to the column RC_4.
3. The basic hole size 2.0000 is written in Fig. 4.9.
4. From the *standard tolerance limits column* under the heading "hole," apply the tolerances as

$$2.0000 \, {}^{+0.0018}_{-0.0000}$$

5. From the column headed "shaft," apply the clearance as indicated.

$$2.0000 - 0.0012 = 1.9988 \text{ in.}$$

6. Apply the tolerances from the "shaft" column 2.4 − 1.2 = 1.2:

$$1.9988 \, {}^{+0.0000}_{-0.0012}$$

Therefore, in Fig. 4.9 (running or sliding fits) a combination of the *smallest hole* and *largest pin* sizes will produce the *minimum clearance*, whereas the *largest hole* and *smallest pin* sizes will produce the *maximum clearance*. The plus sign indicates clearance.

TABLE 4.3

| Nominal Size Range (in.) over–to | Class RC4 | | |
| | Clearance M | Standard Tolerance Limits | |
		Hole	Shaft
1.97–3.15	1.2	+1.8	−1.2
	4.2	0.0	−2.4

TABLE 4.4

Tolerance limits given in body of table are added or subtracted to basic size (as indicated by + or − sign) to obtain maximum and minimum sizes of mating parts.

Values shown below are in thousandths of an inch

Nominal Size Range, Inches Over — To	Class RC 1			Class RC 2			Class RC 3			Class RC 4		
	Clearance	Standard Tolerance Limits Hole H5	Standard Tolerance Limits Shaft g4	Clearance	Standard Tolerance Limits Hole H6	Standard Tolerance Limits Shaft g5	Clearance	Standard Tolerance Limits Hole H7	Standard Tolerance Limits Shaft f6	Clearance	Standard Tolerance Limits Hole H8	Standard Tolerance Limits Shaft f7
0 – 0.12	0.1 / 0.45	+0.2 / 0	−0.1 / −0.25	0.1 / 0.55	+0.25 / 0	−0.1 / −0.3	0.3 / 0.95	+0.4 / 0	−0.3 / −0.55	0.3 / 1.3	+0.6 / 0	−0.3 / −0.7
0.12 – 0.24	0.15 / 0.5	+0.2 / 0	−0.15 / −0.3	0.15 / 0.65	+0.3 / 0	−0.15 / −0.35	0.4 / 1.12	+0.5 / 0	−0.4 / −0.7	0.4 / 1.6	+0.7 / 0	−0.4 / −0.9
0.24 – 0.40	0.2 / 0.6	+0.25 / 0	−0.2 / −0.35	0.2 / 0.85	+0.4 / 0	−0.2 / −0.45	0.5 / 1.5	+0.6 / 0	−0.5 / −0.9	0.5 / 2.0	+0.9 / 0	−0.5 / −1.1
0.40 – 0.71	0.25 / 0.75	+0.3 / 0	−0.25 / −0.45	0.25 / 0.95	+0.4 / 0	−0.25 / −0.55	0.6 / 1.7	+0.7 / 0	−0.6 / −1.0	0.6 / 2.3	+1.0 / 0	−0.6 / −1.3
0.71 – 1.19	0.3 / 0.95	+0.4 / 0	−0.3 / −0.55	0.3 / 1.2	+0.5 / 0	−0.3 / −0.7	0.8 / 2.1	+0.8 / 0	−0.8 / −1.3	0.8 / 2.8	+1.2 / 0	−0.8 / −1.6
1.19 – 1.97	0.4 / 1.1	+0.4 / 0	−0.4 / −0.7	0.4 / 1.4	+0.6 / 0	−0.4 / −0.8	1.0 / 2.6	+1.0 / 0	−1.0 / −1.6	1.0 / 3.6	+1.6 / 0	−1.0 / −2.0
1.97 – 3.15	0.4 / 1.2	+0.5 / 0	−0.4 / −0.7	0.4 / 1.6	+0.7 / 0	−0.4 / −0.9	1.2 / 3.1	+1.2 / 0	−1.2 / −1.9	1.2 / 4.2	+1.8 / 0	−1.2 / −2.4
3.15 – 4.73	0.5 / 1.5	+0.6 / 0	−0.5 / −0.9	0.5 / 2.0	+0.9 / 0	−0.5 / −1.1	1.4 / 3.7	+1.4 / 0	−1.4 / −2.3	1.4 / 5.0	+2.2 / 0	−1.4 / −2.8
4.73 – 7.09	0.6 / 1.8	+0.7 / 0	−0.6 / −1.1	0.6 / 2.3	+1.0 / 0	−0.6 / −1.3	1.6 / 4.2	+1.6 / 0	−1.6 / −2.6	1.6 / 5.7	+2.5 / 0	−1.6 / −3.2
7.09 – 9.85	0.6 / 2.0	+0.8 / 0	−0.6 / −1.2	0.6 / 2.6	+1.2 / 0	−0.6 / −1.4	2.0 / 5.0	+1.8 / 0	−2.0 / −3.2	2.0 / 6.6	+2.8 / 0	−2.0 / −3.8
9.85 – 12.41	0.8 / 2.3	+0.9 / 0	−0.8 / −1.4	0.8 / 2.9	+1.2 / 0	−0.8 / −1.7	2.5 / 5.7	+2.0 / 0	−2.5 / −3.7	2.5 / 7.5	+3.0 / 0	−2.5 / −4.5
12.41 – 15.75	1.0 / 2.7	+1.0 / 0	−1.0 / −1.7	1.0 / 3.4	+1.4 / 0	−1.0 / −2.0	3.0 / 6.6	+2.2 / 0	−3.0 / −4.4	3.0 / 8.7	+3.5 / 0	−3.0 / −5.2
15.75 – 19.69	1.2 / 3.0	+1.0 / 0	−1.2 / −2.0	1.2 / 3.8	+1.6 / 0	−1.2 / −2.2	4.0 / 8.1	+2.5 / 0	−4.0 / −5.6	4.0 / 10.5	+4.0 / 0	−4.0 / −6.5

See footnotes at end of table.

TABLE 4.4 (Continued)

Nominal Size Range, Inches Over — To	Class RC 5 Clearance*	Class RC 5 Standard Tolerance Limits Hole H8	Class RC 5 Standard Tolerance Limits Shaft e7	Class RC 6 Clearance*	Class RC 6 Standard Tolerance Limits Hole H9	Class RC 6 Standard Tolerance Limits Shaft e8	Class RC 7 Clearance*	Class RC 7 Standard Tolerance Limits Hole H9	Class RC 7 Standard Tolerance Limits Shaft d8	Class RC 8 Clearance*	Class RC 8 Standard Tolerance Limits Hole H10	Class RC 8 Standard Tolerance Limits Shaft c9	Class RC 9 Clearance*	Class RC 9 Standard Tolerance Limits Hole H11	Class RC 9 Standard Tolerance Limits Shaft
					Values shown below are in thousandths of an inch										
0 – 0.12	0.6 / 1.6	+0.6 / 0	−0.6 / −1.0	0.6 / 2.2	+1.0 / 0	−0.6 / −1.2	1.0 / 2.6	+1.0 / 0	−1.0 / −1.6	2.5 / 5.1	+1.6 / 0	−2.5 / −3.5	4.0 / 8.1	+2.5 / 0	−4.0 / −5.6
0.12 – 0.24	0.8 / 2.0	+0.7 / 0	−0.8 / −1.3	0.8 / 2.7	+1.2 / 0	−0.8 / −1.5	1.2 / 3.1	+1.2 / 0	−1.2 / −1.9	2.8 / 5.8	+1.8 / 0	−2.8 / −4.0	4.5 / 9.0	+3.0 / 0	−4.5 / −6.0
0.24 – 0.40	1.0 / 2.5	+0.9 / 0	−1.0 / −1.6	1.0 / 3.3	+1.4 / 0	−1.0 / −1.9	1.6 / 3.9	+1.4 / 0	−1.6 / −2.5	3.0 / 6.6	+2.2 / 0	−3.0 / −4.4	5.0 / 10.7	+3.5 / 0	−5.0 / −7.2
0.40 – 0.71	1.2 / 2.9	+1.0 / 0	−1.2 / −1.9	1.2 / 3.8	+1.6 / 0	−1.2 / −2.2	2.0 / 4.6	+1.6 / 0	−2.0 / −3.0	3.5 / 7.9	+2.8 / 0	−3.5 / −5.1	6.0 / 12.8	+4.0 / 0	−6.0 / −8.8
0.71 – 1.19	1.6 / 3.6	+1.2 / 0	−1.6 / −2.4	1.6 / 4.8	+2.0 / 0	−1.6 / −2.8	2.5 / 5.7	+2.0 / 0	−2.5 / −3.7	4.5 / 10.0	+3.5 / 0	−4.5 / −6.5	7.0 / 15.5	+5.0 / 0	−7.0 / −10.5
1.19 – 1.97	2.0 / 4.6	+1.6 / 0	−2.0 / −3.0	2.0 / 6.1	+2.5 / 0	−2.0 / −3.6	3.0 / 7.1	+2.5 / 0	−3.0 / −4.6	5.0 / 11.5	+4.0 / 0	−5.0 / −7.5	8.0 / 18.0	+6.0 / 0	−8.0 / −12.0
1.97 – 3.15	2.5 / 5.5	+1.8 / 0	−2.5 / −3.7	2.5 / 7.3	+3.0 / 0	−2.5 / −4.3	4.0 / 8.8	+3.0 / 0	−4.0 / −5.8	6.0 / 13.5	+4.5 / 0	−6.0 / −9.0	9.0 / 20.5	+7.0 / 0	−9.0 / −13.5
3.15 – 4.73	3.0 / 6.6	+2.2 / 0	−3.0 / −4.4	3.0 / 8.7	+3.5 / 0	−3.0 / −5.2	5.0 / 10.7	+3.5 / 0	−5.0 / −7.2	7.0 / 15.5	+5.0 / 0	−7.0 / −10.5	10.0 / 24.0	+9.0 / 0	−10.0 / −15.0
4.73 – 7.09	3.5 / 7.6	+2.5 / 0	−3.5 / −5.1	3.5 / 10.0	+4.0 / 0	−3.5 / −6.0	6.0 / 12.5	+4.0 / 0	−6.0 / −8.5	8.0 / 18.0	+6.0 / 0	−8.0 / −12.0	12.0 / 28.0	+10.0 / 0	−12.0 / −18.0
7.09 – 9.85	4.0 / 8.6	+2.8 / 0	−4.0 / −5.8	4.0 / 11.3	+4.5 / 0	−4.0 / −6.8	7.0 / 14.3	+4.5 / 0	−7.0 / −9.8	10.0 / 21.5	+7.0 / 0	−10.0 / −14.5	15.0 / 34.0	+12.0 / 0	−15.0 / −22.0
9.85 – 12.41	5.0 / 10.0	+3.0 / 0	−5.0 / −7.0	5.0 / 13.0	+5.0 / 0	−5.0 / −8.0	8.0 / 16.0	+5.0 / 0	−8.0 / −11.0	12.0 / 25.0	+8.0 / 0	−12.0 / −17.0	18.0 / 38.0	+12.0 / 0	−18.0 / −26.0
12.41 – 15.75	6.0 / 11.7	+3.5 / 0	−6.0 / −8.2	6.0 / 15.5	+6.0 / 0	−6.0 / −9.5	10.0 / 19.5	+6.0 / 0	−10.0 / −13.5	14.0 / 29.0	+9.0 / 0	−14.0 / −20.0	22.0 / 45.0	+14.0 / 0	−22.0 / −31.0
15.75 – 19.69	8.0 / 14.5	+4.0 / 0	−8.0 / −10.5	8.0 / 18.0	+6.0 / 0	−8.0 / −12.0	12.0 / 22.0	+6.0 / 0	−12.0 / −16.0	16.0 / 32.0	+10.0 / 0	−16.0 / −22.0	25.0 / 51.0	+16.0 / 0	−25.0 / −35.0

All data above heavy lines are in accord with ABC agreements. Symbols H5, g4, etc. are hole and shaft designations in ABC system. Limits for sizes above 19.69 inches are also given in the USA Standard.

* Pairs of values shown represent minimum and maximum amounts of clearance resulting from application of standard tolerance limits.

TABLE 4.4 *(Continued)*

Tolerance limits given in body of table are added or subtracted to basic size (as indicated by + or − sign) to obtain maximum and minimum sizes of mating parts.

Values shown below are in thousandths of an inch

Nominal Size Range, Inches Over – To	Class LC 1			Class LC 2			Class LC 3			Class LC 4			Class LC 5		
	Clearance*	Standard Tolerance Limits Hole H6	Shaft h5	Clearance*	Standard Tolerance Limits Hole H7	Shaft h6	Clearance*	Standard Tolerance Limits Hole H8	Shaft h7	Clearance*	Standard Tolerance Limits Hole H10	Shaft h9	Clearance*	Standard Tolerance Limits Hole H7	Shaft g6
0 – 0.12	0 / 0.45	+0.25 / 0	0 / −0.2	0 / 0.65	+0.4 / 0	0 / −0.25	0 / 1	+0.6 / 0	0 / −0.4	0 / 2.6	+1.6 / 0	0 / −1.0	0.1 / 0.75	+0.4 / 0	−0.1 / −0.35
0.12 – 0.24	0 / 0.5	+0.3 / 0	0 / −0.2	0 / 0.8	+0.5 / 0	0 / −0.3	0 / 1.2	+0.7 / 0	0 / −0.5	0 / 3.0	+1.8 / 0	0 / −1.2	0.15 / 0.95	+0.5 / 0	−0.15 / −0.45
0.24 – 0.40	0 / 0.65	+0.4 / 0	0 / −0.25	0 / 1.0	+0.6 / 0	0 / −0.4	0 / 1.5	+0.9 / 0	0 / −0.6	0 / 3.6	+2.2 / 0	0 / −1.4	0.2 / 1.2	+0.6 / 0	−0.2 / −0.6
0.40 – 0.71	0 / 0.7	+0.4 / 0	0 / −0.3	0 / 1.1	+0.7 / 0	0 / −0.4	0 / 1.7	+1.0 / 0	0 / −0.7	0 / 4.4	+2.8 / 0	0 / −1.6	0.25 / 1.35	+0.7 / 0	−0.25 / −0.65
0.71 – 1.19	0 / 0.9	+0.5 / 0	0 / −0.4	0 / 1.3	+0.8 / 0	0 / −0.5	0 / 2	+1.2 / 0	0 / −0.8	0 / 5.5	+3.5 / 0	0 / −2.0	0.3 / 1.6	+0.8 / 0	−0.3 / −0.8
1.19 – 1.97	0 / 1.0	+0.6 / 0	0 / −0.4	0 / 1.6	+1.0 / 0	0 / −0.6	0 / 2.6	+1.6 / 0	0 / −1	0 / 6.5	+4.0 / 0	0 / −2.5	0.4 / 2.0	+1.0 / 0	−0.4 / −1.0
1.97 – 3.15	0 / 1.2	+0.7 / 0	0 / −0.5	0 / 1.9	+1.2 / 0	0 / −0.7	0 / 3	+1.8 / 0	0 / −1.2	0 / 7.5	+4.5 / 0	0 / −3	0.4 / 2.3	+1.2 / 0	−0.4 / −1.1
3.15 – 4.73	0 / 1.5	+0.9 / 0	0 / −0.6	0 / 2.3	+1.4 / 0	0 / −0.9	0 / 3.6	+2.2 / 0	0 / −1.4	0 / 8.5	+5.0 / 0	0 / −3.5	0.5 / 2.8	+1.4 / 0	−0.5 / −1.4
4.73 – 7.09	0 / 1.7	+1.0 / 0	0 / −0.7	0 / 2.6	+1.6 / 0	0 / −1.0	0 / 4.1	+2.5 / 0	0 / −1.6	0 / 10.0	+6.0 / 0	0 / −4	0.6 / 3.2	+1.6 / 0	−0.6 / −1.6
7.09 – 9.85	0 / 2.0	+1.2 / 0	0 / −0.8	0 / 3.0	+1.8 / 0	0 / −1.2	0 / 4.6	+2.8 / 0	0 / −1.8	0 / 11.5	+7.0 / 0	0 / −4.5	0.6 / 3.6	+1.8 / 0	−0.6 / −1.8
9.85 – 12.41	0 / 2.1	+1.2 / 0	0 / −0.9	0 / 3.2	+2.0 / 0	0 / −1.2	0 / 5	+3.0 / 0	0 / −2.0	0 / 13.0	+8.0 / 0	0 / −5	0.7 / 3.9	+2.0 / 0	−0.7 / −1.9
12.41 – 15.75	0 / 2.4	+1.4 / 0	0 / −1.0	0 / 3.6	+2.2 / 0	0 / −1.4	0 / 5.7	+3.5 / 0	0 / −2.2	0 / 15.0	+9.0 / 0	0 / −6	0.7 / 4.3	+2.2 / 0	−0.7 / −2.1
15.75 – 19.69	0 / 2.6	+1.6 / 0	0 / −1.0	0 / 4.1	+2.5 / 0	0 / −1.6	0 / 6.5	+4 / 0	0 / −2.5	0 / 16.0	+10.0 / 0	0 / −6	0.8 / 4.9	+2.5 / 0	−0.8 / −2.4

See footnotes at end of table.

TABLE 4.4 (*Continued*)

Values shown below are in thousandths of an inch

Nominal Size Range, Inches Over	To	Class LC 6 Clearance*	Class LC 6 Hole H9	Class LC 6 Shaft f8	Class LC 7 Clearance*	Class LC 7 Hole H10	Class LC 7 Shaft e9	Class LC 8 Clearance*	Class LC 8 Hole H10	Class LC 8 Shaft d9	Class LC 9 Clearance*	Class LC 9 Hole H11	Class LC 9 Shaft c10	Class LC 10 Clearance*	Class LC 10 Hole H12	Class LC 10 Shaft	Class LC 11 Clearance*	Class LC 11 Hole H13	Class LC 11 Shaft
0–	0.12	0.3 / 1.9	+1.0 / 0	−0.3 / −0.9	0.6 / 3.2	+1.6 / 0	−0.6 / −1.6	1.0 / 3.6	+1.6 / 0	−1.0 / −2.0	2.5 / 6.6	+2.5 / 0	−2.5 / −4.1	4 / 12	+4 / 0	−4 / −8	5 / 17	+6 / 0	−5 / −11
0.12–	0.24	0.4 / 2.3	+1.2 / 0	−0.4 / −1.1	0.8 / 3.8	+1.8 / 0	−0.8 / −2.0	1.2 / 4.2	+1.8 / 0	−1.2 / −2.4	2.8 / 7.6	+3.0 / 0	−2.8 / −4.6	4.5 / 14.5	+5 / 0	−4.5 / −9.5	6 / 20	+7 / 0	−6 / −13
0.24–	0.40	0.5 / 2.8	+1.4 / 0	−0.5 / −1.4	1.0 / 4.6	+2.2 / 0	−1.0 / −2.4	1.6 / 5.2	+2.2 / 0	−1.6 / −3.0	3.0 / 8.7	+3.5 / 0	−3.0 / −5.2	5 / 17	+6 / 0	−5 / −11	7 / 25	+9 / 0	−7 / −16
0.40–	0.71	0.6 / 3.2	+1.6 / 0	−0.6 / −1.6	1.2 / 5.6	+2.8 / 0	−1.2 / −2.8	2.0 / 6.4	+2.8 / 0	−2.0 / −3.6	3.5 / 10.3	+4.0 / 0	−3.5 / −6.3	6 / 20	+7 / 0	−6 / −13	8 / 28	+10 / 0	−8 / −18
0.71–	1.19	0.8 / 4.0	+2.0 / 0	−0.8 / −2.0	1.6 / 7.1	+3.5 / 0	−1.6 / −3.6	2.5 / 8.0	+3.5 / 0	−2.5 / −4.5	4.5 / 13.0	+5.0 / 0	−4.5 / −8.0	7 / 23	+8 / 0	−7 / −15	10 / 34	+12 / 0	−10 / −22
1.19–	1.97	1.0 / 5.1	+2.5 / 0	−1.0 / −2.6	2.0 / 8.5	+4.0 / 0	−2.0 / −4.5	3.0 / 9.5	+4.0 / 0	−3.0 / −5.5	5.0 / 15.0	+6 / 0	−5.0 / −9.0	8 / 28	+10 / 0	−8 / −18	12 / 44	+16 / 0	−12 / −28
1.97–	3.15	1.2 / 6.0	+3.0 / 0	−1.0 / −3.0	2.5 / 10.0	+4.5 / 0	−2.5 / −5.5	4.0 / 11.5	+4.5 / 0	−4.0 / −7.0	6.0 / 17.5	+7 / 0	−6.0 / −10.5	10 / 34	+12 / 0	−10 / −22	14 / 50	+18 / 0	−14 / −32
3.15–	4.73	1.4 / 7.1	+3.5 / 0	−1.4 / −3.6	3.0 / 11.5	+5.0 / 0	−3.0 / −6.5	5.0 / 13.5	+5.0 / 0	−5.0 / −8.5	7 / 21	+9 / 0	−7 / −12	11 / 39	+14 / 0	−11 / −25	16 / 60	+22 / 0	−16 / −38
4.73–	7.09	1.6 / 8.1	+4.0 / 0	−1.6 / −4.1	3.5 / 13.5	+6.0 / 0	−3.5 / −7.5	6 / 16	+6 / 0	−6 / −10	8 / 24	+10 / 0	−8 / −14	12 / 44	+16 / 0	−12 / −28	18 / 68	+25 / 0	−18 / −43
7.09–	9.85	2.0 / 9.3	+4.5 / 0	−2.0 / −4.8	4.0 / 15.5	+7.0 / 0	−4.0 / −8.5	7 / 18.5	+7 / 0	−7 / −11.5	10 / 29	+12 / 0	−10 / −17	16 / 52	+18 / 0	−16 / −34	22 / 78	+28 / 0	−22 / −50
9.85–	12.41	2.2 / 10.2	+5.0 / 0	−2.2 / −5.2	4.5 / 17.5	+8.0 / 0	−4.5 / −9.5	7 / 20	+8 / 0	−7 / −12	12 / 32	+12 / 0	−12 / −20	20 / 60	+20 / 0	−20 / −40	28 / 88	+30 / 0	−28 / −58
12.41–	15.75	2.5 / 12.0	+6.0 / 0	−2.5 / −6.0	5.0 / 20.0	+9.0 / 0	−5 / −11	8 / 23	+9 / 0	−8 / −14	14 / 37	+14 / 0	−14 / −23	22 / 66	+22 / 0	−22 / −44	30 / 100	+35 / 0	−30 / −65
15.75–	19.69	2.8 / 12.8	+6.0 / 0	−2.8 / −6.8	5.0 / 21.0	+10.0 / 0	−5 / −11	9 / 25	+10 / 0	−9 / −15	16 / 42	+16 / 0	−16 / −26	25 / 75	+25 / 0	−25 / −50	35 / 115	+40 / 0	−35 / −75

All data above heavy lines are in accordance with American-British-Canadian (ABC) agreements. Symbols H6, H7, s6, etc. are hole and shaft designations in ABC system. Limits for sizes above 19.69 inches are not covered by ABC agreements but are given in the USA standard.

* Pairs of values shown represent minimum and maximum amounts of interference resulting from application of standard tolerance limits.

TABLE 4.4 (Continued)

Values shown below are in thousandths of an inch

Nominal Size Range, Inches Over	To	Class LT 1 Fit*	Class LT 1 Hole H7	Class LT 1 Shaft js6	Class LT 2 Fit*	Class LT 2 Hole H8	Class LT 2 Shaft js7	Class LT 3 Fit*	Class LT 3 Hole H7	Class LT 3 Shaft k6	Class LT 4 Fit*	Class LT 4 Hole H8	Class LT 4 Shaft k7	Class LT 5 Fit*	Class LT 5 Hole H7	Class LT 5 Shaft n6	Class LT 6 Fit*	Class LT 6 Hole H7	Class LT 6 Shaft n7
0	0.12	−0.12 / +0.52	+0.4 / 0	+0.12 / −0.12	−0.2 / +0.8	+0.6 / 0	+0.2 / −0.2							−0.5 / +0.15	+0.4 / 0	+0.5 / +0.25	−0.65 / +0.15	+0.4 / 0	+0.65 / +0.25
0.12	0.24	−0.15 / +0.65	+0.5 / 0	+0.15 / −0.15	−0.25 / +0.95	+0.7 / 0	+0.25 / −0.25							−0.6 / +0.2	+0.5 / 0	+0.6 / +0.3	−0.8 / +0.2	+0.5 / 0	+0.8 / +0.3
0.24	0.40	−0.2 / +0.8	+0.6 / 0	+0.2 / −0.2	−0.3 / +1.2	+0.9 / 0	+0.3 / −0.3	−0.5 / +0.5	+0.6 / 0	+0.5 / +0.1	−0.7 / +0.8	+0.9 / 0	+0.7 / +0.1	−0.8 / +0.2	+0.6 / 0	+0.8 / +0.4	−1.0 / +0.2	+0.6 / 0	+1.0 / +0.4
0.40	0.71	−0.2 / +0.9	+0.7 / 0	+0.2 / −0.2	−0.35 / +1.35	+1.0 / 0	+0.35 / −0.35	−0.5 / +0.6	+0.7 / 0	+0.5 / +0.1	−0.8 / +0.9	+1.0 / 0	+0.8 / +0.1	−0.9 / +0.2	+0.7 / 0	+0.9 / +0.5	−1.2 / +0.2	+0.7 / 0	+1.2 / +0.5
0.71	1.19	−0.25 / +1.05	+0.8 / 0	+0.25 / −0.25	−0.4 / +1.6	+1.2 / 0	+0.4 / −0.4	−0.6 / +0.7	+0.8 / 0	+0.6 / +0.1	−0.9 / +1.1	+1.2 / 0	+0.9 / +0.1	−1.1 / +0.2	+0.8 / 0	+1.1 / +0.6	−1.4 / +0.2	+0.8 / 0	+1.4 / +0.6
1.19	1.97	−0.3 / +1.3	+1.0 / 0	+0.3 / −0.3	−0.5 / +2.1	+1.6 / 0	+0.5 / −0.5	−0.7 / +0.9	+1.0 / 0	+0.7 / +0.1	−1.1 / +1.5	+1.6 / 0	+1.1 / +0.1	−1.3 / +0.3	+1.0 / 0	+1.3 / +0.7	−1.7 / +0.3	+1.0 / 0	+1.7 / +0.7
1.97	3.15	−0.3 / +1.5	+1.2 / 0	+0.3 / −0.3	−0.6 / +2.4	+1.8 / 0	+0.6 / −0.6	−0.8 / +1.1	+1.2 / 0	+0.8 / +0.1	−1.3 / +1.7	+1.8 / 0	+1.3 / +0.1	−1.5 / +0.4	+1.2 / 0	+1.5 / +0.8	−2.0 / +0.4	+1.2 / 0	+2.0 / +0.8
3.15	4.73	−0.4 / +1.8	+1.4 / 0	+0.4 / −0.4	−0.7 / +2.9	+2.2 / 0	+0.7 / −0.7	−1.0 / +1.3	+1.4 / 0	+1.0 / +0.1	−1.5 / +2.1	+2.2 / 0	+1.5 / +0.1	−1.9 / +0.4	+1.4 / 0	+1.9 / +1.0	−2.4 / +0.4	+1.4 / 0	+2.4 / +1.0
4.73	7.09	−0.5 / +2.1	+1.6 / 0	+0.5 / −0.5	−0.8 / +3.3	+2.5 / 0	+0.8 / −0.8	−1.1 / +1.5	+1.6 / 0	+1.1 / +0.1	−1.7 / +2.4	+2.5 / 0	+1.7 / +0.1	−2.2 / +0.4	+1.6 / 0	+2.2 / +1.2	−2.8 / +0.4	+1.6 / 0	+2.8 / +1.2
7.09	9.85	−0.6 / +2.4	+1.8 / 0	+0.6 / −0.6	−0.9 / +3.7	+2.8 / 0	+0.9 / −0.9	−1.4 / +1.6	+1.8 / 0	+1.4 / +0.2	−2.0 / +2.6	+2.8 / 0	+2.0 / +0.2	−2.6 / +0.4	+1.8 / 0	+2.6 / +1.4	−3.2 / +0.4	+1.8 / 0	+3.2 / +1.4
9.85	12.41	−0.6 / +2.6	+2.0 / 0	+0.6 / −0.6	−1.0 / +4.0	+3.0 / 0	+1.0 / −1.0	−1.4 / +1.8	+2.0 / 0	+1.4 / +0.2	−2.2 / +2.8	+3.0 / 0	+2.2 / +0.2	−2.6 / +0.6	+2.0 / 0	+2.6 / +1.4	−3.4 / +0.6	+2.0 / 0	+3.4 / +1.4
12.41	15.75	−0.7 / +2.9	+2.2 / 0	+0.7 / −0.7	−1.0 / +4.5	+3.5 / 0	+1.0 / −1.0	−1.6 / +2.0	+2.2 / 0	+1.6 / +0.2	−2.4 / +3.3	+3.5 / 0	+2.4 / +0.2	−3.0 / +0.6	+2.2 / 0	+3.0 / +1.6	−3.8 / +0.6	+2.2 / 0	+3.8 / +1.6
15.75	19.69	−0.8 / +3.3	+2.5 / 0	+0.8 / −0.8	−1.2 / +5.2	+4.0 / 0	+1.2 / −1.2	−1.8 / +2.3	+2.5 / 0	+1.8 / +0.2	−2.7 / +3.8	+4.0 / 0	+2.7 / +0.2	−3.4 / +0.7	+2.5 / 0	+3.4 / +1.8	−4.3 / +0.7	+2.5 / 0	+4.3 / +1.8

All data above heavy lines are in accord with ABC agreements. Symbols H7, js6, etc. are hole and shaft designations in ABC system.

* Pairs of values shown represent maximum amount of interference (−) and maximum amount of clearance (+) resulting from application of standard tolerance limits.

Tolerance limits given in body of table are added or subtracted to basic size (as indicated by + or − sign) to obtain maximum and minimum sizes of mating parts.

Nominal Size Range, Inches Over To	Class LN 1			Class LN 2			Class LN 3		
	Limits of Interference	Standard Limits		Limits of Interference	Standard Limits		Limits of Interference	Standard Limits	
		Hole H6	Shaft n5		Hole H7	Shaft p6		Hole H7	Shaft r6
	Values shown below are given in thousandths of an inch								
0 – 0.12	0 / 0.45	+0.25 / 0	+0.45 / +0.25	0 / 0.65	+0.4 / 0	+0.65 / +0.4	0.1 / 0.75	+0.4 / 0	+0.75 / +0.5
0.12– 0.24	0 / 0.5	+0.3 / 0	+0.5 / +0.3	0 / 0.8	+0.5 / 0	+0.8 / +0.5	0.1 / 0.9	+0.5 / 0	+0.9 / +0.6
0.24– 0.40	0 / 0.65	+0.4 / 0	+0.65 / +0.4	0 / 1.0	+0.6 / 0	+1.0 / +0.6	0.2 / 1.2	+0.6 / 0	+1.2 / +0.8
0.40– 0.71	0 / 0.8	+0.4 / 0	+0.8 / +0.4	0 / 1.1	+0.7 / 0	+1.1 / +0.7	0.3 / 1.4	+0.7 / 0	+1.4 / +1.0
0.71– 1.19	0 / 1.0	+0.5 / 0	+1.0 / +0.5	0 / 1.3	+0.8 / 0	+1.3 / +0.8	0.4 / 1.7	+0.8 / 0	+1.7 / +1.2
1.19– 1.97	0 / 1.1	+0.6 / 0	+1.1 / +0.6	0 / 1.6	+1.0 / 0	+1.6 / +1.0	0.4 / 2.0	+1.0 / 0	+2.0 / +1.4
1.97– 3.15	0.1 / 1.3	+0.7 / 0	+1.3 / +0.7	0.2 / 2.1	+1.2 / 0	+2.1 / +1.4	0.4 / 2.3	+1.2 / 0	+2.3 / +1.6
3.15– 4.73	0.1 / 1.6	+0.9 / 0	+1.6 / +1.0	0.2 / 2.5	+1.4 / 0	+2.5 / +1.6	0.6 / 2.9	+1.4 / 0	+2.9 / +2.0
4.73– 7.09	0.2 / 1.9	+1.0 / 0	+1.9 / +1.2	0.2 / 2.8	+1.6 / 0	+2.8 / +1.8	0.9 / 3.5	+1.6 / 0	+3.5 / +2.5
7.09– 9.85	0.2 / 2.2	+1.2 / 0	+2.2 / +1.4	0.2 / 3.2	+1.8 / 0	+3.2 / +2.0	1.2 / 4.2	+1.8 / 0	+4.2 / +3.0
9.85–12.41	0.2 / 2.3	+1.2 / 0	+2.3 / +1.4	0.2 / 3.4	+2.0 / 0	+3.4 / +2.2	1.5 / 4.7	+2.0 / 0	+4.7 / +3.5
12.41–15.75	0.2 / 2.6	+1.4 / 0	+2.6 / +1.6	0.3 / 3.9	+2.2 / 0	+3.9 / +2.5	2.3 / 5.9	+2.2 / 0	+5.9 / +4.5
15.75–19.69	0.2 / 2.8	+1.6 / 0	+2.8 / +1.8	0.3 / 4.4	+2.5 / 0	+4.4 / +2.8	2.5 / 6.6	+2.5 / 0	+6.6 / +5.0

All data in this table are in accordance with American-British-Canadian (ABC) agreements.

Limits for sizes above 19.69 inches are not covered by ABC agreements but are given in the USA Standard.

Symbols H7, p6, etc. are hole and shaft designations in ABC system.

*Pairs of values shown represent minimum and maximum amounts of interference resulting from application of standard tolerance limits.

TABLE 4.4 (Continued)

Values shown below are in thousandths of an inch

Nominal Size Range, Inches (Over–To)	Class FN 1			Class FN 2			Class FN 3			Class FN 4			Class FN 5		
	Interference	Hole H6	Shaft	Interference	Hole H7	Shaft s6	Interference	Hole H7	Shaft t6	Interference	Hole H7	Shaft u6	Interference	Hole H8	Shaft x7
0–0.12	0.05 / 0.5	+0.25 / 0	+0.5 / +0.3	0.2 / 0.85	+0.4 / 0	+0.85 / +0.6				0.3 / 0.95	+0.4 / 0	+0.95 / +0.7	0.3 / 1.3	+0.6 / 0	+1.3 / +0.9
0.12–0.24	0.1 / 0.6	+0.3 / 0	+0.6 / +0.4	0.2 / 1.0	+0.5 / 0	+1.0 / +0.7				0.4 / 1.2	+0.5 / 0	+1.2 / +0.9	0.5 / 1.7	+0.7 / 0	+1.7 / +1.2
0.24–0.40	0.1 / 0.75	+0.4 / 0	+0.75 / +0.5	0.4 / 1.4	+0.6 / 0	+1.4 / +1.0				0.6 / 1.6	+0.6 / 0	+1.6 / +1.2	0.5 / 2.0	+0.9 / 0	+2.0 / +1.4
0.40–0.56	0.1 / 0.8	+0.4 / 0	+0.8 / +0.5	0.5 / 1.6	+0.7 / 0	+1.6 / +1.2				0.7 / 1.8	+0.7 / 0	+1.8 / +1.4	0.6 / 2.3	+1.0 / 0	+2.3 / +1.6
0.56–0.71	0.2 / 0.9	+0.4 / 0	+0.9 / +0.6	0.5 / 1.6	+0.7 / 0	+1.6 / +1.2				0.7 / 1.8	+0.7 / 0	+1.8 / +1.4	0.8 / 2.5	+1.0 / 0	+2.5 / +1.8
0.71–0.95	0.2 / 1.1	+0.5 / 0	+1.1 / +0.7	0.6 / 1.9	+0.8 / 0	+1.9 / +1.4				0.8 / 2.1	+0.8 / 0	+2.1 / +1.6	1.0 / 3.0	+1.2 / 0	+3.0 / +2.2
0.95–1.19	0.3 / 1.2	+0.5 / 0	+1.2 / +0.8	0.6 / 1.9	+0.8 / 0	+1.9 / +1.4	0.8 / 2.1	+0.8 / 0	+2.1 / +1.6	1.0 / 2.3	+0.8 / 0	+2.3 / +1.8	1.3 / 3.3	+1.2 / 0	+3.3 / +2.5
1.19–1.58	0.3 / 1.3	+0.6 / 0	+1.3 / +0.9	0.8 / 2.4	+1.0 / 0	+2.4 / +1.8	1.0 / 2.6	+1.0 / 0	+2.6 / +2.0	1.5 / 3.1	+1.0 / 0	+3.1 / +2.5	1.4 / 4.0	+1.6 / 0	+4.0 / +3.0
1.58–1.97	0.4 / 1.4	+0.6 / 0	+1.4 / +1.0	0.8 / 2.4	+1.0 / 0	+2.4 / +1.8	1.2 / 2.8	+1.0 / 0	+2.8 / +2.2	1.8 / 3.4	+1.0 / 0	+3.4 / +2.8	2.4 / 5.0	+1.6 / 0	+5.0 / +4.0
1.97–2.56	0.6 / 1.8	+0.7 / 0	+1.8 / +1.3	0.8 / 2.7	+1.2 / 0	+2.7 / +2.0	1.3 / 3.2	+1.2 / 0	+3.2 / +2.5	2.3 / 4.2	+1.2 / 0	+4.2 / +3.5	3.2 / 6.2	+1.8 / 0	+6.2 / +5.0
2.56–3.15	0.7 / 1.9	+0.7 / 0	+1.9 / +1.4	1.0 / 2.9	+1.2 / 0	+2.9 / +2.2	1.8 / 3.7	+1.2 / 0	+3.7 / +3.0	2.8 / 4.7	+1.2 / 0	+4.7 / +4.0	4.2 / 7.2	+1.8 / 0	+7.2 / +6.0
3.15–3.94	0.9 / 2.4	+0.9 / 0	+2.4 / +1.8	1.4 / 3.7	+1.4 / 0	+3.7 / +2.8	2.1 / 4.4	+1.4 / 0	+4.4 / +3.5	3.6 / 5.9	+1.4 / 0	+5.9 / +5.0	4.8 / 8.4	+2.2 / 0	+8.4 / +7.0
3.94–4.73	1.1 / 2.6	+0.9 / 0	+2.6 / +2.0	1.6 / 3.9	+1.4 / 0	+3.9 / +3.0	2.6 / 4.9	+1.4 / 0	+4.9 / +4.0	4.6 / 6.9	+1.4 / 0	+6.9 / +6.0	5.8 / 9.4	+2.2 / 0	+9.4 / +8.0

See footnotes at end of table.

TABLE 4.4 (Continued)

Values shown below are in thousandths of an inch

Nominal Size Range, Inches (Over–To)	Class FN 1 Interference*	Class FN 1 Hole H6	Class FN 1 Shaft	Class FN 2 Interference*	Class FN 2 Hole H7	Class FN 2 Shaft s6	Class FN 3 Interference*	Class FN 3 Hole H7	Class FN 3 Shaft t6	Class FN 4 Interference*	Class FN 4 Hole H7	Class FN 4 Shaft u6	Class FN 5 Interference*	Class FN 5 Hole H8	Class FN 5 Shaft x7
4.73–5.52	1.2 / 2.9	+1.0 / 0	+2.9 / +2.2	1.9 / 4.5	+1.6 / 0	+4.5 / +3.5	3.4 / 6.0	+1.6 / 0	+6.0 / +5.0	5.4 / 8.0	+1.6 / 0	+8.0 / +7.0	7.5 / 11.6	+2.5 / 0	+11.6 / +10.0
5.52–6.30	1.5 / 3.2	+1.0 / 0	+3.2 / +2.5	2.4 / 5.0	+1.6 / 0	+5.0 / +4.0	3.4 / 6.0	+1.6 / 0	+6.0 / +5.0	5.4 / 8.0	+1.6 / 0	+8.0 / +7.0	9.5 / 13.6	+2.5 / 0	+13.6 / +12.0
6.30–7.09	1.8 / 3.5	+1.0 / 0	+3.5 / +2.8	2.9 / 5.5	+1.6 / 0	+5.5 / +4.5	4.4 / 7.0	+1.6 / 0	+7.0 / +6.0	6.4 / 9.0	+1.6 / 0	+9.0 / +8.0	9.5 / 13.6	+2.5 / 0	+13.6 / +12.0
7.09–7.88	1.8 / 3.8	+1.2 / 0	+3.8 / +3.0	3.2 / 6.2	+1.8 / 0	+6.2 / +5.0	5.2 / 8.2	+1.8 / 0	+8.2 / +7.0	7.2 / 10.2	+1.8 / 0	+10.2 / +9.0	11.2 / 15.8	+2.8 / 0	+15.8 / +14.0
7.88–8.86	2.3 / 4.3	+1.2 / 0	+4.3 / +3.5	3.2 / 6.2	+1.8 / 0	+6.2 / +5.0	5.2 / 8.2	+1.8 / 0	+8.2 / +7.0	8.2 / 11.2	+1.8 / 0	+11.2 / +10.0	13.2 / 17.8	+2.8 / 0	+17.8 / +16.0
8.86–9.85	2.3 / 4.3	+1.2 / 0	+4.3 / +3.5	4.2 / 7.2	+1.8 / 0	+7.2 / +6.0	6.2 / 9.2	+1.8 / 0	+9.2 / +8.0	10.2 / 13.2	+1.8 / 0	+13.2 / +12.0	13.2 / 17.8	+2.8 / 0	+17.8 / +16.0
9.85–11.03	2.8 / 4.9	+1.2 / 0	+4.9 / +4.0	4.0 / 7.2	+2.0 / 0	+7.2 / +6.0	7.0 / 10.2	+2.0 / 0	+10.2 / +9.0	10.0 / 13.2	+2.0 / 0	+13.2 / +12.0	15.0 / 20.0	+3.0 / 0	+20.0 / +18.0
11.03–12.41	2.8 / 4.9	+1.2 / 0	+4.9 / +4.0	5.0 / 8.2	+2.0 / 0	+8.2 / +7.0	7.0 / 10.2	+2.0 / 0	+10.2 / +9.0	12.0 / 15.2	+2.0 / 0	+15.2 / +14.0	17.0 / 22.0	+3.0 / 0	+22.0 / +20.0
12.41–13.98	3.1 / 5.5	+1.4 / 0	+5.5 / +4.5	5.8 / 9.4	+2.2 / 0	+9.4 / +8.0	7.8 / 11.4	+2.2 / 0	+11.4 / +10.0	13.8 / 17.4	+2.2 / 0	+17.4 / +16.0	18.5 / 24.2	+3.5 / 0	+24.2 / +22.0
13.98–15.75	3.6 / 6.1	+1.4 / 0	+6.1 / +5.0	5.8 / 9.4	+2.2 / 0	+9.4 / +8.0	9.8 / 13.4	+2.2 / 0	+13.4 / +12.0	15.8 / 19.4	+2.2 / 0	+19.4 / +18.0	21.5 / 27.2	+3.5 / 0	+27.2 / +25.0
15.75–17.72	4.4 / 7.0	+1.6 / 0	+7.0 / +6.0	6.5 / 10.6	+2.5 / 0	+10.6 / +9.0	9.5 / 13.6	+2.5 / 0	+13.6 / +12.0	17.5 / 21.6	+2.5 / 0	+21.6 / +20.0	24.0 / 30.5	+4.0 / 0	+30.5 / +28.0
17.72–19.69	4.4 / 7.0	+1.6 / 0	+7.0 / +6.0	7.5 / 11.6	+2.5 / 0	+11.6 / +10.0	11.5 / 15.6	+2.5 / 0	+15.6 / +14.0	19.5 / 23.6	+2.5 / 0	+23.6 / +22.0	26.0 / 32.5	+4.0 / 0	+32.5 / +30.0

All data above heavy lines are in accordance with American-British-Canadian (ABC) agreements. Symbols H6, H7, s6, etc. are hole and shaft designations in ABC system. Limits for sizes above 19.69 inches are not covered by ABC agreements but are given in the USA standard.
* Pairs of values shown represent minimum and maximum amounts of interference resulting from application of standard tolerance limits.

Minimum clearance		*Maximum clearance*	
Smallest hole	2.0000	Largest hole	2.0018
Largest pin	1.9988	Smallest pin	1.9976
Minimum clearance	+0.0012	Maximum clearance	+0.0042

L-location fits.   These are used primarily for stationary assemblies. *Clearance fits* (LC) are those which have their tolerances applied to the hole and shaft sizes in such a way that any combination of mating sizes produces a clearance allowance between the two parts. They are intended for stationary assemblies which can be freely assembled or disassembled.

Example 4.2

A 2-in. bushing and locator pin are to be assembled so that two parts are in alignment. Draw the bushing and pin, insert the appropriate dimensions and tolerances, and do an analysis of the allowances.

Solution   Table 4.5 has been extracted from Table 4.4 for class LC fits. Figure 4.10 shows the tolerances and the allowance applied to the hole and shaft. The analysis of this class of fits produces the following. A combination of the *smallest hole* and the *largest shaft* will produce *no allowance*, whereas the *largest hole* in combination with the *smallest shaft* will create a *maximum clearance*.

Minimum clearance		*Maximum clearance*	
Smallest hole	2.000	Largest hole	2.0045
Largest shaft	2.000	Smallest hole	1.9970
Allowance	0.000	Maximum clearance	0.0075

TABLE 4.5

Nominal Size Range (in.) over–to	Class LC4		
	Clearance M	Standard Tolerance Limits	
		Hole	Shaft
1.97–3.15	0.0	+4.5	0.0
	7.5	0.0	−3.0

Figure 4.10

nominal size range inches	Class LC 4		
	clearance M	Standard tolerance limits	
over - to		Hole	Shaft
1.97– 3.15	0	+4.5	0
	7.5	0	−3

2.000 +0.0045 / -0.0000   2.000 +0.000 / -0.003

Transition fits (LT) are those which result after the tolerances have been applied to the mating parts and which produce either a clearance or an interference between them upon assembly. This class is used where greater accuracy in assembly is needed.

Example 4.3

> A 2-in. shaft and bushing are to provide for selective assembly. The fit is a class 4 transition. Draw the shaft and the bushing, insert the dimensions and tolerances, and do an analysis.

Solution   Table 4.6 has been extracted from Table 4.4 for the LT class. Figure 4.11 shows the tolerances and the allowances applied to the hole and the shaft. Figure 4.11 (transition location fit) a combination of the *smallest hole* and the *largest shaft* sizes will produce an *interference*, whereas the combination of the *largest hole* and the *smallest shaft* sizes will produce a *clearance fit*. The negative sign indicates an interference.

Interference allowance		*Clearance allowance*	
Smallest hole	2.0000	Largest hole	2.0018
Largest shaft	2.0013	Smallest shaft	2.0001
Interference	−0.0013	Clearance	+0.0017

TABLE 4.6

Nominal Size Range (in.) over–to	Class LT4		
		Standard Tolerance Limits	
	Fit M	Hole	Shaft
1.97–3.15	−1.3 +1.7	+1.8 0.0	+1.3 +0.1

Figure 4.11

nominal size range inches over - to	Class LT4			
	fit M	Standard tolerance limits		
		Hole	Shaft	
1.97– 3.15	−1.3 +1.7	+1.8 0	+1.3 +0.1	

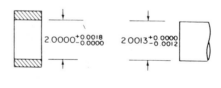

Interference fits (LN) are those which result after the tolerances have been applied to mating parts and which always produce an interference when they are assembled no matter which combination of hole and shaft sizes is selected. These are used where accuracy of locating one part with another is of primary importance. Again, it is not intended to be a fit that will transmit motion.

Example 4.4

A 2-in. key is to fit a slot in a fixture in order to locate a positioning block. Under no conditions is the block to transmit motion, yet it must be assembled with a slight interference. A class 2 fit is prescribed. Draw the slot and key and dimension them. Make an analysis of the allowances.

Solution   Table 4.7 has been extracted from Table 4.4 for the LN classes. Figure 4.12 shows the tolerances and the allowance applied to a slot and a key. An analysis shows the following. A combination of the *smallest slot* and the *largest key* sizes will produce the *tightest interference allowance,* whereas the combination of the *largest slot* and the *smallest key* sizes will also produce the *smallest interference allowance.*

Maximum interference		*Minimum interference*	
Smallest slot	2.0000	Largest slot	2.0012
Largest key	2.0021	Smallest key	2.0014
Max. interference	−0.0021	Min. interference	−0.0002

TABLE 4.7

Nominal Size Range (in.) over–to	Class LN2		
	Interference M	Standard Tolerance Limits	
		Hole	Shaft
1.97–3.15	−0.2	+1.2	+2.1
	−2.1	0.0	+1.4

Figure 4.12

nominal size range inches	Class LN2		
	interference M	Standard tolerance limits	
over - to		Hole	Shaft
1.97– 3.15	−0.2	+1.2	+2.1
	−2.1	0	+1.4

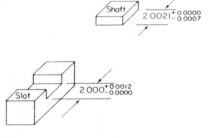

FN-force and shrink fits (Fig. 4.13).   FN1 through FN5 produce a series of force or shrink fits. Each series produces its own characteristic pressures throughout a particular series.

Example 4.5

A 2-in. rod is to be assembled with a bushing by heating the bushing. A class 4 fit is recommended. Draw the rod and bushing, dimension them, and do an analysis of the allowances.

Figure 4.13

nominal size range inches	Class FN4		
	interference M	Standard tolerance limits	
over - to		Hole	Shaft
1.97–2.56	−2.3 −4.2	+1.2 0	+4.2 +3.5

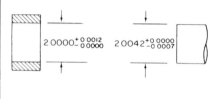

Solution   Table 4.8 has been extracted from Table 4.4 for the FN classes. Figure 4.13 shows the application of these tolerances and the allowance. From Fig. 4.13 an analysis will produce the following results. A combination of the *smallest hole* and the *largest rod* sizes will produce the *maximum interference*, whereas a combination of the *largest hole* and the *smallest rod* sizes will produce *minimum interference*.

Maximum interference		Minimum interference	
Smallest hole	2.0000	Largest hole	2.0012
Largest rod	2.0042	Smallest rod	2.0035
Max. interference	−0.0042	Min. interference	−0.0023

TABLE 4.8

Nominal Size Range (in.) over–to	Class FN4		
	Interference M	Standard Tolerance Limits	
		Hole	Shaft
1.97–2.56	−2.3 −4.2	+1.2 0.0	+4.2 +3.5

4.7. PRECISION IN MANUFACTURING

The degree to which tolerances can be applied is a function of the fabricating process used. It is true that good-quality fabrication can be achieved by a very highly trained and skilled toolmaker. But it is also true that the cost of production increases rapidly. The use of tools, jigs, fixtures, and gages will reduce the cost of production. However, when a specific tool is relied upon to do a job, the limitations of that tool must be considered.

In general, the precision that can be expected from a fabricating process is shown in Table 4.9. This is correlated with some machining processes, shown by the bar graph chart below the table. It should be emphasized that these are only guidelines.

TABLE 4.9 APPLICATION OF TOLERANCES TO MACHINING

Nominal Size Range (in.) over–to	Grade									
	4	5	6	7	8	9	10	11	12	13
	Tolerance (thousandths)									
0.04–0.12	0.15	0.20	0.25	0.40	0.60	1.00	1.6	2.5	4.0	6.0
0.12–0.24	0.15	0.20	0.30	0.50	0.70	1.20	1.8	3.0	5.0	7.0
0.24–0.40	0.15	0.25	0.40	0.60	0.90	1.40	2.2	3.5	6.0	9.0
0.40–0.71	0.20	0.30	0.40	0.70	1.00	1.60	2.8	4.0	7.0	10.0
0.71–0.19	0.25	0.40	0.50	0.80	1.20	2.00	3.5	5.0	8.0	12.0
1.19–1.97	0.30	0.40	0.60	1.00	1.60	2.50	4.0	6.0	10.0	16.0
1.97–3.15	0.30	0.50	0.70	1.20	1.80	3.00	4.5	7.0	12.0	18.0
3.15–4.73	0.40	0.60	0.90	1.40	2.20	3.50	5.0	9.0	14.0	22.0
4.73–7.09	0.50	0.70	1.00	1.60	2.50	4.00	6.0	10.0	16.0	25.0
7.09–9.85	0.60	0.80	1.20	1.80	2.80	4.50	7.0	12.0	18.0	28.0
9.85–12.41	0.60	0.90	1.20	2.00	3.00	5.00	8.0	12.0	20.0	30.0
12.41–15.75	0.70	1.00	1.40	2.20	3.50	6.00	9.0	14.0	22.0	35.0
15.75–19.69	0.80	1.00	1.60	2.50	4.00	6.00	10.0	16.0	25.0	40.0

Machining Operation										
Lapping-honing										
Cyl. grinding										
Surface grind										
Diamond turning										
Diamond boring										
Broaching										
Reaming										
Turning										
Boring										
Milling										
Shaping-planing										
Drilling										

In general, the expectations from the various operations need not be as rigorous as shown in Table 4.9. They are generalizations. These generalizations are related to production setups.

The lathe, or turret lathe, operation of finish turning for sizes from about $\frac{1}{64}$ in. to 3 in. in diameter can be held to the range 0.002 to 0.007 in. Finish boring can be held to the range 0.002 to 0.003 in. Lengths can be held to the range 0.003 to 0.007 in. for the same range of sizes.

Milling may generally be held to 0.005 in., although it is not uncommon to machine to 0.003 in. if the machine is of good quality.

Drilling over a range of sizes of $\frac{1}{64}$ to 3 in. may range from 0.002 to 0.015 in. The reaming operation is more reliable and the tolerances may range from 0.005 to 0.0015 in.

Planing and shaping may produce variations of sizes of up to 0.005 in.

Broaching may be accomplished within the range of 0.001 in. per inch of surface.

Grinding may be used to produce workpieces that can be held to 0.0002 to 0.0005 thousands of an inch, honing to 0.0001 to 0.0003 thousands of an inch.

Lapping, which is extremely accurate, may be done to tolerances of 10 to 25 millionths of an inch (0.000010 to 0.000025 in.).

Punches and dies used for blanking or piercing, on a punch press, can produce variations of sizes of from 0.005 to 0.015 in. for a stock up to $\frac{1}{4}$ in. Shear cuts on a brake will operate to within $\pm\frac{1}{32}$ in. A secondary operation of shaving can produce parts to within 0.001 in.

Sand castings generally are considered to vary by about plus or minus $\frac{1}{16}$ in. for 6 in. of length with $\pm\frac{1}{32}$ in. added for each additional 6 in. of length.

Die castings made from aluminum and magnesium may vary about ±0.001 in. per inch of length, but not less than ±0.006 in. Lead and tin can be held to 50% that of aluminum, whereas zinc is held to about 65% that of aluminum. Movable cores may vary as much as $\frac{1}{64}$ in. off their location.

In permanent mold castings such as the lost wax process, the variation could be ±0.003 in. but not less than $\pm\frac{1}{64}$ in.

The powdered metallurgy process may produce variations of ±0.005 in. per inch of length, and ±0.001 in. on a diameter.

Plastics that are molded may vary as much as ±0.005 in. per inch of length, whereas holes can be held to ±0.002 in.

Forgings are produced with two halves of a die. Shrinkage, die wear, and mismatch may produce error, which, of course, affects the tolerances that are produced. Every 5 lb of forging may produce an error across the forging, parallel to the parting plane, of ±0.003 in. in addition to a minimum of $\pm\frac{1}{64}$ in. Shrinkage and die wear may require ±0.003 in. per inch of length be added to $\pm\frac{1}{32}$ in. for forgings that weigh up to 5 lb; from 5 to 25 lb, add $\pm\frac{1}{16}$ in.; from 25 to 50 lb, add $\pm\frac{1}{8}$ in. Thickness tolerances in a direction perpendicular to the parting plane may be held to $+\frac{1}{16}$ to $-\frac{1}{64}$ in. for a 5lb forging; $+\frac{3}{32}$ to $-\frac{1}{32}$ in. for castings that range from 5 to 25 lb; and $+\frac{1}{8}$ to $-\frac{3}{64}$ in. for forgings that have a weight range of 25 to 50 lb.

4.8 SELECTIVE ASSEMBLY

Where design requirements call for very close tolerances to be machined, the cost of producing these parts may be very high. To cope with this situation, the tolerances are opened up for production purposes. Once the parts are produced, they are sorted in pairs to maintain the allowance, and then assembled. Because the allowance is preserved, the performance of the assembly is preserved. In the case of a ball bearing the performance allowances of the inner and outer races, and the balls are preserved.

In Fig. 4.14a it becomes necessary to machine each part to a tolerance of 0.001 thousandth of an inch to a *minimum allowance* of 3.000 − 2.998 = *0.002 in.* and a *maximum allowance* of 3.001 − 2.997 = *0.004 in.*

The drawing is first redimensioned as shown in Fig. 4.14b. The assumption

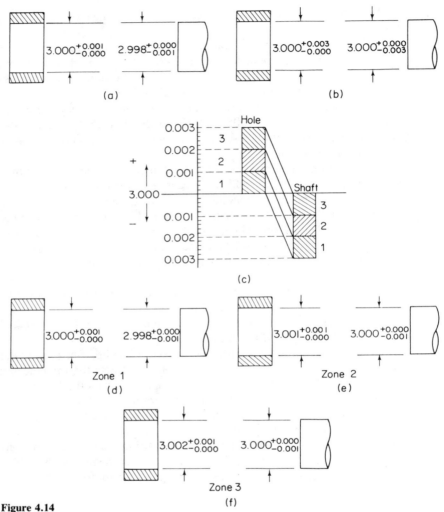

Figure 4.14

here is that there will result three groups of dimensional combinations which will preserve the allowance of 0.002 in.

These tolerance groups are plotted as shown in Fig. 4.14c and parallel lines drawn as shown connecting the various zones. Group matching of zone 1 hole with zone 1 shaft; zone 2 hole with zone 2 shaft; zone 3 hole with zone 3 shaft will produce the three sets of assemblies shown in Fig. 4.14d, e, and f.

Thus in zone 1 (Fig. 4.14d)

Smallest hole	3.000	Largest hole	3.001
Largest shaft	2.998	Smallest shaft	2.997
Min. allowance	0.002	Max. allowance	0.004

and in zone 2 (Fig. 4.14e)

Smallest hole	3.001	Largest hole	3.002
Largest shaft	2.999	Smallest shaft	2.998
Min. allowance	0.002	Max allowance	0.004

and finally, in zone 3 (Fig. 4.14f)

Smallest hole	3.002	Largest hole	3.003
Largest shaft	3.000	Smallest shaft	2.999
Min. allowance	0.002	Max. allowance	0.004

This process is known as *selective assembly.*

4.9 COMPUTER DRAFTING AND DESIGN

Computer-aided design or drafting (CAD) has revolutionized the approach to and the processes for designing mechanisms, tools, and fixtures. Drawings may be viewed on a video terminal Fig. 4.15(a) and changes in design made very quickly. The erasures and additions to new designs or old designs are made on the screen of a video terminal through a keyboard or digitizing tablet and cursor (Fig. 4.15b).

Once all changes have been made, the computer can be directed to develop a three-dimensional drawing, or picture of the object. It can be rotated through any desired angle. In addition, it can be sectioned at any point. This section can be rotated to expose the inside of the object under consideration. If changes are to be made, they are made and the section portion rotated and joined with its mating part.

The next step is to direct the computer to make a blueprint (hard copy) of the object that has been designed. Plotters may be tabletop size, or drum-type capable of generating full-size blue prints.

The use of CAD is frequently coupled with computer-aided manufacturing, CAM, and referred to as CAD/CAM. This became possible when computers were applied to numerical control (NC) machines to create a generation of computer numerical control (CNC) machine tools. The programs developed with CAD were applied directly, or indirectly, to CNC machines. Thus programs could be designed and changes made on the terminal without the need to make hard prototypes of the objects to be manufactured.

New developments indicate the use of CAD/CAM in conjunction with CNC machine tools and transfer systems to move materials into position automatically so that sequenced operations may be completed without human intervention at the processing station. Computer-integrated manufacturing (CIM) and or flexible machining systems (FMS) are concepts currently in the developmental stage.

In general, FMS requires the development of transporters, workstations, and computer controllers. Transporters are used to move materials from storage to machine, or from machine to machine, where the part processing is to take place. They may be powered tow lines, powered carts, or conveyors moved by chains and guided

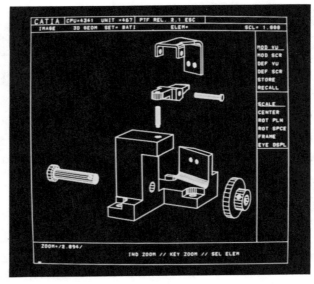

(a)

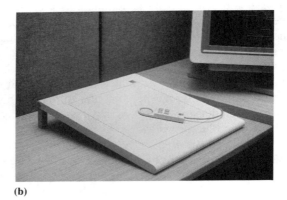

(b)

Figure 4.15   Courtesy: IBM:5080 Graphics System: pp's 6 and 13. Dept-805 White Plains NY 10604

by floor or overhead rails, guided along electronic paths, or moved and positioned by robots.

The control of the operations of NC machines from a central processor, known as *downloading,* makes possible the use of these machines from remote stations. The central processor loads the memory of the computer at the machine tool, which then operates independently of the central processor. This independence allows for easier editing of programs at the machine tool. It also allows the central processor to service many more machine tool, or production lines. The latter is known as distributed numerical control, a new use for the acronym DNC. With this system of controls an FMS may control the processes at the workstation, such as tool life, fixture control and maintenance, record keeping, performance monitoring, and so on.

4.10. CHECKING THE DRAWING

Another important aspect of engineering design is the checking of drawings. The importance of this function of the engineering department cannot be overstated. It is the checker's job to verify all aspects of the project and inform the draftsman, who makes the changes. He must be familiar with drawing practices, standards parts and machining practices, and so on. The job is so all-encompassing that checkers will usually seek answers to the following questions:

1. What is the general appearance of the drawings? Do they meet drafting and design standards?
2. Are the strength requirements met so that the part will not fail while in use?
3. Can the parts be machined economically? Is there enough room for a tool so that all machined surfaces can be machined? Can the parts be easily assembled and disassembled? Can it readily be serviced? Is it possible to use purchased cutting tools, or existing tools?
4. Are all views to scale? Are they correct? Are there enough sectional views? If parts are to be bent, are the developed lengths correct?
5. Are all necessary dimensions shown? Are they correct? Do they conform to the layout? Are there any dimensions that have been repeated? Does the machinist or toolmaker need to do any calculations?
6. Are all allowances and tolerances correct? What about the accumulation of tolerances? Are all fillets and drafts shown?
7. Is there sufficient clearance between parts? Are all finish signs included? Are all symbols, such as welding, included?
8. What about notes? Are notes such as concentricity, parallelism, flatness, and so on, included, and if so, are they correct? Are the heat treatment, plating, and material specifications included and correct?
9. Are all surfaces that need to be machined provided with enough material to ensure cleanup? If banking pads are needed, are they provided?
10. Does the bill of materials have all the parts necessary for complete assembly? Does the rough stock listed have enough material allowance?
11. Is the title block complete: signed, dated, and so on?
12. Have all revisions been entered in the change block? Have they been made? Have they been cross-checked?
13. If trademarks, part numbers, or identification numbers must be included, are they shown properly?

PROBLEMS

4.1. How does a tool drawing differ from a working shop drawing?

4.2. How does a set of tool drawings used in a large toolroom differ from those used in a one-person toolroom?

4.3. List the 11 rules set forth in this chapter that will conserve design time and aid in clarifying drawings.

4.4. Discuss the use of descriptive notes instead of additional views on drawings.

4.5. Why is it only necessary to draw the outlines of a standard part such as a screw or bolt on a tool drawing?

4.6. Make a hand sketch of a simple standard part that needs to be altered and show how this part should be dimensioned.

4.7. Explain the use of sections when making a drawing of a punch and die.

4.8. **(a)** What is a partial view?
(b) How does it conserve design time?
(c) Sketch a partial view of an object in the room.

4.9. Why should a dimension be shown only in one view and not repeated in another?

4.10. Check your drafting book. Find an assembly drawing and indicate the critical dimensions and the location dimensions. Discuss your reasons for choosing the dimensions that you have selected. Check your reasons with your instructor.

4.11. Assume several parts that have one of a large number of like dimensions.
(a) How should the like dimensions appear on the drawing?
(b) How should unlike dimensions appear on the drawing?

4.12. List as many advantages as you can for outlining the workpiece in red when designing a jig, fixture, or a die.

4.13. Make freehand sketches for the details for making the vise in Fig. 4.16. You should balloon Fig. 4.16 first.

4.14. Assign balloons and part numbers to the vise in Fig. 4.16 and develop a bill of materials.

4.15. What is the sequence used when ballooning an assembly?

4.16. What information should be included in a change block?

4.17. Explain the rationale for the preference for sectional drawings over details when making tool drawings.

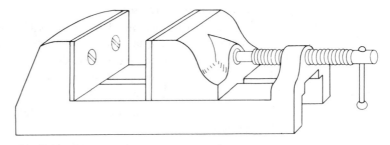

Figure 4.16

4.18. (a) Explain the views used when drawing a punch and die.
(b) Why is a die combination shown as a "shut height"?

4.19. Explain the advantages that you see when a die is shown in open and shut positions, as in Fig. 4.4b.

4.20. What are the penalties of poor dimensioning? Explain each and its effect on the management effort.

4.21. How is the lead dimension of one that has a tolerance applied determined?

4.22. What is the difference in interpretation of the three types of dimensions shown below?

$$2.005\begin{smallmatrix}+0.000\\-0.005\end{smallmatrix} \qquad 2.000\begin{smallmatrix}+0.005\\-0.000\end{smallmatrix} \qquad \frac{2.005}{2.000}$$

4.23. Figure 4.17 shows a dimensioned drawing. Which of the dimensions are superfluous?

4.24. List the four dimensions that may be used to dimension tapers. How are they used?

4.25. Define a datum point, line, or surface.

4.26. (a) Why should a dimension be shown in one view only?
(b) How may a dimension be repeated in another view?

4.27. Define and illustrate the four types of surfaces on a workpiece.

4.28. Sketch Fig. 4.16, letter each surface, and identify whether the surfaces are datum, functional, clearance, or atmospheric.

4.29. List the items that the tool designer must consider when selecting tolerances and allowances.

4.30. Write the dimension *three and three-quarters* as
(a) a nominal size,
(b) a basic size, and a
(c) design size. Explain your answers.

4.31. Define and illustrate the following terms as they relate to dimensioning:
(a) nominal size,
(b) basic size,
(c) design size,
(d) allowance,
(e) limits,
(f) tolerances.

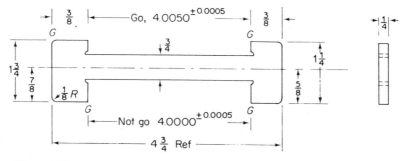

Figure 4.17

4.32. How should limits be written when dimensioning
   (a) a shaft and
   (b) a hole? Illustrate.

4.33. Define
   (a) bilateral and
   (b) unilateral tolerances. Illustrate.

4.34. Define
   (a) basic hole and
   (b) basic shaft systems.
   (c) When is it appropriate to use the latter?

4.35. In Fig. 4.18 calculate the allowances.

4.36. In Fig. 4.19 calculate the allowances.

4.37. A 4-in. shaft and bushing are to operate with a class 2 running fit. Make a drawing of the bushing and shaft and do an analysis of the allowances.

4.38. A $2\frac{1}{4}$-in. keyway and key are to operate with a class 5 sliding fit. Make a drawing of both and do an analysis.

4.39. A $\frac{1}{2}$-in. pin is to be used to position a locator pad in a milling fixture. The pin is to have a class 5 clearance with the reamed hole into which it fits. Make a sketch of the pin and the hole, dimension it, and do an allowance analysis.

4.40. Same as Prob. 4.39 except that the fit is a transition class 6 which is to provide for selective assembly.

4.41. A $\frac{3}{4}$-in. pin is to be pressed into a reamed hole with a class 1 fit. Draw the pin and hole, dimension them, and do an analysis of the allowances.

4.42. In general, what is the precision that may be applied to
   (a) machining,
   (b) milling,
   (c) drilling, and
   (d) reaming a diameter, bore, or length of work?

4.43. What are the ranges of machining accuracies for
   (a) grinding
   (b) lapping, and
   (c) honing?

4.44. What accuracies can be expected from punches and dies?

4.45. What is the accuracy size reliability of
   (a) sand casting,
   (b) die casting, and
   (c) permanent mold casting?

4.46. What is the accuracy of powdered metallurgy parts?

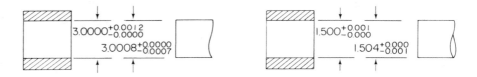

Figure 4.18   **Figure 4.19**

4.47. What limits of accuracy can be expected from forgings?

4.48. Discuss the production of mating parts where selective assembly principles must be used.

4.49. A shaft and a bushing are to be machined to the tolerance shown in Fig. 4.20a. Figure 4.20b is the same drawing redimensioned to produce the same functionality. Apply the principle of selective assembly. Do a complete analysis supporting selective assembly in groups of two zones.

4.50. Figure 4.21a shows a 1-in. slot and a key which is to slide with 0.001 in. allowance. Since the cost of producing this combination was excessive, the company decided to produce the assembly to the dimensions shown in Fig. 4.21b and according to the principles of selective assembly. Do a complete analysis supporting a four-zone grouping of dimensions.

4.51. **(a)** Describe the design process known as CAD.
(b) What are its unique features that make it so desirable when designing machine parts?
(c) List the advantages that make it more efficient than the use of drafting instruments.

4.52. How are computers and CNC machine tools used in CIM (FMS)?

4.53. What is downloading?

4.54. What is the effect of downloading on the use of central processing?

4.55. List the questions to which a checker will seek answers when he reviews a set of drawings.

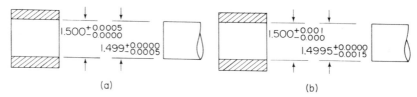

Figure 4.20

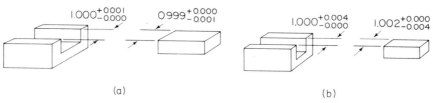

Figure 4.21

Chapter 5

Processing and Production

5.1 PROJECT ANALYSIS

It is not the purpose of this book to study the intricacies of process planning. However, since the design of tools, jigs, and fixtures depends directly on the processes planned for a job, it becomes important to take a look at how a process evolves and the reasons for the procedure and sequence of operations chosen.

The first step that must be established is the production feasibility of the product or part. This includes the suitability of the project to the type of industry that is to produce the project. That is, how adaptable is the present equipment to the proposed project? What new equipment will be needed? Do the present personnel have the know-how to do the proposed job?

The next series of considerations that must be evaluated include an estimate of the length of time it will take to run the job, and the cost of the project. This should include the cost of development and design as well as the cost of production.

If the product is one that is to be developed and sold by the company producing it, a market analysis should be done. These steps are followed by a company: design, process planning, production drawings, material procurement, and tooling. The steps followed in the processing of a job must follow a logical procedure which leads to the best output at the lowest cost. One procedure that can be followed when processing a new job is the following:

1. Analyze the blueprints, bill of materials, and specifications.
2. Redesign parts wherever necessary without destroying functionality.
3. Select reference or critical surfaces.

4. List the operations without regard to their sequence.

5. Combine or group the operations for quality and economic production.

6. Rearrange those operations in proper sequence.

7. Specify the equipment and tooling that will be needed.

8. Estimate the time to machine the components and assemble the end product.

If the manufacturing company has been updated to NC and CNC equipment, the feasibility of using that equipment, or a part of it, must be seriously considered in the project analysis. One-piece, or short-run, production may dictate the use of this equipment. The use of computer control machine tools will make a substantial difference in the sequencing of operations, the tools and fixtures required, the caliber of the personnel used, and the length of time it takes to produce the components and assemble them. These considerations can have a very profound effect on the cost of production and how competitive the company is. The eight steps listed above can be applied to the processing of a job when using NC or CNC machine tools.

5.2 SPECIFICATIONS, BLUEPRINTS, AND BILL OF MATERIALS

The key to a determination of what may be done is found in the specifications, bill of materials, and prints. It is important that the process engineer understand the operation of the mechanism and the relationship of the part to the operation of the assembly for which he intends to write the process sheets. Thus it is important that the part be studied carefully in relation to the assembly, bill of materials, notes, and special specification sheets. Below are listed some of the items that may be a source of help to the designer or process sheet writer.

A. The assembly drawing should reveal:

1. The functional surfaces and those surfaces that are not quite as important

2. Important tie-in dimensions

3. Concentricity, parallelism, squareness, and the angular requirements that will ensure proper functioning

4. Which surfaces need to be processed to withstand wear, abrasion, shock, and so on

5. Possible redesign

B. The working detail drawing should reveal:

1. Tolerances and allowances

2. Relationship of surfaces

3. Finishes required

4. Hole sizes, shoulder and step sizes, and so on

5. Types of machining operations that might be used
6. Whether or not the parts are to be heat treated

C. The bill of materials should reveal:

1. The number of parts needed, which could influence the type of machines, jigs, fixtures, and gages used, and will, no doubt, influence the sequence of operations
2. The type of raw materials to be used: whether a casting, forging, stamping, bar or flat stock, and so on, is to be used

D. General specification sheets should reveal:

1. The overall functional requirements of the assembly or specific parts
2. The overall precision of standard parts, such as bearings, pillar blocks, and so on
3. The minimum physical requirements of the raw materials, castings, forgings, and so on
4. The requirements of precision, such as tolerances and allowances, when not specifically stated
5. The general requirements of parallelism, squareness, concentricity, etc., when not specifically stated
6. The general requirements of finish

There are other sources that the process sheet writer may investigate. He can gather information from other departments, meetings with customers, suppliers of standard parts, subcontractors (foundry, plating company, etc.), toolmakers and foremen in his own shop, and so on.

5.3 REDESIGN OF PARTS

One of the areas of concern of the process writer is the ease of producing a part. It is therefore important for him always to be on the alert for making redesign suggestions to management. Obviously, if the redesign adversely affects the functioning of the part or assembly, the redesign should be discarded. By the same token, designs may be improved and the cost substantially reduced when they are reworked. To the process writer this is important because it will change his approach to the problem he faces and certainly will produce a completely different sequence of processes. He should ask himself the questions: What does the part do? Does it do the job? What else will do the job as effectively? How does the cost of one compare with the other?

Following are some design changes that could be made to reduce cost without materially affecting the functionality of the part:

1. *Combination of several separate parts into one.* This will eliminate parts that require separate processing, and separate or additional machining operations. A good place to start is to be critical of all secondary operations. Questioning the value of secondary operations will sometimes lead to a solution that eliminates them. Figure 5.1a shows a bearing flange that is machined as a separate piece of work fitted and bolted to an accurately bored hole in the main housing. In Fig. 5.1b the same end result is accomplished with a piece of round stock welded to the main housing and machined in position.

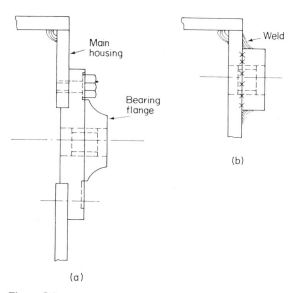

(b)

(a)

Figure 5.1

Figure 5.2a shows a bearing support block which is made from four separate parts to be bolted together. Figure 5.2b shows the same block made from welded pieces. Figure 5.2c is the same support block made as a sand casting. Certainly, the casting is the easiest to process and the cheapest to produce. The welded structure is intermediate between Figure 5.2a and c in terms of ease of production and cost. Figure 5.2a is the most difficult to produce and costs the most.

2. *Selection of the fabricating method.* Some operations are more expensive when done with one type of machine than when done with another type. If one piece is to be made, it may not be feasible to sand cast it. The cost of the pattern alone may be prohibitive. Yet if the part is complicated and requires a great deal of hand work and a very expensive experienced toolmaker to do the job, it may be cheaper and easier to make a wood pattern and cast the part. It may even be better to redesign the part into several parts and provide for their assembly.

However, in most instances, and especially if the quantities permit, a turret

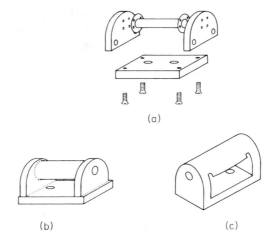

Figure 5.2

lathe is faster and more economical to produce a run than an engine lathe would be; an automatic screw machine is more economical than a turret lathe; and so on. At other times it may be more economical to stamp parts rather than trying to machine them. The selection of the method of fabrication is a consideration when redesigning parts.

3. *Substitution of standard parts.* It is almost always cheaper to buy and use standard parts than it is to try to make these parts. Companies that manufacture the parts have researched the materials, tools, dies, production methods, and so on. Thus bolts, nuts, washers, springs, bushings, spacers, clamps, support buttons, and so on, should all be purchased whenever possible.

5.4 SELECTION OF A REFERENCE SURFACE

Once the design has been established, the next step is to start the processing. The processing of a part begins with the identification of the critical planes, surfaces, or points. These *critical areas* may be identified in various ways.

In some instances the *assembly drawing* will reveal a surface, or surfaces, which are dimensioned to some other part in the assembly. Such a *tie-in* dimension would generally indicate a critical area that could be used as a reference surface in the design of jigs, or fixtures, and certainly in processing.

In other instances, special requirements, such as surface finishes, squareness, flatness, parallelism, and so on, may reveal that the related surface could be used as a critical area from which to process a part.

In still other instances, a dimension that has *close tolerances* may reveal critical areas. It may be that to preserve these tolerances, such a surface would have to be used as a critical area for processing.

Functional surfaces generally are used as critical areas since other dimensions are related to these surfaces. This may occur even though the tolerances that apply to these dimensions are not close.

In Fig. 5.3 the dimension *a* need *not* be dimensioned accurately, whereas dimensions *b* and *c* probably should be. The presence of the *baseline* would indicate that it is important that the two holes should be particular heights above the baseline. Dimension *a* need not have close tolerances because surface x need not be machined.

In the other plane of the front view of the workpiece, Fig. 5.3, the centerline becomes critical. This would seem to indicate that the keyway would be a logical point from which to machine the distances *d* and *e*.

In the right-side elevation, Fig. 5.3, surface y would appear to be a good reference surface. Apparently, the bored shoulder distances *g* and *h* are critical. The distance *f*, although not dimensioned accurately, is also taken from y.

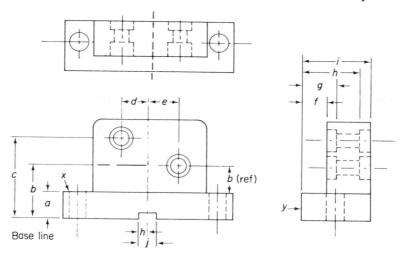

Figure 5.3

5.5 NONSEQUENTIAL LISTING OF OPERATIONS

Once the critical areas, usually in three planes, have been identified, the part may then be analyzed for the operations that will be necessary to complete the job of processing. These operations may be separated into and listed under one of four major headings:

1. Vendor

2. Critical surface

3. Auxiliary

4. Assembly

All categories need not apply to all jobs. The processing of a single part may require only critical surface and auxiliary operations.

It should be understood that whether or not an operation is to be placed under the heading of vendor or auxiliary is determined by the decision of management to subcontract, or to perform the operation itself. It is therefore important that the process engineer understand fully the capabilities of his own company when analyzing a job for processing.

Vendor operations.   The functions that fall into this category are those which the company subcontracts. They may be separated into two categories: *nonmachined* operations and *machined* operations.

Under *vendor-nonmachined operations* are included raw stock, patterns, castings, forgings, molded shapes, platings, protective coatings, heat treating, and so on. These are the operations that may be done at the "home" plant of a large manufacturing facility if the equipment is available. Some of these operations may or may not require further processing. A casting that needs further machining is processed by the process engineer only for those operations that will be machined at the "home" plant. He does not need to process the pattern or the steps needed to make the mold and cores. If the casting is to be machined at the "home" plant, he must know the draft applied, the allowances, where the parting line will be, the type of finish that sand castings have, and so on. If the operations are to be done at the "home" plant, these operations will not appear under the vendor category.

Under *vendor-machined operations* are included standard parts and those machined parts which the company "farms out" for one reason or another. These are the standard parts, such as screws, bolts, nuts, washers, pins, and so on. Most manufacturing plants have the capability of producing these parts even if it means buying new machinery. However, if they can be purchased as standard parts, the probability is that their cost will be less than it would be when machined "in-house." In some instances it may mean that the company might not have, or want to commit, the personnel to manufacturing these parts; or it might not want to buy new equipment.

Under certain circumstances, parts that need machining may be farmed out. Another manufacturer may have more appropriate equipment and personnel. In other instances, even though a company has the capabilities for producing a part, it may farm out the part because the work schedule may be too overloaded to justify breaking into existing production schedules that might endanger production deadlines. Under these conditions management might make the decision to farm out a particular part.

If the decision is made to farm out a part to be produced in part or as a whole, the part need not be processed in detail by the home plant. The operation should be placed in the *vendor* category.

Critical surface operations.   Section 5.4 dealt with the importance of selecting critical surfaces. Once these critical surfaces are selected, they may be the surfaces which require machining, so that this machined surface may become the banking or reference surface for subsequent operations. It should be pointed out that this does *not* imply that as a general rule the critical surface should be machined. *The* *rule is that as many surfaces as possible should be machined in each setting of the part, and that as many subsequent operations as possible should be done from the same banking surfaces.* Each time new banking surfaces are used, the ability to hold the stated tolerances becomes more difficult, and the chances of error become greater.

Usually, because the *initial banking surface* is not machined, its position in the fixture will vary from workpiece to workpiece. Figure 5.4a shows an exaggerated version of what could happen if pad *a* is longer than pad *b*. Using pads *a* and *b* as banking surfaces causes the cutter to machine more material from pad *d* than from *c*. If pads *c* and *d*, Fig. 5.4b, had been chosen as the reference pads, the excess material on pad *a* would have been removed and all surfaces on the workpiece would now be more nearly parallel.

Figure 5.4c shows all surfaces cleaning up. It is entirely possible that all subsequent surfaces that need to be machined may not clean up because of a poor choice of *initial* banking points. Of course, the choice of banking points will have to be changed if subsequent surfaces do not clean up.

The discussion above points up the need to study and to be completely familiar with the variations of the surfaces of the rough casting.

Auxiliary operations. These operations may be auxiliary-nonmachined or auxiliary-machined operations. They deal with those operations that are done in-house and therefore need processing. They are the operations other than the critical surface or assembly operations. They may be done in separate setups, or in the same setup as the operation in which the critical surface is machined. As an example, in Fig. 5.4b the rough pads *c* and *d* are auxiliary surfaces used to machine the critical surface pads *a* and *b*. If subsequent operations are to be done on the angle plate, the machined pads *a* and *b* should be used as banking pads.

Auxiliary-nonmachined operations are receiving and shipping, visual inspection, all inspection operations done during and after manufacturing, heat treating, burring, etc. If the plant makes its own pattern for casting, or does its own plating, these operations will need processing and therefore should be classified as auxiliary.

Auxiliary-machined operations are all those operations that require machining as an in-house effort. They are all the lathe, milling, drilling, grinding, sawing,

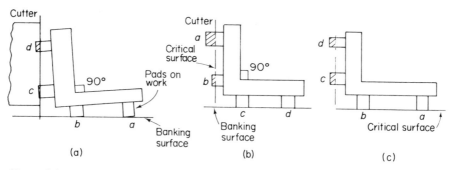

Figure 5.4

stamping, drawing, welding, and so on, operations encountered in a machine shop. They are the operations that use the critical surfaces as banking surfaces as a reference for further machining. As indicated earlier, in Fig. 5.4b, after pads *a* and *b* have been machined, the part is reset to rest on these pads and pads *c* and *d* are machined as shown in Fig. 5.4c. Pads *a* and *b* are critical surfaces. Pads *c* and *d* are auxiliary surfaces.

Assembly operations. These are the operations done by the company at assembly. They include final functional inspection, or any machining that is required as part of the assembly operation. At times fitting of bearings to shafts, or turning babbit bearings at assembly to match bored holes, or selective assembly of bushings and shafts, drilling and pinning of gears, cams, and so on, to shafts at assembly and other such operations are required. These should be written into the process sheet. Correction of machining mistakes or reworked parts at assembly are not included

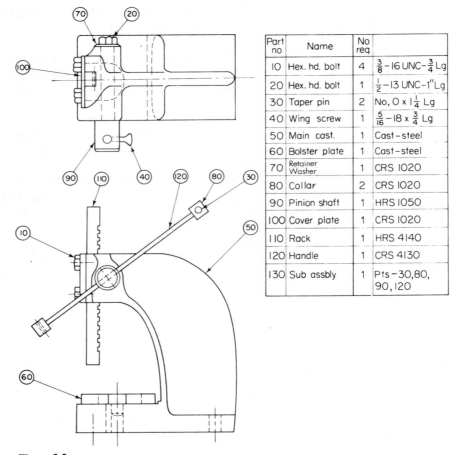

Part no	Name	No req	
10	Hex. hd. bolt	4	$\frac{3}{8}$–16 UNC–$\frac{3}{4}$ Lg
20	Hex. hd. bolt	1	$\frac{1}{2}$–13 UNC–1"Lg
30	Taper pin	2	No, 0 x 1$\frac{1}{4}$ Lg
40	Wing screw	1	$\frac{5}{16}$–18 x $\frac{3}{4}$ Lg
50	Main cast.	1	Cast–steel
60	Bolster plate	1	Cast–steel
70	Retainer Washer	1	CRS 1020
80	Collar	2	CRS 1020
90	Pinion shaft	1	HRS 1050
100	Cover plate	1	CRS 1020
110	Rack	1	HRS 4140
120	Handle	1	CRS 4130
130	Sub assbly	1	Pts–30,80, 90,120

Figure 5.5a

under this category. Allowances for manufacturing of scrap or the correction of mistakes in machining are provided for, within limits, when the job is bid. It should not be processed into the manufacture of the assembly.

5.6 NONSEQUENTIAL OPERATIONS ANALYSIS

Figure 5.5a shows an assembly drawing and a bill of materials for an arbor press. Tables 5.1 through 5.9 show the *nonsequential operations* and categories of all parts.

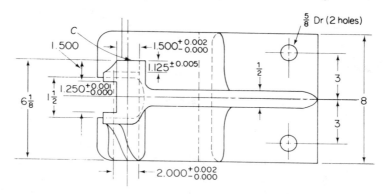

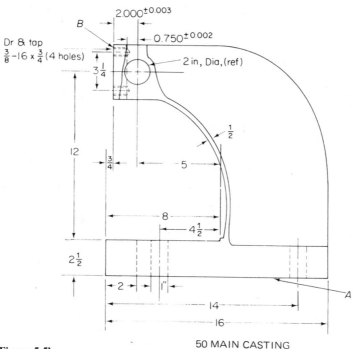

50 MAIN CASTING

Figure 5.5b

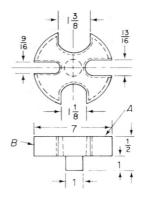

60 Bolster plate

Figure 5.5c

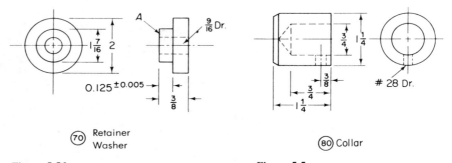

70 Retainer Washer

Figure 5.5d

80 Collar

Figure 5.5e

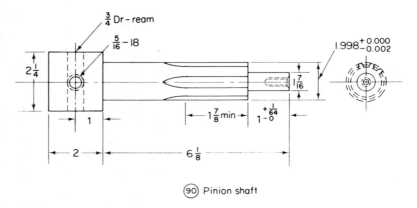

90 Pinion shaft

Figure 5.5f

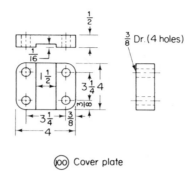

(100) Cover plate

Figure 5.5g

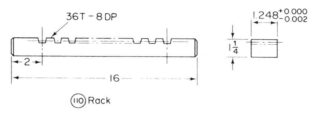

(110) Rack

Figure 5.5h

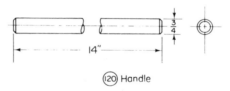

(120) Handle

Figure 5.5i

Parts 10 through 40 are standard and need not be processed. They are all in the vendor category. Part 50 is the main housing. It will be made from cast steel, which will be farmed out because this company does not do its own casting. The detail drawing is shown in Fig. 5.5b. However, in some instances this company will need to make the drawing for the wood pattern, subcontract the making of the pattern, check all dimensions, draft, and allowances, and supply it to the foundry. In almost all cases, the foundry will assume responsibility only for the quality of the material from which the casting is made. The operations and their categories for part 50 are shown in Table 5.1. They are not in sequence.

Part 60, Fig. 5.5c, is the bolster plate. It is to be a steel casting and therefore needs a pattern. The same procedure of supplying the pattern will be followed as in part 50. The operations are as shown in Table 5.2.

Part 70, Fig. 5.5d, is a retainer washer that secures the pinion shaft in the main

TABLE 5.1   PART 50—MAIN CASTING—
FIG. 5.5b

Description	Category
Pattern	Vendor
Casting	Vendor
Mill base—face *A*	Critical
Mill face *B*	Critical
Mill face *C*	Critical
Mill 1.250 slot	Auxiliary
Shaper—upper face of base ($2\frac{1}{2}$ in.)	Auxiliary
Drill, bore, ream (2.000 in.) hole	Auxiliary
Drill, ream (1 in.) hole in base	Critical
Drill (2) $\frac{3}{8}$-dia. holes in base	Auxiliary
Drill and tape (4) holes $\frac{3}{8}$–16	Auxiliary

TABLE 5.2   PART 60—BOLSTER
PLATE—FIG. 5.5c

Description	Category
Pattern	Vendor
Casting	Vendor
Turn face *A*	Critical
Turn diameter *B*	Critical
Turn $\frac{1}{2}$ dimension	Auxiliary
Turn 1-in. dia.	Auxiliary

housing but leaves it free to rotate. It is made from 1-in. cold-rolled bar stock. Since the 2-in. dimension has no function, it need not be machined. The parts will, therefore, be machined from a 2-in.-diameter cold-finished bar. The operations are shown in Table 5.3.

Part 80, Fig. 5.5e, is a collar that is assembled with the handle. Its outside diameter need not be machined and since it is purchased as cold-rolled steel, the

TABLE 5.3   PART 70—RETAINER
WASHER—FIG. 5.5d

Description	Category
2-in. bar stock	Vendor
Turn face *A*	Auxiliary
Turn diameter $1\frac{7}{16}$	Critical
Turn $\frac{1}{8}$ face	Critical
Cut off on lathe	Auxiliary
Turn overall $\frac{3}{8}$ dimension	Auxiliary
Drill $\frac{9}{16}$-in.-dia. hole	Auxiliary

finish is acceptable. The part will be machined from $1\frac{1}{4}$-in. bar stock. The operations are shown in Table 5.4.

TABLE 5.4   PART 80—COLLAR—
FIG. 5.5e

Description	Category
$1\frac{1}{4}$ CRS bar stock	Vendor
Turn face A	Critical
Drill $\frac{3}{4}$ hole to $\frac{47}{64}$-in. dia.	Auxiliary
Ream drill hole to $\frac{3}{4}$-in. dia.	Critical
Cut off to $1\frac{3}{8}$ in. long	Auxiliary
Turn overall length $1\frac{1}{4}$ in.	Auxiliary
Chamfer	Auxiliary
Drill No. 28 hole	Auxiliary

Part 90, Fig. 5.5f, is the pinion shaft and will be heat treated. It will need to be ground after heat treatment. Therefore, the bar stock will be $2\frac{3}{8}$ in.-diameter hot-rolled steel. This material has a scale surface that needs machining. The operations are shown in Table 5.5.

TABLE 5.5   PART 90—PINION SHAFT—FIG. 5.5f

Description	Category
$2\frac{3}{8}$ HRS bar stock	Vendor
Saw cut to $6\frac{1}{4}$	Auxiliary
Face end A	Critical
Center drill end A	Critical
Drill, tap, chamfer $\frac{1}{2}$–13 thread	Auxiliary
Face overall length	Auxiliary
Center drill	Critical
Turn $2\frac{1}{4}$ OD	Auxiliary
Turn OD between centers to 2.010 in.	Auxiliary
Turn OD to rough dimension $1\frac{1}{2}$ in.	Auxiliary
Turn OD to $1\frac{7}{16}$ in.	Auxiliary
Face 1-in. shoulder	Critical
Drill $\frac{49}{64}$-in. hole	Auxiliary
Ream $\frac{3}{4}$-in. hole	Auxiliary
Drill and tap $\frac{5}{16}$–18 NC	Auxiliary
Cut teeth	Auxiliary
Heat treat	Vendor
Lap centers	Critical
Grind 1.998 dia.	Auxiliary

Part 100, Fig. 5.5g, the cover plate, is made from 4-in.-wide by $\frac{5}{8}$-in.-thick CRS flat stock. Its sides and top need not be finished. The operations are shown in Table 5.6.

TABLE 5.6 PART 100—COVER
PLATE—FIG. 5.5g

Description	Category
4 × $\frac{3}{8}$-in. CRS flat stock	Vendor
Saw cut to $4\frac{1}{4}$-in. lg.	Auxiliary
Mill ends to 4 in.	Critical
Mill $\frac{1}{2}$-in. thickness	Critical
Mill $1\frac{1}{2}$-in. clearance slot	Auxiliary
Drill 4 holes $\frac{3}{8}$-in.	Auxiliary
Mill 4 radii	Auxiliary

Part 110, Fig. 5.5h, the rack, is made from HRS, $1\frac{3}{8} \times 1\frac{3}{8}$ square stock. It should be heat treated for toughness. The process is shown in Table 5.7.

TABLE 5.7 PART 110—RACK—
FIG. 5.5h

Description	Category
$1\frac{3}{8} \times 1\frac{3}{8}$ sq. bar stock	Vendor
Cut to $16\frac{1}{2}$-in.	Auxiliary
Face ends to 16 in.	Critical
Mill 4 sides to $1.260\,{}^{+0.000}_{-0.002}$	Critical
Cut teeth	Auxiliary
Chamfer one end	Auxiliary
Chamfer opposite end	Auxiliary
Heat treat	Vendor
Grind 4 sides	Auxiliary

Part 120, Fig. 5.5i, is the handle and is made from CRS round stock, and therefore the outside diameter need not be machined. This is shown in Table 5.8.

TABLE 5.8 PART 120—HANDLE—FIG. 5.5i

Description	Category
$\frac{3}{4}$ CRS (4130) bar stock	Vendor
Saw cut to length $14\frac{1}{8}$ in.	Auxiliary
Face one end and chamfer	Critical
Face overall to 14 in. and chamfer	Auxiliary

Part 130 is a subassembly of parts 30, 80, 90, and 120. This series of operations needs to be performed at assembly. The process is shown in Table 5.9. The subassembly is shown in Fig. 5.5a.

TABLE 5.9   PART 130—SUBASSEMBLY—FIG. 5.5a

Description	Category
At one end parts 120 and 80 are assembled	Assembly
No. 28 hole is drilled through	Assembly
Ream for No. 0 taper pin	Assembly
Insert No. 0 taper pin	Assembly
Assemble rod with pinion shaft part 90	Assembly
Insert screw 40 into pinion shaft	Assembly
Assemble part 80 at other end of rod	Assembly
Drill and taper ream through	Assembly
Insert No. 0 taper pin	Assembly

The remainder of the assembly operations require no machining. Subassembly 130 is assembled with part 50, using the washer, part 70. The rack, part 110, is inserted into the slot, part 50. The cover plate, part 100, is assembled to the main casting. The plate, part 60, is assembled with the casting, part 50.

5.7 GROUPING OF OPERATIONS

The operations, as listed in Sec. 5.6, are now ready for grouping into those that can be performed in one setup (CIM). This section will group the operations, but will *not* arrange them into the sequence in which one setup follows another.

There are many criteria to be considered when grouping the machining that is to be done on each part.

1. The *size of the order* will determine to a large extent how much can be spent on tools, fixtures, and special machinery. This, in turn, will be one of the factors that will determine how large the batch will be which is to run. It may very well be that the decision is made to run the entire job as one continuous batch. Thus the processor must always keep in mind the cost of labor, tooling, setup and tear down, maintenance, and the equipment if new equipment is to be purchased.

2. The *rate of production* is another of the factors that must be considered when writing process sheets. Production schedules and commitments to the customer or vendor must be met. The machines selected, the cost of labor, the batch sizes, and the tooling all must be aimed at some target date and cost schedule.

3. The *size of the workpiece* is another factor that must be considered when combining operations. Small pieces that have holes drilled on two sides may be enclosed in a box fixture and both the fixture and workpiece rotated on the drill press table. It might be impossible to invert a very large casting, and therefore it might not be practical to combine drilling operations on such a workpiece.

4. The *machinery available* is another consideration. The processor must be familiar with the limitations of the equipment in the plant. He must be aware of what the machines can do, their power rating, range, the size of the work which they can handle, their precision, and their dependability.

Availability, in another sense, deals with the amount of free time that can be applied to the job. It may be that a particular machine is suitable in every respect but is overloaded. Or it may be that the machine is at one end of the plant but is needed at the other end. Under these conditions, availability may revolve around the feasibility of transporting the workpieces to the machine, or the machine to the workpieces.

5. The *nature of the surfaces to be machined* will also enter into decisions when a part is processed. Adjacent surfaces may be machined in one setup. A surface that is controlled by close tolerances may need to be machined independently of all others. All holes that have one size or are in close proximity to each other may be drilled in one setup. It is generally not good practice to machine a finishing cut of one surface simultaneously with a roughing cut of another surface. The vibrations from the roughing cut will usually produce an unsatisfactory surface on the finished surface.

With these items as background, Fig. 5.5 and the nonsequential operations will now be studied to determine how many can be combined.

From Table 5.1, part 50, main casting, and Fig. 5.5b, and still *nonsequential*, the operations shown in Table 5.10 result.

TABLE 5.10   PART 50—MAIN CASTING—FIG. 5.5b

No.	Operation	Description
1	Vendor	Pattern
2	Vendor	Steel casing
3	Mill base—face *A*—drill and ream 1-in.-dia. hole	Use milling machine and face milling cutter, drills, and reamer
4	Mill face *B*, slot 1.250	Combine operation; locate off 1-in. hole
5	Mill face *C*	Locate off 1-in. hole
6	Drill 1.500 and 2.000 holes to $1\frac{3}{8}$ and $1\frac{7}{8}$ in.; bore to 1.409 and 1.990; line ream to size	Use gun drills, bore, and line reamer on milling machine; locate off 1-in. hole
7	Shape upper face to $2\frac{1}{2}$ in.	Locate off 1-in. hole and base
8	Drill and tap $\frac{3}{8}$–16 (4 holes) in face *B*; invert fixture and drill two $\frac{3}{8}$ holes	Drill head and 3 spindle drill press; use drill fixture

From Table 5.2, part 60, bolster plate, Fig. 5.5c, the nonsequential *combined* operations are shown in Table 5.11.

TABLE 5.11   PART 60—BOLSTER PLATE—FIG. 5.5c

No.	Operation	Description
1	Vendor	Pattern
2	Vendor	Steel casing
3	Machine face *A*	Lathe, hold on 7-in. OD
4	Machine diameter *B*, turn $\frac{1}{2}$ in. thickness, turn 1-in. boss, and turn 1-in. OD	Lathe, fixture hold in 4 slots

From Table 5.3, part 70, washer, Fig. 5.5d, the nonsequential *combined* operations are shown in Table 5.12.

TABLE 5.12 PART 70—RETAINER WASHER—FIG. 5.5d

No.	Operation	Description
1	Vendor	2-in.-dia. 1020 CRS bar stock
2	Turn and face; turn $1\frac{7}{16}$-in. OD × $\frac{1}{8}$ shoulder; drill $\frac{9}{16}$-in.-dia. hole; cut off to $\frac{7}{16}$ in. overall	Hand screw machine
3	Face overall to $\frac{3}{8}$ in.	Lathe, fixture

From Table 5.4, part 80, collar, Fig. 5.5e, the nonsequential *combined* operations are shown in Table 5.13.

TABLE 5.13 PART 80—COLLAR—FIG. 5.5e

No.	Operation	Description
1	Vendor	$1\frac{1}{4}$-in.-dia. CRS bar stock
2	Turn face A; drill $\frac{47}{64}$-in.-dia.; ream to $\frac{3}{4}$-in. dia.; cut off to $1\frac{3}{8}$ lg.	Hand screw machine
3	Face overall to $1\frac{1}{4}$ and chamfer	Lathe and fixture
4	Drill No. 28 hole	Drill press and fixture

From Table 5.5, part 90, pinion shaft, Fig. 5.5f, the nonsequential *combined* operations are shown in Table 5.14.

TABLE 5.14 PART 90—PINION SHAFT—FIG. 5.5f

No.	Operation	Description
1	Vendor	$2\frac{3}{8}$ HRS bar stock
2	Saw cut to $8\frac{1}{4}$ in. lg.	Hacksaw
3	Face end A; center drill; drill for $\frac{1}{2}$–13 thread; tap; chamfer thread 60°; rough turn $1\frac{7}{16}$ to $1\frac{1}{2}$ in.	Chuck in turret lathe
4	Face overall length to $8\frac{1}{8}$ in.; center drill turn $2\frac{1}{4}$ OD × $2\frac{1}{8}$ in. lg.	Lathe / Lathe
5	Turn OD to 2.010 in.; turn OD to $1\frac{7}{16}$ in.; face 1-in. shoulder × 1 in. lg.	Lathe; turn between centers
6	Drill $\frac{49}{64}$-in. hole; ream $\frac{3}{4}$ in.; drill and tap $\frac{5}{16}$–18 UNC	Drill press; fixture
7	Cut teeth	Gear hobber
8	Vendor	Heat treat
9	Lap centers	Drill press and 60° lap
10	Grind 1.998 dia.	Cylindrical grinder

From Table 5.6, part 100, cover plate, Fig. 5.5g, the nonsequential *combined* operations are shown in Table 5.15.

TABLE 5.15   PART 100—COVER PLATE—FIG. 5.5g

No.	Operation	Description
1	Vendor	$4 \times \frac{3}{8}$ thick CRS flat steel
2	Saw cut to $4\frac{1}{4}$ in. lg.	Power hacksaw
3	Mill $4\frac{1}{8}$ in. and 2 radii	Straddle mill; fixture
4	Mill 4 in. and 2 radii	Straddle mill; fixture
5	Mill $\frac{1}{2}$-in. thick, and $1\frac{1}{2}$ wide clearance slot	Form cutter; fixture
6	Drill 4 holes $\frac{3}{8}$-in. dia.	Fixture and $\frac{3}{8}$ drill

From Table 5.7, part 110, rack, Fig. 5.5h, the nonsequential *combined* operations are shown in Table 5.16.

TABLE 5.16   PART 110—RACK—FIG. 5.5h

No.	Operation	Description
1	Vendor	$1\frac{3}{8} \times 1\frac{3}{8}$ sq. HRS (4140) bar stock
2	Saw to length $16\frac{1}{2}$ in.	Power saw
3	Face ends to 16 in. lg.	Lathe
4	Mill 4 sides to $1.260 \, {}^{+.000}_{-.002}$	Milling machine, shell cutter, fixture
5	Chamfer both ends	Belt sander, fixture
6	Cut teeth	Gear generator, fixture
7	Vendor	Heat treat
8	Grind 4 sides	Surface grinder

From Table 5.8, part 120, handle, Fig. 5.5i, the nonsequential *combined* operations are shown in Table 5.17.

TABLE 5.17   PART 120—HANDLE—FIG. 5.5i

No.	Operation	Description
1	Vendor	$\frac{3}{4}$-dia. CRS (4130) bar stock
2	Saw cut to $14\frac{1}{8}$ lg.	Power saw
3	Face both ends and chamfer to 14 in. lg.	Lathe

From Table 5.9, part 130, subassembly, Fig. 5.5a, the nonsequential *combined* operations are shown in Table 5.18.

TABLE 5.18   PART 130—SUBASSEMBLY—FIG. 5.5a

No.	Operation	Description
1	Assemble Parts 80, 120, 40, and 90	Wing screw (40), pinion shaft (90) and collars (80)
2	Drill No. 28 hole (2) and taper ream No. 0	No. 28 drill, No. 0 taper reamer
3	Insert taper pins (2)	Pins (30)

5.8 OPERATIONS ARRANGED IN SEQUENCE

The next step is to rearrange the combined operations into the *sequence* in which they will be done. Thus in Table 5.10, for Fig. 5.5b, the eight operations will be rearranged as shown in the operation sheet—part 50, Table 5.19.

TABLE 5.19   OPERATION SHEET—PART 50—MAIN CASTING

No. from Table 5.10	Operation number	Operation	Machine	Tools and fixtures
1	10	Pattern	Vendor	
2	20	Steel casting	Vendor	
—	30	Visual inspection	Insp. dept.	
3	40	Mill base—face A—drill and ream 1-in. hole	Horizontal miller	Face mill cutter, $\frac{63}{64}$-in. drill, 1-in. reamer, fixture
5	50	Mill face C	Horizontal miller	Fixture locate off 1-in. ream hole and facemill cutter
6	60	Drill 1.500 and 2.000 hole to $1\frac{3}{8}$ and $1\frac{7}{8}$ in; bore to 1.490 and 1.990; line ream to size	Horizontal miller	$1\frac{3}{8}$ and $1\frac{7}{8}$ gun drills; boring bar with blocks; line reamer; fixture with change bushings; locate off 1-in. hole
4	70	Mill face B, and 1.250-in. slot	Horizontal miller	$1\frac{1}{4}$ shell mill; fixture located off 1-in. hole
—	80	Inspect all milling	Insp. dept.	
7	90	Shape upper face of $2\frac{1}{2}$-in. dimension	Shaper	Fixture
8	100	Drill and tap $\frac{3}{8}$–16 (4) holes in face B, invert fixture and drill (2) $\frac{5}{8}$ holes	Radial drill press	Drill fixture tap drill $\frac{5}{16}$ dia. $\frac{3}{8}$—16 tap $\frac{5}{8}$-in. drill
	110	Final inspection	Insp. dept.	

It should be noted that the operation sheet does not include the various plug, thread, and distance gages which will be used at each operation and in the inspection department. These will be covered in a subsequent chapter dealing with gages. The design of the various fixtures is also covered later.

The rearrangement of the operations from Table 5.11 is shown in operation sheet—part 60, Table 5.20.

TABLE 5.20   OPERATION SHEET—PART 60—BOLSTER PLATE

No. from Table 5.11	Operation number	Operation	Machine	Tools and fixtures
1	10	Pattern	Vendor	
2	20	Steel casting	Vendor	
	30	Visual inspection	Insp. dept.	
3	40	Turn face *A*	Lathe	Grip in chuck on 7-in. OD
4	50	Turn dia. *B*, turn $\frac{1}{2}$-in. thickness, turn 1-in. boss, and turn 1-in. OD	Lathe	Special jaws to grip plate in 4 slots; square turret
	60	Final inspection	Insp. dept.	

The rearrangement of the operation from Table 5.12 is shown in the operation sheet—part 70, Table 5.21. Since this part is rather simple there is no rearrangement necessary. Table 5.12, operation, will be repeated here for continuity.

TABLE 5.21   OPERATION SHEET—PART 70—RETAINER WASHER

No. from Table 5.12	Operation number	Operation	Machine	Tools and fixtures
1	10	2-in.-dia. CRS 1020 bar stock	Vendor	
2	20	Turn and face; turn $1\frac{7}{16}$ in. OD $\times$ $\frac{1}{8}$ shoulder; drill $\frac{9}{16}$ hole; cut off to $\frac{7}{16}$ in.	Hand screw machine	$\frac{9}{16}$ drill; parting tool
3	30	Face overall to $\frac{3}{8}$ in.	Lathe	Fixture to grip $1\frac{7}{16}$ in.
	40	Final inspection	Insp. dept	

The collar in Fig. 5.5e and in Table 5.13 has been repeated in operation sheet—part 80, Table 5.22, for continuity of presentation.

The rearrangement of the operations from Table 5.14 is shown in the operation sheet—part 90, Table 5.23.

TABLE 5.22 OPERATION SHEET—PART 80—COLLAR

No. from Table 5.13	Operation number	Operation	Machine	Tools and fixtures
1	10	$1\frac{1}{4}$-in.-dia. CRS bar stock	Vendor	
2	20	Turn face A; drill $\frac{47}{64}$-in.-dia. hole; ream to $\frac{3}{4}$-in. dia.; cut off to $1\frac{3}{8}$ in. lg.	Hand screw machine	$\frac{47}{64}$ in. drill; $\frac{3}{4}$ reamer; parting tool
3	30	Inspection	Insp. dept.	
3	40	Face overall to $1\frac{1}{4}$-in. chamfer	Lathe	Fixture
4	50	Drill No. 28 hole	Drill press	Fixture

TABLE 5.23 OPERATION SHEET—PART 90—PINION SHAFT

No. from Table 5.14	Operation number	Operation	Machine	Tools and fixtures
1	10	$2\frac{3}{8}$ HRS bar stock	Vendor	
2	20	Cut to $8\frac{1}{4}$ in. lg.	Power hack saw	
3	30	Face end A; center drill; drill $\frac{27}{64}$ in. hole; tap $\frac{1}{2}$–13 UNC; chamfer 60°; rough turn $1\frac{7}{16}$ to $1\frac{1}{2}$ in.	Turret lake	Chuck; drills; taps; 60°C', sink
4	40	Turn overall to $8\frac{1}{8}$ in. center drill; turn $2\frac{1}{4}$ OD × $2\frac{1}{8}$ in. lg.	Lathe	Center drill
	50	Inspection	Insp. dept.	
5	60	Turn OD to 2.010 in. turn OD to $1\frac{7}{16}$ in.; face 1-in., shoulder 1 in. lg.	Lathe	Turn between centers
6	70	Drill $\frac{49}{64}$ hole; ream $\frac{3}{4}$ in.; drill F hole; tap $\frac{5}{16}$–18 UNC	Drill press	Fixture; drills and taps, reamer
	80	Inspection	Insp. dept.	
7	90	Cut teeth	Gear hobber	Hob
8	100	Heat treat	Vendor	
	120	Inspect heat treatment	Insp. dept.	
9	130	Lap centers	Drill press	60° lap
10	140	Grind 1.998 OD	Cylindrical grinder	
	150	Final inspection	Insp. dept.	

The rearrangement of the operations from Table 5.15 is shown in the operation sheet—part 100, Table 5.24.

TABLE 5.24 OPERATION SHEET—PART 100—COVER PLATE

No. from Table 5.15	Operation number	Operation	Machine	Tools and fixtures
1	10	$4 \times \frac{3}{8}$-thick CRS flat stock	Vendor	
2	20	Mill $\frac{1}{2}$-in. thick; $1\frac{1}{2}$-in. wide clearance	Milling machine	Fixture; form cutter
	30	Inspection	Insp. dept.	
3	40	Saw cut to $4\frac{1}{4}$ in. lg.	Power saw	
4	50	Mill $4\frac{1}{8}$in.; 2 radii	Miller	Straddle mill fixture
5	60	Mill 4 in.; 2 radii	Miller	Straddle mill fixture
6	70	Drill 4 holes $\frac{3}{8}$-in. dia.	Drill press	Fixture and $\frac{3}{8}$ drill
	80	Final inspection	Insp. dept.	

The rearrangement of the operations from Table 5.16 is shown in the operation sheet—part 110, Table 5.25.

TABLE 5.25 OPERATION SHEET—PART 110—RACK

No. from Table 5.16	Operation number	Operation	Machine	Tools and fixtures
1	10	$1\frac{3}{8} \times 1\frac{3}{8}$ sq. bar stock HRS-4140	Vendor	
2	20	Saw to $16\frac{1}{2}$ in. lg.	Power saw	
3	30	Face ends to 16 in. lg.	Lathe	
4	40	Mill four sides to $1.260^{+.000}_{-.002}$	Miller	Fixture; shell cutter
5	50	Chamfer both ends	Belt sander	Fixture
	60	Inspection	Insp. dept.	
6	70	Cut teeth	Gear cutter	Fixture; hob
	80	Inspection	Insp. dept.	
7	90	Heat treat	Vendor	
	100	Inspection	Insp. dept.	
8	110	Grind four sides	Surface grinder	Fixture
	120	Final inspection	Insp. dept.	

The rearrangement of the operations from Table 5.17 is shown in the operation sheet—part 120, Table 5.26.

TABLE 5.26 OPERATION SHEET—PART 120—HANDLE

No. from Table 5.17	Operation number	Operation	Machine	Tools and fixtures
1	10	¾-dia. CRS bar stock 4130	Vendor	
2	20	Saw cut to 14⅛ lg.	Power saw	
3	30	Face both ends to 14 in. and chamfer	Lathe	
4	40	Final inspection	Insp. Dept.	

The rearrangement of the operations from Table 5.18 is shown in the subassembly operation sheet—part 130, Table 5.27.

TABLE 5.27 OPERATION SHEET—PART 130—SUB ASSEMBLY

No. from Table 5.18	Operation number	Operation	Machine	Tools and fixtures
1	10	Assemble parts 80, 120, 40, and 90	Assembly dept.	
2	20	Drill No. 28 holes (2); ream No. 0 taper	Drill press	No. 28 drill; No. 0 taper reamer
3	30	Insert taper pins		
	40	Final inspection	Insp. dept.	

Table 5.28 is the final assembly operation sheet.

It is now possible to determine the cost of producing the assembly and the types of machines, fixtures, gages, and cutters needed.

TABLE 5.28 OPERATION SHEET—ASSEMBLY

Operation Number	Operation	Tools and fixtures
10	Assemble part 130 (subassembly) with main casting part 50; and washer part 70 and bolt part 20	
20	Assemble rack, part 110; main casting, part 50; cover plate, part 100, and bolts, part 10	
30	Plate, part 60, is assembled with the main casting, part 50	
40	Final inspection	

PROBLEMS

5.1. Write a short explanation of the significance of the following questions as related to processing.

(a) How adaptable is the present equipment to the proposed job?

(b) Do the present personnel have the know-how to do the proposed job?

(c) Is this type of industry suitable for the proposed project?

5.2. A company is to sell its own product. List the activities that must be figured into the cost of the product.

5.3. What do you think a market analysis accomplishes for a company preparing to sell a new product?

5.4. List the eight steps that may be followed when processing a new job.

5.5. A company converts its machine tools from conventional to CNC. What is the impact on project analysis?

5.6. The process sheet writer will find help by studying the assembly drawing. List the items that will be revealed.

5.7. Repeat Prob. 5.6 for the working detail drawing.

5.8. Repeat Prob. 5.6 for the bill of materials.

5.9. Repeat Prob. 5.6 for the general specifications sheets.

5.10. Of what value to the process sheet writer are

(a) the customer,

(b) supplier of standard parts,

(c) subcontractors,

(d) shop personnel, and

(e) other departments? Write a short statement about each.

5.11. What questions should the designer attempt to answer when he is thinking about redesigning a part?

5.12. **(a)** Explain the advantages gained when combining several parts into one part.

(b) Given an example of such a redesign and, referring to part (a), explain the advantages.

5.13. What role should the selection of the method of fabrication have in the decision to redesign a part?

5.14. Why is it generally advisable to substitute a standard part for one that must be machined?

5.15. **(a)** List the five indicators of critical surfaces.

(b) Why are they important as critical surfaces?

5.16. List and explain each of the four major categories used when processing for manufacture.

5.17. A casting may be "farmed out" or produced in the "home" plant. What effect will the decision to do one or the other have on the process categories to which an operation is assigned?

5.18. Discuss and support with examples vendor-nonmachined operations and vendor-machined operations.

5.19. Explain, and use an example, why the same banking surface should be used for as many operations as possible.

5.20. Explain the importance of the choice of the initial banking surface.

5.21. Discuss and illustrate

(a) auxiliary-nonmachined surfaces, and

(b) auxiliary-machined surfaces.

5.22. Which machining operations are processed under the category of assembly?

5.23. When grouping operations into the machining sequence, which criteria should be considered?

5.24. Discuss the effect of the size of an order upon the various factors that must be considered when processing a job.

5.25. Repeat Prob. 5.24 for the rate of production.

5.26. Repeat Prob. 5.24 for the size of the workpiece.

5.27. Repeat Prob. 5.24 for the machinery that is available.

5.28. Repeat Prob. 5.24 when related to the nature of the surfaces to be machined.

5.29. Using Fig. 5.6,

    (a) select the critical surfaces,

    (b) list the operations without reference to sequence to be done,

    (c) combine the operations that you intend to machine in one setup, and

    (d) rearrange the operations for quality and economical production.

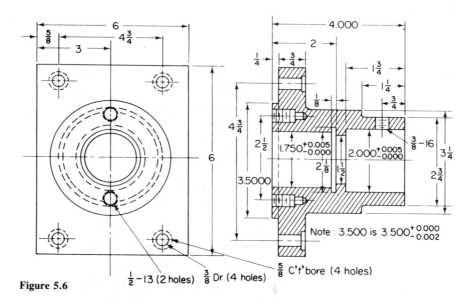

Figure 5.6

5.30. Repeat Prob. 5.29 for Fig. 5.7.

5.31. Process the assembly in Fig. 5.8 for production using the procedures discussed in this chapter. Apply your own dimensions.

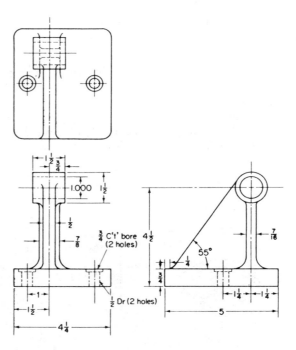

Figure 5.7

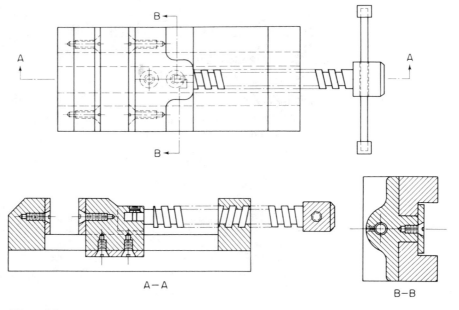

A—A   B—B

Figure 5.8

5.32. Process the hand tapping assembly, Fig. 5.9, for production. Apply your own dimensions.

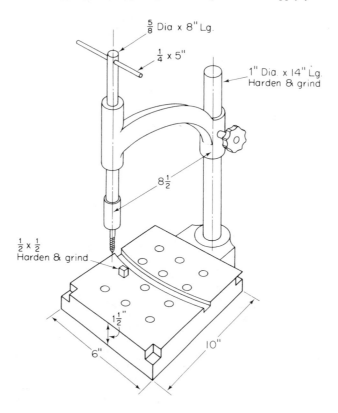

$\frac{5}{8}$ Dia x 8" Lg.

$\frac{1}{4}$ x 5"

1" Dia. x 14" Lg.
Harden & grind

$8\frac{1}{2}$

$\frac{1}{2}$ x $\frac{1}{2}$
Harden & grind

$1\frac{1}{2}$"

6"

10"

Figure 5.9

5.33. Process the magnetic V-blocks shown in Fig. 5.10 for production. Note that each of the magnetic and nonmagnetic plates must be absolutely flat before they are riveted.

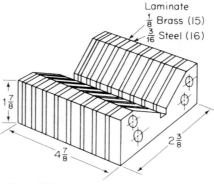

Laminate
$\frac{1}{8}$ Brass (15)
$\frac{3}{16}$ Steel (16)

$1\frac{7}{8}$

$4\frac{7}{8}$

$2\frac{3}{8}$

Figure 5.10

5.34. Process the compound die, Fig. 5.11. Apply your own dimensions.

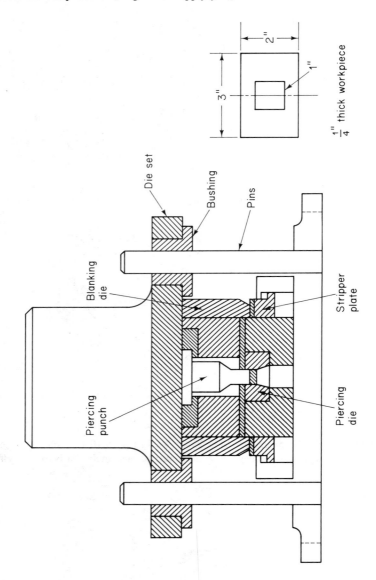

Figure 5.11

Chapter 6

Cost Analysis

6.1 COST METHODS

There are many methods available which can be used to calculate the cost of manufacturing a product by a company. In general, the cost is defined as the expenses charged to an account for items purchased, for the cost of distribution of items sold, or for the distribution or production of services. Specifically, the cost of manufacturing refers to direct and overhead costs. It is the cost of those products that are ready for shipment.

Direct costs are those that enter directly into the manufacturing of the product. They are those costs of *direct labor* expended for work done directly on the workpiece. The cost of the raw material that becomes part of the product is referred to as *direct material* cost.

Overhead costs are the costs of those activities that do not enter directly into the manufacture of the product. They may be divided into *indirect labor, indirect material,* and *fixed* costs. Examples of indirect labor are inspection, setup, maintenance of machines and tools, safety devices on machines, chip removal and parts cleanup, and material and parts routing and moving. Indirect materials are those materials that are needed to build jigs, fixtures, gages, and standard parts such as nuts, bolts, rivets, and pins. Examples of fixed costs are insurance, taxes, depreciation, and interest on investment.

It is not the intention of this book to investigate the various methods used to calculate cost. It is of the utmost importance that a tool designer understand that whatever he designs will need to be justified, among other things, in terms of the cost of producing that item. There are two major methods used for calculating costs: based

141

on the *actual cost* of production; and the *"standard" cost* method, which relies on estimates based on standard data generated from past performances.

Actual cost includes such items as the cost of materials, direct labor, and an equitable distribution of the other manufacturing expenses incurred in support of the enterprise. The actual cost method, however, relies on calculating the cost of production after the production effort has started.

The *standard cost method* is based on standards and tries to determine the cost of a product before it is put into production. Basically, this method of calculating costs is based on (1) a determination of the quantities of materials used and their cost based on standard pricing; (2) a determination of the standard labor costs from standard wage rates, time studies of the cost, or from established standard timetables; and (3) standard overhead costs, fairly apportioned to the job at hand. More specifically, there are many other ways of calculating the costs of production.

Probably the most widely used method of figuring costs is the *job-order method*. In this method, an order number is assigned to the job, a lot, or an item. All raw materials, standard parts, tools, and so on, are charged to that job number. Labor charges are collected from time tickets and charged directly to the job number. Overhead charges are distributed according to some preestablished rate. The summation of all of these charges divided by the total number of units produced yields the unit price. The cost of production of tools, jigs, fixtures, dies, and so on, may be calculated using this method.

6.2 ESTIMATING DIRECT LABOR COSTS

As already mentioned, *direct labor* is that effort which is expended directly on the product during the time that it is being manufactured. It may be divided into four major efforts: (1) setup, (2) nonmachining, (3) machining, and (4) down times.

The *setup time* includes the time required to collect all the tools and materials needed for a particular operation. It also includes the time required to set these into the machine and to make the necessary trial runs and adjustments. Also included in setup time is the time to tear down a setup. A setup *or* tear-down is charged to the job as a unit every time that it is made. Thus, if there are five job lots, and a setup is required for each lot, the entire job should be charged with five setup and teardown times.

Nonmachining time includes those activities that are necessary to the production of the workpiece but are not reflected directly in it. They constitute activities that take place at times other than during the actual machining. These activities are cutter or tool sharpening, inspection, personal needs, and fatigue. These activities are a part of the real time that must be considered but which are auxiliary and not part of the actual working hours.

In addition to the activities above, peripheral to the actual machining, there are other activities associated with the actual machining. If these activities are missing, the machining could not take place. Such activities are: advancing or retracting the cutting tool, tightening a chuck, a trial cut, trial measurements, changing the tool in the post, deburring in the machine, cleaning a fixture, loading and clamping the part, and others. It should be noted that the time required to perform these activities has been studied extensively and may be obtained from good time-and-motion study books.

Machine time is that time which deals with the actual cutting of the material, and two additional quantities—*overtravel* and *approach*.

The *overtravel* is a small amount that needs to be added to the actual length of cut at the points where the automatic feed is engaged or disengaged. This is a small amount, typically $\frac{1}{8}$ to $\frac{3}{8}$ in. It is quite different from the ''advancing or retracting of the tool'' as stated under nonmachining time. These two overtravel quantities are necessary because it is not possible to advance the tool by hand to the exact point of cutting. In the same way, it is impossible to stop the feed at the very end of the cutting cycle.

The *approach* is the additional feed time that must be added to the actual time required to cut a given length of work. This is necessary because of the contour of the cutting tool, such as the radius of a lathe tool or the diameter of a cutter.

Another of the direct labor costs is *downtime*. This includes any downtime that results from unpredictable circumstances. If the machinery in a plant is comparatively new, the downtime that results from breakdown is less than if the machinery is old. Mishaps will cause some downtime. In other instances, the machinist may have to wait for materials or tools to be delivered to his machine. Usually, a case history of such events will permit the establishing of a multiplication factor that will take care of the frequency of such events at a particular plant.

6.3 SETUP TIME AND ECONOMIC LOT SIZE

The *setup time* is allocated to a particular lot size on the basis of past experience. It is possible to break down the setup procedure into elements small enough so that the time required to perform an element may be estimated, checked, or obtained from a time study book.*

The *number of setups* needed for a job is a function of the *economic lot size*. The economic lot size will yield the lowest cost for each unit to be produced. It is a reflection of the cost of setup, tooling, handling, and carrying charge, when related to the cost of storing finished parts. Thus, the following equation is one that will provide the economic lot size needs for most plants. It is, however, only a guide and should be used to provide an approximation and starting point for estimating costs.

The economic lot size is

$$L = 5 \sqrt{\frac{MS}{KC}}$$

$L =$ lot size, pieces
$M =$ pieces used per month
$C =$ cost per piece, dollars
$K =$ carrying charge factor, %* (5 to 30%)

*W. A. Nordhoff, *Machine Shop Estimating* (New York: McGraw-Hill, 1960).

*Includes interest, taxes, insurance, storage, obsolescence, and so on.

$$S = \text{setup cost per log, dollars}$$
$$s = \text{setup, hours}$$

The hours permitted for each setup is

$$s = \frac{LS}{M}$$

Example 6.1

Assume that a company uses 2000 pieces per month in its assembly department. The cost of each piece is \$25.00 and the cost of each setup is \$50.00 per lot. If a carrying charge factor of 30% is used, calculate (a) the economic lot size and (b) the setup hours permitted for each lot.

Solution

(a) The economic lot size is

$$L = 5\sqrt{\frac{MS}{KC}} = 5\sqrt{\frac{2000 \times 50}{0.30 \times 25}}$$
$$= 578 \text{ pieces}$$

$M = 2000$ pieces per month
$S = \$50.00$ per lot
$C = \$25.00$ per piece
$K = 30\%$

(b) The hours permitted for each setup is

$$s = \frac{LS}{M} = \frac{578 \times 50}{2000}$$
$$= 14.45 \text{ hr}$$

6.4 NONMACHINING TIMES

Nonmachining, as indicated, deals with the time consumed while the machine is not in operation. The easiest way to deal with these costs is to develop percentages of the machining and nonmachining times. Thus if the total estimated time required to handle and machine a workpiece is 26 min and 5% is allowed for *personal needs,* (26 × 0.05) or 1.3 min is added to 26 min as an additional allowance to be charged against the workpiece.

The various percentage ranges that are generally used for *fatigue* depends on the activity and the manner in which it is carried on in the manufacturing enterprise. Close tolerance and very heavy work will require a 25 to 30% allowance for fatigue. For short cutting cycles the range is from 15 to 20%, and for very long cuts the fatigue range allowance is about 10%. Automatic machinery working to close tolerances may require 15 to 20% and normal work tolerances as little as 5%.

Tool sharpening is another of the nonmachining times that must be included in the cost of production. Tool sharpening is related to the amount of cutting which the tool is doing and the tool life. Included in this calculation is the amount of time needed to change the tool. Thus the sum of the length of time (T_k) a tool can be used between sharpenings and the time (T') required to remove and reset a sharp tool divided by the length of the life (T) of the tool is the portion (t) of the total life of the tool added to each cycle. Thus

$$t = \frac{T_k T'}{T}$$

T_k = cutting time, min
T' = tool change time, min
T = life of tool, min
t = time added to cost, min

Example 6.2

A tool will cut for 4 hr before it needs sharpening. It takes 18 min to change the tool. If the tool can be sharpened 15 times before it is discarded, how much time is charged to one cycle?

Solution

$$t = \frac{T_k T'}{T} = \frac{240 \times 18}{3600} \qquad \begin{aligned} T' &= 18 \text{ min} \\ T_k &= 4 \times 60 = 240 \text{ min} \end{aligned}$$

$$= 1.2 \text{ min} \qquad T = 15 \times 4 \times 60 = 3600 \text{ min}$$

Measuring and *gaging* are other intangibles that depend on the type of operation performed, when the check is made, the type of measuring or checking device used, the precision and contour of the work surface, and so on. If the operation is performed in an automatic machine, the parts may require spot checking only while the operation is running. This would require no cost allowance for measuring and gaging. If the operation is precision, it may require checking each piece. In any event, the allowances may range from 0 to 30 sec for each measurement or gage check made.

An equation that is sometimes used relies on a knowledge of the time required to make each check and the frequency of checking. The equation is

$$t_c = \frac{T_c}{N + 1}$$

t_c = measuring time allowance, min
T_c = checking each dimension, min
N = number of parts
F = frequency, parts/min

$$\text{or} \quad t_c = FT_c$$

t_c is the time for checking a specific dimension. If multiplied by the number of parts in a lot, the time will be the total time required to check that dimension. If the value T_c is computed for other dimensions, their sum will produce the total time required to measure and gage all dimensions for all the parts in a lot. This can be seen by solving the equation above for T_c:

$$T_c = t_c (N + 1)$$

Example 6.3

It takes a machinist 12 sec to check a hole that he is boring on a lathe with both ends of a GO/NO GO plug gage. He uses the gage twice for each piece and he has 300 pieces to bore. How much time must he charge to the job to pay for this activity?

Solution   The time allotted for each piece is

$$t_c = T_c \times F = 12(2/60) \qquad \begin{aligned} T_c &= 12 \text{ sec} \\ F &= 2 \end{aligned}$$
$$= 0.40 \text{ min}$$

The total time required to check all pieces for this dimension is

$$T_c = t_c(N + 1) = 0.40(300 + 1) \qquad \text{where } t_c = 0.40 \text{ min}$$
$$= 121 \text{ min} \qquad\qquad\qquad\qquad\qquad N = 300 \text{ pcs}$$

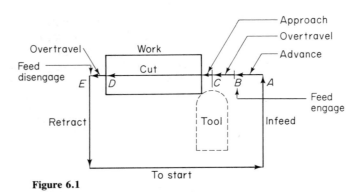

Figure 6.1

6.5 MACHINING TIME: CUTTING SPEED

This element of direct cost refers to the actual time that the tool cuts. Elements of feed that take place just prior to the time when the tool starts to cut, and after the tool stops cutting, are to be added to the length of cut. This is called *overtravel*. The allowance is generally between ⅛ and ⅜ in. Overtravel may be added to the length of the workpiece and the approach as part of direct labor.

Distances A to B and E to A are values that are added to the noncutting time. They are referred to as *advancing* and *retracting time*. They are found in tables of standard times. Figure 6.1 shows these tool motions as they relate to a lathe. A complete cycle in Fig. 6.1 consists of advancing the tool from A to B. At B the feed is engaged. It travels to point C, where it contacts the work and begins to cut. It cuts until it reaches D. Notice that distance CD includes the approach. The tool is allowed to feed past point D to E, where it is disengaged. The length of cut is the sum of the length of the workpiece, the approach, and the two overtravels. The tool is retracted and returned to point A to complete the cycle. This section deals with the methods used to calculate the actual cutting time and the overtravel.

Machine tools may be classified as *rotary* or *reciprocating*. Tables have been developed which state the ideal relationship between the movement of the work and the tool material. This relative motion is called *cutting speed*. The cutting speed is the length of cut (lathe) taken in 1 min. Since the largest diameter cut has the greatest surface speed, it is the removal of the material at the largest diameter that will generate the most heat energy. It is this diameter that must be considered when calculating tool breakdown which results from the development of this heat and which determines the cutting speed.

From the cutting speed definition, if the circumference of a workpiece is 2 in. and a tool cuts one revolution of material in 1 min from its surface, the cutting speed is 2 in. per minute. If the work turns at 2 revolutions per minute, the cutting speed is 4 in. per minute. That is, a length of 4 in. of material is removed every minute that the tool cuts.

In Fig. 6.2 the tool has a point and travels (feeds) from right to left. In one revolution of the work it will cut one circumference. Thus

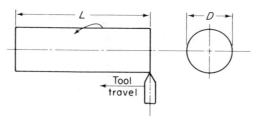

Figure 6.2

$$C = \pi D \qquad \begin{array}{l} D = \text{diameter, in.} \\ C = \text{circumference, in.} \end{array}$$

If the work revolves at N revolutions per minute, the length of the chip cut in 1 minute will be

$$\text{chip length} = \pi DN \qquad N = \text{revolutions per minute (rpm)}$$

Since the cutting speed is the length in units of feet of the chip removed every minute, the cutting speed equation is

$$S = \frac{\pi DN}{12} \qquad \begin{array}{l} S = \text{cutting speed, ft/min} \\ D = \text{diameter in.} \\ N = \text{rpm} \end{array}$$

This equation is the same for the lathe, drill press, and milling machine. Cutting speeds for selected materials are given in Table 6.1. More extensive tables may be found in any good reference book.

TABLE 6.1   CUTTING SPEEDS[a] (FT/MIN)

Work material	Cutter material			
	Carbon	HSS[b]	Cast Steel	Carbide
Aluminum	300	700	900	1000
Brass	50	200	500	700
Bronze	40	120	200	300
Copper	40	120	250	300
Iron				
Gray	30	50	125	150
Malleable	50	100	175	250
Magnesium	250	700	1000	1200
Steel				
Soft	40	90	250	400
Medium	30	70	200	250
Hard	—	40	100	150
Stainless	30	70	150	250

[a]Values are approximate.
[b]High-speed steel.

Example 6.4

A lathe is to take one cut on the diameter of a soft-steel shaft 2.5 in. in diameter. Calculate the rpm if a high-speed steel tool is used.

Solution

$$N = \frac{12S}{\pi D} = \frac{12 \times 90}{\pi \times 2.5} \qquad \begin{aligned} D &= 2.5 \text{ in.} \\ S &= 90 \text{ ft/min (Table 6.1)} \end{aligned}$$
$$= 138 \text{ rpm}$$

If the machine used is a drill press and the operation performed is drilling, the same formula is used and D is the diameter of the drill.

Example 6.5

A drill press is used to drill six holes into a cast iron block. The drill bit is $\frac{5}{8}$ in. in diameter and is made from high-speed steel. Calculate the rpm of the drill bit.

Solution

$$N = \frac{12S}{\pi D} = \frac{12 \times 50}{\pi \times 5/8} \qquad \begin{aligned} D &= \tfrac{5}{8} \text{ in.} \\ S &= 50 \text{ ft/min} \end{aligned}$$
$$= 306 \text{ rpm}$$

If the machine used is a milling machine, the same formula is used and D is the diameter of the cutter.

Example 6.6

A milling machine uses a carbide-tooth cutter to mill a copper block. The diameter of the cutter is 10 in. Calculate the rpm of the cutter.

Solution

$$N = \frac{12S}{\pi D} = \frac{12 \times 300}{\pi \times 10} \qquad \begin{aligned} D &= 10 \text{ in.} \\ S &= 300 \text{ ft/min} \end{aligned}$$
$$= 115 \text{ rpm}$$

The cutting speed of reciprocating machines must include the cutting as well as the noncutting part of the stroke.

The planer table oscillates under the tool bit. With each reverse motion of the table, the tool feeds a predetermined distance (feed) so that on the forward stroke, the tool will remove material from the work. The return stroke is generally faster than the cutting stroke by a quantity that is designated as the *ratio*. The equation for cutting speed on a planer is

$$S = N\left(L + \frac{L}{r}\right) \qquad \begin{aligned} L &= \text{length of stroke, ft} \\ r &= \text{ratio} \\ N &= \text{rpm} \\ S &= \text{cutting speed, ft/min} \end{aligned}$$

The return speed of the planer is

$$S_r = Sr \qquad \begin{aligned} S &= \text{cutting speed, ft/min} \\ S_r &= \text{return speed, ft/min} \\ r &= \text{ratio} \end{aligned}$$

Example 6.7

A planer, ratio 2, is to cut the surface of a large block of cast iron. The block is 4 ft long. It is to be machined with a high-speed tool bit. (a) How many strokes per minute should be used? (b) Calculate the return speed of the table.

Solution

(a) The number of strokes per minute is

$$N = \frac{S}{L + \dfrac{L}{r}} = \frac{50}{4 + \dfrac{4}{2}} \qquad \begin{aligned} L &= 4 \text{ ft} \\ r &= 2 \\ S &= 50 \text{ ft/min} \\ S_r &= \text{return speed} \end{aligned}$$

$$= 8.3 \text{ strokes/min}$$

(b) The return speed of the planer table is

$$S_r = 50 \times 2 = 100 \text{ ft/min}$$

The shaper is a smaller reciprocating machine than the planer. Its tool is carried by a ram which oscillates, and the table (work) feeds with each stroke.

Crank shapers have a ratio of 3:2 cutting to return stroke. That is, three-fifths of the circumference of the bull wheel is used to do the cutting; whereas two-fifths of the circumference of the bull wheel is used to return the tool to its starting position. The equations are:

Cutting speed:

$$S = \frac{L}{12}(N)\frac{5}{3} = \frac{LN}{7.2} \qquad \begin{aligned} S &= \text{cutting speed, ft/min} \\ S_r &= \text{return speed, ft/min} \\ L &= \text{length of stroke, in.} \\ N &= \text{strokes/min} \end{aligned}$$

Return speed:

$$S_r = \frac{L}{12}(N)\frac{5}{2} = \frac{LN}{4.8}$$

Example 6.8

A crank shaper is to machine a block of mild steel, 3 in. long, with a high-speed-steel tool bit. Calculate (a) the strokes per minute used and (b) the return speed.

Solution

(a) The number of strokes per minute is

$$N = \frac{7.2 \times S}{L} = \frac{7.2 \times 70}{3} \qquad \begin{aligned} L &= 3 \text{ in.} \\ S &= 70 \text{ ft/min} \end{aligned}$$

$$= 168 \text{ st/min}$$

(b) The return speed is

$$S_r = \frac{LN}{4.8} = \frac{3 \times 168}{4.8}$$
$$= 105 \text{ ft/min}$$

For *hydraulic shapers* with a ratio of 2:1, the equation is

$$S = \frac{LN}{8} \qquad \begin{aligned} S &= \text{cutting speed, ft/min} \\ L &= \text{length of stroke, in.} \\ N &= \text{strokes/min} \end{aligned}$$

Example 6.9

Repeat Ex. 6.8 for a hydraulic shaper.

Solution

(a) The number of stokes per minute is

$$N = \frac{8S}{L} = \frac{8 \times 70}{3} \qquad \begin{aligned} S &= 70 \text{ ft/min} \\ L &= 3 \text{ in.} \end{aligned}$$
$$= 187 \text{ strokes/min}$$

(b) The return stroke is

$$S_c = S \times 2 = 70 \times 2 = 140 \text{ ft/min}$$

6.6 MACHINING TIME: TIME TO MACHINE

The machining time is calculated by dividing the length of the cut, or the length of the stroke, by the feed of the tool in inches per minute. Thus

$$T = \frac{L \text{ in.}}{F \text{ in./min}} = \text{min} \qquad \begin{aligned} L &= \text{length of cut, in.} \\ F &= \text{feed, in./min} \\ T &= \text{time to machine, min} \\ f &= \text{feed, in./rev.} \\ N &= \text{rpm} \end{aligned}$$

Feed (F) is defined as the distance a tool, or table, moves every minute. The feed (f) may also be defined as the distance a tool moves for every revolution of the work (or tool). Therefore,

$$f = \frac{F}{N} = \frac{\text{in./\cancel{min}}}{\text{rev/\cancel{min}}} = \frac{\text{in.}}{\text{rev}}$$

The time to machine may also be stated as

$$T = \frac{L}{fN} = \frac{\cancel{\text{in.}}}{\cancel{\dfrac{\text{in.}}{\text{rev}}} \times \dfrac{\cancel{\text{rev}}}{\text{min}}} = \text{min}$$

Example 6.10

A lathe, using a sharp-pointed tool bit, is to take one cut off the 3.5-in. diameter of a soft-steel shaft. The length of the shaft is 7 in. The tool bit is made from high-speed steel. Calculate the time to machine one cut, using a feed of 0.015 in. per revolution. Assume a total overtravel of $\frac{1}{2}$ in.

Solution   The rpm to be used is

$$N = \frac{12S}{\pi D} = \frac{12 \times 90}{\pi \times 3.5}$$
$$= 98 \text{ rpm}$$

$D = 3.5$ in.
$L = 7$ in.
$S = 90$ ft/min
$f = 0.015$ in./rev
$o = \frac{1}{2}$ in.

The time needed to machine one cut is

$$T = \frac{L + o}{fN} = \frac{7.0 + 0.5}{0.015 \times 98}$$
$$= 5.1 \text{ min.}$$

The time required to drill through a thickness of material must also consider the time it takes the point cone of the drill to cut through the material. This is called the *approach* and is shown in Fig. 6.3. The general grind of a drill point is 118°. Thus, if D is the diameter of the drill, the approach A_d may be found from

$$\tan 59° = \frac{D/2}{A_d}$$

Solving for A_d gives

$$A_d = \frac{D/2}{\tan 59°}$$

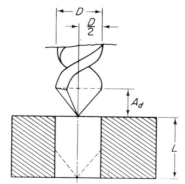

Figure 6.3

It should be noted that to drill harder materials, the included drill point angle could increase to as much as 135°, and to drill materials softer than mild steel, the included angle could be reduced to 90° or less.

Example 6.11

A $\frac{7}{8}$-in. high-speed drill is used to drill a hole into a cast iron plate 3 in. thick. The feed to be used is 0.012 in./rev. Calculate the time required to drill the hole. Assume an overtravel of $\frac{1}{4}$ in. at each end of the cut.

Solution   The rpm is

$$N = \frac{12S}{\pi D} = \frac{12 \times 50}{\pi \times 0.875}$$
$$= 218 \text{ rpm}$$

$D = \frac{7}{8}$ in.
$S = 50$ ft/min
$f = 0.012$ in./rev
$L = 3$ in.
$o = 0.250$ in.

The approach is

$$A_d = \frac{D/2}{\tan 59°} = \frac{0.875/2}{1.67}$$
$$= 0.262 \text{ in.}$$

The time to drill one hole is

$$T = \frac{L + A_d + 2o}{fN} = \frac{3.0 + 0.262 + 2(0.250)}{0.012 \times 218}$$
$$= 1.44 \text{ min}$$

The calculation of approach for a milling cutter depends on the type of cutter used and the size of the cutter as related to the width of the material being removed.

If a face milling cutter is the *exact diameter of the width of the work*, the approach is shown in Fig. 6.4a. The approach is one-half the diameter of the cutter.

$$A_1 = \frac{D}{2}$$

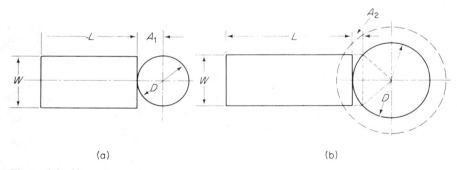

(a)   (b)

Figure 6.4(a,b)

Example 6.12

A 10-in.-diameter carbide face milling cutter is used to take one cut across the face of a block of aluminum which is 8 in. wide. The length of the block is 20 in. The feed is 0.025 in./rev. How long will it take to machine one cut on the block? The total overtravel is $\frac{1}{2}$ in.

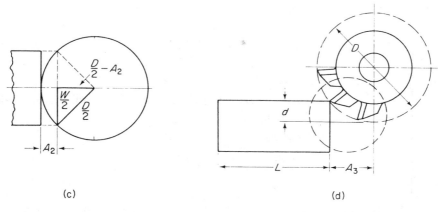

(c)   (d)

Figure 6.4(c,d)

Solution   The rpm of the cutter is

$$N = \frac{12S}{\pi D} = \frac{12 \times 1000}{\pi \times 10}$$

$$= 382 \text{ rpm}$$

$D = 10$ in.
$W = 10$ in.
$L = 20$ in.
$f = 0.025$ in./rev
$o = \frac{1}{2}$ in.
$S = 1000$ ft/min

The approach is

$$A_1 = \frac{D}{2} = \frac{10}{2}$$

$$= 5 \text{ in.}$$

The time to take one cut is

$$T = \frac{L + A + o}{fN} = \frac{20 + 5 + 0.5}{0.025 \times 382}$$

$$= 2.7 \text{ min.}$$

If a face milling cutter is *less than* the width of the work, at least one additional cut must be taken to finish the surface.

If a face milling cutter has a *greater diameter* than the width of the work, as shown in Fig. 6.4b, the approach (A_2) from Fig. 6.4c is

$$\left(\frac{D}{2}\right)^2 = \left(\frac{W}{2}\right)^2 + \left(\frac{D}{2} - A_2\right)^2$$

D = diameter of cutter
W = width of work
A_2 = approach

Solving for A_2 yields

$$A_2 = \tfrac{1}{2}(D - \sqrt{D^2 - W^2})$$

Example 6.13

Assume the same data as in Ex. 6.12 with the exception that a 12-in.-diameter cutter is used. Calculate the time to take one cut.

Solution   The rpm is

$$N = \frac{12S}{\pi D} = \frac{12 \times 1000}{\pi \times 12}$$

$$= 318$$

The approach is

$$A_2 = \tfrac{1}{2}(D - \sqrt{D^2 - W^2}) = \tfrac{1}{2}(12 - \sqrt{12^2 - 10^2})$$

$$= 2.683 \text{ in.}$$

The time to take one cut

$$T = \frac{L + A_2}{fN} = \frac{20 + 2.683}{0.025 \times 318}$$

$$= 2.85 \text{ min.}$$

$D = 12$ in.
$W = 10$ in.
$L = 20$ in.
$f = 0.025$ in.
$o = \tfrac{1}{2}$ in.
$S = 1000$ ft/min

If a *plain milling cutter* is used, the approach would be as shown in Fig. 6.4d. The approach (A_3) from Fig. 6.4d is

$$\left(\frac{D}{2}\right)^2 = \left(\frac{D}{2} - d\right)^2 + A_3{}^2$$

Solving for A_3, we obtain

$$A_3 = \sqrt{d(D - d)}$$

D = diameter of cutter, in.
d = depth of cut, in.
A_3 = approach, in.

Example 6.14

A 6-in.-diameter carbide cutter is used to cut medium-hard steel with a plain milling cutter. The depth of cut is 0.040 in., the feed is 0.015 in./rev, and the length of the work is 28 in. Calculate the time required to take one cut. The total overtravel is $\tfrac{1}{2}$ in.

Solution   The rpm is

$$N = \frac{12S}{\pi D} = \frac{12 \times 250}{\pi \times 6}$$

$$= 159 \text{ rpm}$$

The approach is

$$A_3 = \sqrt{d(D - d)} = \sqrt{0.040(6 - 0.040)}$$

$$= 0.488 \text{ in.}$$

The time to take one cut

$$T = \frac{L + A_3 + o}{fN} = \frac{28 + 0.488 + 0.5}{0.015 \times 159}$$

$$= 12.36 \text{ min}$$

$D = 6$ in.
$S = 250$ ft/min
$d = 0.040$ in.
$f = 0.015$ in./rev
$L = 28$ in.
$o = \tfrac{1}{2}$ in.

The time to machine the surface of a workpiece with a planer or a shaper is related to the width of the work, the feed, and number of strokes per minute. It should be noted that the overtravel should be added to the width of the workpiece. The amount allowed ranges from $\tfrac{1}{2}$ to $\tfrac{3}{4}$ in.

$$T = \frac{W + o}{fN}$$

W = width of work
f = feed, in./stroke
N = strokes/min
T = time, min
o = overtravel

The crank shaper allowance for overtravel should be $\frac{1}{4}$ to $\frac{1}{2}$ in.

Example 6.15

A planer has a ratio of 2 and is set to feed the tool bit at 0.030 in./stroke. The workpiece is 8 ft long and 18 in. wide. The overtravel is 0.200 in. at each end of the cut. If the workpiece is cast iron and the tool bit is cast steel, how long will it take to machine the surface?

Solution   The strokes per minute are

$$N = \frac{S}{L + L/2} = \frac{125}{8 + 8/2}$$
$$= 10.42 \text{ strokes/min}$$

The time to machine the surface is

$$T = \frac{W + o}{fN} = \frac{18 + 2(0.200)}{0.030 \times 10.42}$$
$$= 58.22 \text{ min}$$

L = 8 ft
r = 2
f = 0.030 in./stroke
W = 18 in.
S = 125 ft/min
o = 0.200 in.

Example 6.16

A crank shaper is used to machine the surface of a block of medium steel. The dimensions of the block are 24 in. long by 5 in. wide. A high-speed steel tool bit is set to overtravel the work by $\frac{1}{2}$ in. at each end of the cut. (a) How long will it take to machine the surface of the work if the feed is 0.010 in./stroke? (b) How long will it take if a hydraulic shaper is used? (c) What are the return speeds on both shapers?

Solution

(a) For the crank shaper

$$N = \frac{7.2S}{L} = \frac{7.2 \times 70}{24} = 21 \text{ st/min}$$

The time to machine the surface of the block is

$$T = \frac{W + 2o}{fN} = \frac{5 + 2(0.5)}{0.010 \times 21}$$
$$= 28.6 \text{ min}$$

S = 70 ft/min
W = 5 in.
f = 0.010 in./stroke
o = $\frac{1}{2}$ in.
L = 24 in.

(b) For the hydraulic shaper, the strokes per minute are

$$N = \frac{8S}{L} = \frac{8 \times 70}{24}$$
$$= 23.3 \text{ st/min}$$

The time required to machine the surface is

$$T = \frac{W + 2o}{fN} = \frac{5 + 2(0.5)}{0.010 \times 23.3}$$
$$= 25.7 \text{ min}$$

(c) The return speed for a crank shaper is

$$S_r = \frac{LN}{4.8} = \frac{24 \times 21}{4.8}$$
$$= 105 \text{ ft/min}$$

The hydraulic shaper is

$$S_r = S \times 2 = 70 \times 2$$
$$= 140 \text{ ft/min}$$

6.7 PERFORMANCE FACTOR

As indicated earlier, *downtime* factors are based on many series of case histories taken over a long period of time. They are factors that relate the *actual* time it takes to do a series of specific operations to the *estimate* of the time it takes to do these jobs. Since *downtime* relates to any and all causes that result in nonproduction, the ratio will be greater than 1. This ratio is termed a *performance factor*. It is estimated that performance factors will have values of 1.2 to 2.0. The equation that may be used to establish these performance factors is

$$P = \frac{T_a}{T_e} \qquad \begin{array}{l} T_a = \text{actual time} \\ T_e = \text{estimated time} \\ P \ = \text{performance factor} \end{array}$$

Example 6.17

Assume that you are to machine 30 castings in two setups. What is the cost of production if the following data are used?

Setup time	45 min per setup
Nonmachining time	20.6 min per casting
Machining time	12 min per casting
Tool sharpening	4 min per casting
Fatigue	20%
Personal needs	4%
Tool change time	8 min
Tool life	10 hrs
Checking or gaging time	15 sec and 5 checks per casting
Performance factor	1.3
Direct labor cost	$2.50 per hour

Solution The calculation is as follows:

1. Machining time   12.0 min

2. Nonmachining time   20.6 min

3. Fatigue   (12.0 + 20.6) 20% = 32.6 × 0.20

6.52 min

4. Personal needs   32.6 × 0.04 = 1.3 min

5. Tool sharpening:

$$t = \frac{T_k T'}{T} \times \frac{20.6 \times 8}{600} = 0.27 \text{ min/casting}$$

$T_k = 20.6$ min
$T' = 8$ min
$T = 10$ hr = 600 min

6. Measuring and checking:

$$t_c = \frac{15 \times 5}{60} = 1.25 \text{ min/casting}$$

7. The time is:

Machining time	12.00 min
Nonmachining time	20.60
Tool sharpening	0.27
Measuring and checking	1.25
Fatigue	6.52
Personal needs	1.30
Time per casting	41.94 min

8. The time for 30 castings is

$$41.94 \times 30 = 1258.2 \text{ min}$$

9. The setup time is

$$45 \text{ (min/setup)} \times 2 \text{ setups} = 90 \text{ min}$$

10. The direct labor time for 30 castings is

$$1258.2 + 90 = 1348.2 \text{ min}$$

11. The performance factor is 1.3; therefore, the total hours charged against the 30 castings is

$$\frac{1348.2 \times 1.3}{60} = 29.21 \text{ hr}$$

12. At $2.50 per hour, the direct labor cost is

$$29.21 \text{ hr} \times 2.5 \text{ per hour} = \$73.00$$

6.8 DIRECT MATERIAL COST

Material costs must reflect the material that finally becomes the finished product. It must also include the scrap material that results from spoilage, ends that are not long enough to be machined into the finished product and waste that becomes the chips machined from the raw stock. The cost of raw material is affected by the manner in which the material is ordered and the quantities ordered. Cutting charges may be added to the cost of material if other than standard sizes are ordered.

When tools, jigs, fixtures, or dies are to be manufactured by the home plant, the material cost includes the raw material, the standard items such as die sets, clamps, bushings, and so on.

Material may be ordered by weight, or by the linear foot. If material is ordered by weight, an estimate is made and the weight in pounds or tons is multiplied by the price per pound, or ton.

Castings are generally purchased by weight. Two methods may be used to find the weight of a casting from the pattern.

The weight of a casting may be estimated from the volume (cubic inches) of the pattern and multiplying the volume by the specific weight (sp wt) of the cast material. These multipliers are found in Table 6.2.

TABLE 6.2 CASTING MATERIAL FACTORS

Pattern mat.	Cast Iron	Aluminum	Brass	Zinc
White pine[a]	16.5	5.8	19.0	15.0
Yellow pine	14.0	4.2	16.0	12.0
Cherry	10.4	3.8	13.0	10.0
Mahogany	12.5	4.5	14.0	12.0
Aluminum	2.8	1.0	3.3	2.7
Cast iron	1.0	0.4	1.2	0.9
Dry core sand	4.0	1.4	4.8	3.6
Sp wt (lb/in.^{3})	0.2778	0.09375	0.3112	0.2532

[a]Sp wt of white pine = 0.017 lb/in.^{3}.

Example 6.18

Figure 6.5 is a pattern to be used to cast a brass part. Calculate the weight of the casting using its volume.

Solution The net volume is the casting less the cored hole. Thus

$$V = \left(\frac{\pi D^2}{4} \times L \right) - \left(\frac{\pi d^2}{4} \times L \right) = \left(\frac{\pi 6^2}{4} \times 4 \right) - \left(\frac{\pi 2^2}{4} \times 4 \right)$$

$$= 113.04 - 12.56 = 100.5 \text{ in.}^3 \qquad D = 6 \text{ in.}$$
$$d = 2 \text{ in.}$$
$$L = 4$$
$$\text{sp wt} = 0.3112 \text{ lb/in.}^3$$

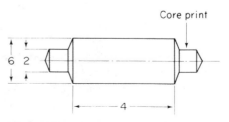

Figure 6.5

The weight of the casting is

$$W = V \times \text{sp wt} = 100.5 \times 0.3112 = 31.28 \text{ lb}$$

The weight of the cast material may also be determined if the weight of the pattern is known and if the pattern material is known.

Example 6.19

Calculate the weight of an aluminum casting made from yellow pine pattern that weighs 14 lb.

Solution

$$W = K_p W_p = 4.2(14)$$
$$= 58.8 \text{ lb}$$

K_p = constant (Table 6.2)
W_p = pattern weight = 14 lb
W = weight of casting

If the pattern is cored, the core box should be packed with sand. The sand core is weighed, multiplied by the appropriate constant from Table 6.2, and subtracted from the casting weight.

Example 6.20

Assume a sand core in Ex. 6.19 that weighs 4 lb. What is the weight of the cored casting?

Solution

$$W = K_p W_p - K_c W_c = 4.2(14) - 1.4(4)$$
$$= 53.2 \text{ lb}$$

K_p = 4.2 (from Table 6.2)
W_p = 14 lbs
K_c = 1.4 (Table 6.2)
W_c = 4 lb

Stampings are made from sheet or strip stock. Much saving can be gained by the way in which the blanks are laid out on the material strips. It should be remembered that the cost of making a die is warranted only when a large number of workpieces are to be made. Therefore, even a saving of a small amount of material could result in large overall savings. In any event, 50 to 70% of the cost of the parts are material costs.

Thus, in Fig. 6.6a the metal strip shows the position of a simple equilateral triangle. It is obvious that there is a great deal of waste material in this scrap strip. This is a *single-pass* layout.

If the triangular pieces are positioned as shown in Fig. 6.6b, a considerable savings of material is obtained. This is a *single-row two-pass* scrap strip layout. The material strip is passed through the die and the shaded triangular parts are produced. The strip is then turned around and the nonshaded triangular parts are produced.

One blank area in Fig. 6.6a is

where W = width of strip
           L = length of blank
$A = WL$ A = area of blank

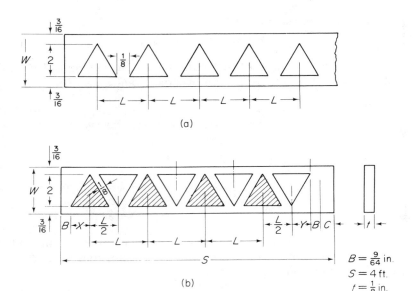

Figure 6.6(a,b)

The blank area for one blank in Fig. 6.6b is

$$A = \frac{WL}{2}$$

The allowance between blanks and between the edges of the scrap strip is material that is part of the cost of the material used.

A good general rule for calculating the material *allowance between blanks* is shown in Table 6.3.

The *allowance between the edge of the strip and the blank* may be approximated to be equal to $1\frac{1}{2}$ times the material thickness but not less than $\frac{1}{32}$ in. It is important to understand that these rules produce approximate values. Good reference handbooks give specific allowance values based on the configuration of the blank, the type of material, and the thickness of the material.

TABLE 6.3

Material thickness, t (in.)	Allowance (in.)
less than $\frac{1}{32}$	$\frac{1}{32}$
$\frac{1}{32}-\frac{3}{16}$	Stock thickness
greater than $\frac{3}{16}$	$\frac{3}{16}$

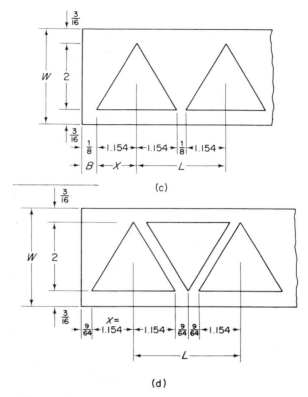

Figure 6.6(c,d)

Example 6.21

Refer to Fig. 6.6a and b. Calculate the area of each blank given the data in the drawings.

Solution   For Fig. 6.6c (single-row single-pass)

1. The width of the strip is

$$W = 2 + 2 (0.187) = 2.375 \text{ in.}$$

2. The length of each blank is

$$L = 2(1.154) + 0.125 = 2.433 \text{ in.}$$

3. The area is

$$A = WL = 2.375 \times 2.433 = 5.778 \text{ in.}^2$$

For Fig. 6.6d (single-row double-pass)

1. The width of the strip is

$$W = 2 + 2 (0.1875) = 2.375 \text{ in.}$$

2. The length of the strip is (see Fig. 6.6d)

$$L = 2(1.154) + 2\left(\frac{9}{64}\right) = 2.590 \text{ in.}$$

3. The area of each blank for a single-row double-pass scrap strip is

$$A = \frac{WL}{2} = \frac{2.375 \times 2.590}{2} = 3.075 \text{ in.}^2$$

A = area, in.^{2}
$W = 2.375$ in.
$L = 2.590$ in.

The scrap end, C, is calculated by determining the number of blanks which can be cut from a given length of material.

If the run is a *single-row single-pass* operation as shown in Fig. 6.6a, the equation which yields the number of blanks is

$$N = \frac{S - (X + Y + 2B)}{L} + 1$$

The waste end is

$$C = S - [L(N - 1) + (X + Y + 2B)]$$

S = length of strip
X = reference line first blank
Y = reference line last blank
B = allowances
L = distance between blanks per run
C = scrap end

If the run is a *single-row double-pass* operation as shown in Fig. 6.6b, the equations are the same except that $\frac{1}{2} L$ is substituted in both equations for L. Thus

$$N = \frac{S - (X + Y + 2B)}{0.5L} + 1$$

and the waste end is

$$C = S - [0.5L(N - 1) + (X + Y + 2B)]$$

Example 6.22

Use the data from Ex. 6.21 and Fig. 6.6b. (a) Calculate the number of blanks that can be machined from a 4-ft-long strip. (b) How much waste scrap remains at the end of the strip? (c) Compare your answer in part (b) with the number of blanks produced in Fig. 6.6a for a 4-ft strip. (d) What is the loss from the waste end?

Solution

(a) The number of blanks is

$$N = \frac{S - (X + Y + 2B)}{0.5L} + 1$$
$$= \frac{48 - [1.154 + 1.154 + 2(\frac{9}{64})]}{0.5(2.590)} + 1$$
$$= 36.06 \sim 36 \text{ blanks}$$

$S = 4 \text{ ft} \times 12 = 48$ in.
$X = 1.154$ in.
$Y = 1.154$ in.
$B = \frac{9}{64}$ in.
$L = 2.590$ in.

(b) The waste material is

$$C = S - [0.5L(N - 1) + (X + Y + 2B)]$$
$$= 48 - [0.5(2.590)(36 - 1) + 1.154 + 1.154 + 2(\tfrac{9}{64})]$$
$$= 48.000 - 47.915$$
$$= 0.085 \text{ in.}$$

(c) The number of possible blanks produced in a single-row single-pass (Fig. 6.6a) is

$$N = \frac{S - (X + Y + 2B)}{L} + 1 = \frac{48 - [1.154 + 1.154) + 2(\tfrac{1}{8})]}{2.433} + 1$$
$$= 19.67 \sim 19 \text{ blanks}$$

$$L = 2.433 \text{ in.}$$
$$B = \tfrac{1}{8} \text{ in.}$$

Thus the single-row double-pass scrap strip shows that this layout produces almost 90% more parts from the same quantity of material. The area per blank also reveals this saving.

(d) The waste end is

$$C = S - L(N - 1) + (X + Y + 2B)$$
$$= 48 - 2.433 (19 - 1) + [1.154 + 1.154 + 2(\tfrac{1}{8})]$$
$$= 48 - 46.352 = 1.648 \text{ in.}$$

Another type of calculation which needs to be made to determine material costs is that for lathe turning or any of the machine tools in a machine shop. If material is cut into lengths with a power hacksaw, and if this material is to be machined after cutting, an amount must be allowed for the material removed by the hacksaw. When a hacksaw is used, the allowance can be as little as $\tfrac{1}{32}$ in. to as much as $\tfrac{1}{8}$ in.

If the bar stock is inserted into the spindle of a turret, or automatic lathe, the allowances for cutoff are shown in Table 6.4. An additional allowance for scrap of 5% is added. This allowance will take care of scrap ends and spoilage.

The weight of the raw material may be calculated by using the specific weight values in Table 6.5. Multiply the volume of the material by the specific weight to get the weight of the material.

TABLE 6.4 CUTOFF

Depth of Cut	Width of Cut
Up to $\tfrac{5}{8}$	$\tfrac{5}{32}$
$\tfrac{1}{2}$–$\tfrac{3}{4}$	$\tfrac{3}{16}$
$\tfrac{5}{8}$–1	$\tfrac{1}{4}$
$\tfrac{3}{4}$–$1\tfrac{1}{4}$	$\tfrac{5}{16}$
Over $1\tfrac{1}{4}$	$\tfrac{5}{16}$

TABLE 6.5 SPECIFIC
WEIGHT

Material	Sp Wt (lb/in.^{3})
Aluminum	0.09375
Brass	0.31120
Copper	0.32118
Iron	0.27777
Lead	0.41012
Steel	0.28332
Tin	0.26562
Zinc	0.25318

Example 6.23

If $\frac{1}{8}$ in. of material is added to the diameter for finishing the part in Fig. 6.7, how much will it cost to buy the material necessary to make 500 parts? The cost of steel is 90 cents per pound. The finished part is $2\frac{1}{2}$ in. long.

Solution In Fig. 6.7 the largest diameter is $1\frac{3}{4}$ in. The stock needed to finish this diameter should not be less than $1\frac{7}{8}$ in. if it is to be machined.

The amount of material added for cutoff is $\frac{1}{4}$ in. Thus

$$L + C = 2.5 + 0.250$$
$$= 2.750 \text{ in.}$$

$L = 2.5$ in.
$D = 1.875$ in.
$C = \frac{1}{4}$ in.

The addition of 5% will give a length of

cost/lb = \$0.90

$$S = 2.750 \times 1.05$$
$$= 2.887 \text{ in.}$$

$N = 500$
sp wt = 0.28332

The weight of 500 pieces is

$$W = V(N)(\text{sp wt}) = \frac{\pi D^2}{4} (S)(N)(\text{sp wt})$$

$$= \frac{\pi 1.875^2}{4}(2.887)(500)(0.28332)$$

$$= 1129 \text{ lb}$$

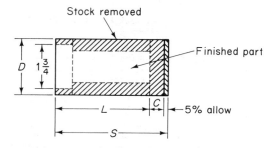

Figure 6.7

$$\text{cost} = 1129 \times 0.90$$
$$= \$1016.32$$

6.9 OVERHEAD COSTS

As indicated, overhead costs arise from peripheral activities that have no relationship to direct production output but which affect production costs. Management, engineering, office, plant, indirect labor, and material are all part of overhead. In essence they are the costs that support the enterprise without which there could be no efficient production.

Since it is difficult to assess which portion of the overhead costs is to be charged to a particular job, several methods have been developed for assessing costs. They may be related to the *actual hours of operation* of a machine. The charges are made against a machine, are totaled, and divided by the actual hours used by the machine. Therefore, the charges are allocated to each department according to a judgment made by management. A study of the machines in a particular department reveals the allocation of a portion of the departmental overhead to be charged against each machine.

The *space allocation method* is determined based on the space occupied by a machine. If the cost bears a relationship to the space, the allocation of a portion of the cost according to some portion of the space makes sense. Heat and electricity costs may be allocated this way.

The overhead costs may be distributed in relationship to the *direct labor dollar, direct labor hour,* or *direct material costs.* The apportioning of overhead according to the direct labor dollar or hour costs make sense if applied where labor is the main item of cost. The direct labor dollar or hour methods are flexible and allow for variations in quality and quantity of labor expended in the production effort. The allocation according to the direct material costs makes sense only where the materials processed are uniform. It is also appropriate for allocating costs for materials, handling, storing, purchasing and shipping, and so on. Where *mass production of one part* is involved, the overhead costs may be distributed over these parts. The charge is then applied to each piece, or lot. Whichever method is used, the choice is arbitrary and will not meet all the requirements of a manufacturing plant.

Maintenance of equipment is often considered an indirect cost and charged to overhead. The daily, or periodic, adjustments and lubrication of machines are forms of maintenance. Repairs and part replacement are still another form of maintenance which are usually charged to overhead. Again a case history associated with a particular machine is used to establish the rate to be charged to that machine. If a machine is completely rebuilt, the degree to which it is rebuilt will determine whether or not it should be charged to overhead, or considered a new machine and treated as an improvement.

Machinery does *deteriorate* with time and maintenance costs will increase with increased use. Or the machine may be come *obsolete.* Deterioration refers to the physical condition of the machine, whereas obsolescence refers to the outdating of a machine with reference to a particular job. On the basis of these definitions, deterioration is determined from the records. Obsolescence is determined by considering the performance of a machine in terms of other new and improved machines, change in product design, a change in manufacturing method, or a change in the production requirements.

There are many ways in which a replacement analysis may be done. The aim of this book is to deal with the economics of today. In general, the use of a fixture in production has the same purpose as the use of a machine tool. That is, it is used to make a profit by paying for itself from savings that result because of its use.

The equation that may be used to determine the number of pieces that need to be produced in order to break even is*

$$N = \frac{C[i + u + t + (1/a)] + Si}{s(1 + L)}$$

If f equals the fixed charges

$$f = i + u + t + \frac{1}{a}$$

then

$$N = \frac{Cf + Si}{s(1 + L)}$$

N = no. of pieces produced per year
C = cost of fixture, dollars
S = difference between unamortized and scrap value of old fixture, dollars
i = interest on capital investment
u = yearly percentage of cost—upkeep
t = yearly percentage of cost—taxes, insurance, etc.
a = yearly cost of amortization
s = savings in direct operating cost per piece
L = percentage of overhead

Example 6.24

A fixture is to cost $600. This investment is expected to save $0.05 per piece. The old fixture, which originally cost $450, has a scrap value of $150. The percentage of overhead charged to this fixture is 30%. The following percentages are to be charged as fixed costs: interest 8%, upkeep 3%, taxes and insurance 12%, amortization $1\frac{1}{2}$ years. Calculate the number of pieces that must be produced to break even so that the fixture may be paid for in 1 year.

Solution   The fixed charges are

$$f = i + u + t + \frac{1}{a} \qquad\qquad\qquad C = \$600$$
$$s = \$0.05 \text{ per piece}$$

*From Cincinnati Milling Machine Co., *A Treatise on Milling and Milling Machines,* 3rd ed., 1951, p.

$$= 0.08 + 0.03 + 0.12 + \frac{1}{1.5}$$

$$= 0.90$$

The break-even number of pieces is

$$N = \frac{Cf + Si}{s(1 + L)} = \frac{600(0.90) + 300(0.08)}{0.05(1 + 0.30)}$$

$$= 8677 \text{ pieces/year}$$

$S = \$450-\$150 = \$300$
$L = 0.30$
$i = 0.08$
$u = 0.03$
$t = 0.12$
$a = 1.5$ years
$N = ?$

The number of dollars that may be spent for a fixture may be calculated using the same equation, but solving it for C. Thus

$$C = \frac{Ns(1 + L) - Si}{f}$$

Example 6.25

Using the data from Ex. 6.24 and changing the number of pieces to be manufactured to 10,000, how much can be spent for a new fixture?

Solution

$$C = \frac{Ns(1 + L) - Si}{f}$$

$$= \frac{10,000(0.05)(1 + 0.30) - 300(0.08)}{0.90}$$

$$= \$695.56$$

$N = 10,000$ pieces
$s = 0.05$
$L = 0.30$
$S = \$300$
$i = 0.08$
$f = 0.90$

The number of years that would be required to recoup an investment in a fixture may be calculated using this equation, but solving it for a.

$$a = \frac{C}{Ns(1 + L) - Si - C(i + u + t)}$$

Example 6.26

Assume the cost of a new fixture to be \$850 when the yearly production is 7000 pieces. Using the same data as in Ex. 6.24, how long will it take to amortize the fixture?

Solution The number of years required to amortize the fixture is

$$a = \frac{C}{Ns(1 + L) - Si - C(i + u + t)}$$

$$= \frac{850}{7000(0.05)(1 + 0.30) - 300(0.08) - 850(0.08 + 0.03 + 0.12)}$$

$$= 3.6 \text{ years}$$

$C = \$850 \quad S = \$300$
$N = 7000$ pieces $\quad i = 0.08$
$s = 0.05 \quad u = 0.03$
$L = 0.30 \quad t = 0.12$

The net savings that are realized when a new fixture is used may be obtained from

$$P = Ns(1 + L) - Si - Cf$$

Example 6.27

In Ex. 6.26, if the fixture results in a production of 10,000 pieces per year and the fixture costs $600, what is the saving?

Solution   The saving is

$$P = Ns\ (1 + L) - Si - Cf$$
$$= 10,000(0.05)(1 + 0.30) - 300(0.08) - 600(0.90)$$
$$= \$86 \text{ saved per year}$$

$$N = 10,000 \text{ pieces}$$
$$s = \$0.05 \text{ per piece}$$
$$L = 0.30$$
$$S = \$300$$
$$i = 0.08$$
$$C = \$600$$
$$f = 0.90$$

PROBLEMS

6.1. Define
   (a) direct,
   (b) overhead, and
   (c) fixed costs.
   Give examples of each.

6.2. Explain and illustrate
   (a) the actual cost and
   (b) the standard cost methods used to calculate the costs of production.

6.3. **(a)** Describe the job-order method of calculating costs.
   (b) Explain how and why it is useful when estimating the cost of jigs, fixtures, and so on.

6.4. In estimating direct labor, one of the items included is setup time. What does this include?

6.5. Assume that a company uses 1000 pieces per month in its assembly department. The cost of each piece is $5 and the cost of each setup is $30 per lot. If a carrying charge factor of 25% is used, what are
   (a) the economic lot size and
   (b) the setup hours permitted for each lot?

6.6. Another item of direct labor is nonmachining time.
   (a) Define nonmachining time.
   (b) Which items are included as nonmachining time?

6.7. List the three elements that constitute machine time in direct labor. Explain each and illustrate.

6.8. **(a)** Define downtime as a direct cost.
   (b) Give several examples.
   (c) How is the cost factor determined?

6.9. **(a)** Define economic lot size.
   (b) Explain how it relates to setup time.

6.10. A tool will cut for 5 hr before it needs sharpening. It take 15 min to change the tool. If the tool can be sharpened 12 times before it is discarded, how much time is charged to one cycle?

6.11. It takes a machinist 8 sec to check a hole which he is boring on a lathe with both ends of a go/no-go plug gage. He uses the gage *twice* for each piece and he has 150 pieces to bore. How much time must be charged to the job to pay for this activity?

6.12. A company produces 15,000 workpieces per month at a cost of $30 each. The setup cost is $65 for each setup. Assume the carrying charge factor to be 20%. Calculate
    (a) the economic lot size and
    (b) the setup hours permitted for each lot.

6.13. A certain plant manufactures L number of pieces of work per lot for a production contract of 5000 pieces per month. The carrying charge is 20% and the setup cost is $50. It takes 5 hr for each setup. Calculate
    (a) the number of pieces per lot and
    (b) the cost per piece.

6.14. The pieces used each month by a plant are 500. The carrying charges are 15% and the cost for each setup is $25. Determine
    (a) the size of each lot and
    (b) the hours allotted to each setup if the cost per piece to the company to manufacture these parts is $40.

6.15. **(a)** List the cost items which are classified in the nonmachining category.
    (b) Explain each and give an example.

6.16. The time required to handle and machine a workpiece is about 70 min. If allowances of 18% must be made for fatigue, and 5% for personal needs, calculate the total time allotted to this job for each workpiece.

6.17. The cutting time required to machine a piece of work is 15 min. It takes 6 min to change the tool for each sharpening. If the tool can be sharpened 20 times, calculate the time that is to be added to the cost of each part.

6.18. It takes a total of 1.5 min to measure a hole that is being bored on a lathe. If there are 60 holes to be bored, how much time should be allotted to the cost of the job?

6.19. The actual cutting time usually includes four elements. Illustrate and name them.

6.20. Define
    (a) cutting speed and
    (b) feed.

6.21. A lathe is to take a cut off a $2\frac{1}{2}$-in.-diameter stainless steel shaft with a HSS tool bit. Calculate the rpm to be used.

6.22. A lathe is to take one cut on the diameter of a soft-steel shaft 3 in. in diameter. Calculate the rpm if a high-speed steel tool is used.

6.23. A $1\frac{1}{2}$-in.-diameter HSS drill is used to cut a hole into a copper block. What is the rpm to be used?

6.24. A drill press is used to drill four holes into a block of cast iron. The drill bit is $\frac{3}{4}$ in. in diameter and is made from HSS. Calculate the rpm of the drill bit.

6.25. Assume a carbide plain milling cutter 9 in. in diameter which is to machine a brass block. What rpm should be used?

6.26. A milling machine uses a carbide-tooth cutter to mill a copper block. The diameter of the cutter is 8 in. Calculate the rpm.

6.27. A planer is set up to machine a large soft-steel block with a HSS tool bit. The block is 6 ft long.
  (a) How many strokes per minute should be used to machine this block?
  (b) What is the return speed of this planer if the ratio is 3?

6.28. A crank shaper is set to machine an 8-in.-long cast iron block with a high-speed steel cutter. Calculate
  (a) the strokes per minute to be used,
  (b) the return speed, and
  (c) the strokes per minute if the shaper is hydraulically powered.

6.29. A planer, ratio 2, is to cut the surface of a large block of cast iron. The block is $3\frac{1}{2}$ ft long. It is to be machined with a HSS tool bit.
  (a) How many strokes per minute should be used?
  (b) Calculate the return speed of the planer.

6.30. A crank shaper is to machine a block of mild steel, 4 in. long, with a HSS tool bit. Calculate
  (a) the strokes per minute used and
  (b) the return speed.

6.31. Repeat Prob. 6.30 for a hydraulic shaper.

6.32. A lathe, using a sharp-pointed tool bit, is to take one cut off the 2-in. diameter of a soft-steel shaft. The length of the shaft is 8 in. The tool bit is made from HSS. Calculate the time to machine one cut using a feed of 0.012 in./rev. Assume a total overtravel of $\frac{3}{8}$ in.

6.33. In Prob. 6.21 the length of the work is $7\frac{1}{2}$ in. and the feed is 0.0085 in./rev. Calculate the time required to machine one cut on this lathe. The overtravel is $\frac{3}{8}$ in. at each end of the cut.

6.34. The thickness of the block in Prob. 6.23 is 2 in. Calculate the time required to drill four holes if the feed is 0.15 in./rev. The total overtravel is $\frac{5}{8}$ in.

6.35. In Prob. 6.25, if the block is 10 in. long, the feed is $4\frac{1}{2}$ in. per minute and the depth of cut is $\frac{1}{16}$ in. Calculate the time required to take one cut across the block. The overtravel at each end of the block is $\frac{1}{4}$ in.

6.36. A $1\frac{1}{4}$-in. HSS drill is used to drill a hole in a cast iron plate 4 in. thick. The feed to be used is 0.015 in./rev. Calculate the time required to drill the hole. Assume an overtravel of $\frac{1}{8}$ in. at each end of the cut.

6.37. An 8-in.-diameter carbide face milling cutter is used to take one cut across the face of a block of aluminum which is 8 in. wide. The length of the block is 18 in. If a feed of 0.030 in./rev is used, how long will it take to machine one cut on the block? The total overtravel is $\frac{1}{2}$ in.

6.38. If the data are the same as the data in Prob. 6.37 with the exception that a 12-in.-diameter milling center is used, calculate the time to take one cut.

6.39. An 8-in.-diameter carbide cutter is used to cut medium-hardness steel with a plain cutter. The depth of cut is 0.050 in., the feed is 0.012 in./rev, and the length of the work is 18 in. Calculate the time required to take one cut. The total overtravel is $\frac{3}{8}$ in.

6.40. A milling job calls for a 12-in. face cutter, HSS, to take a face cut across the face of a workpiece 12 in. wide. The workpiece is 14 in. long and is made from aluminum. If the feed is 0.032 in./rev., calculate the time to take one cut. The overtravel is $\frac{3}{8}$ in. at each end of the cut.

6.41. The data are the same as in Prob. 6.40 except that the cutter is a 16-in. fly cutter which has carbide insert teeth. Calculate the time to take one cut.

6.42. Refer to Prob. 6.27 and calculate the time to machine the surface of the block if the surface is 8 in. wide and the feed is $\frac{3}{64}$ in./stroke. The total overtravel is $\frac{1}{2}$ in.

6.43. A planer has a ratio of 2 and is set to feed the tool bit at 0.015 in./stroke. The workpiece is 4 ft long and 16 in. wide. The overtravel is $\frac{1}{4}$ in. at each end of the cut. If the workpiece is cast iron and the tool bit is cast steel, how long will it take to traverse the surface of the work once?

6.44. **(a)** Calculate the time to machine the block in Prob. 6.28 if the feed used is $\frac{1}{32}$ in./stroke and the surface to be machined is 6 in. wide. The overtravel is $\frac{3}{16}$ in. at each end of the cut.
   (b) Repeat this calculation for a hydraulic shaper.

6.45. A crank shaper is used to machine the surface of a block of medium steel. The dimensions of the block are 15 in. long by $4\frac{1}{2}$ in. wide. A HSS tool bit is set to overtravel the work by $\frac{3}{8}$ in. at each end of the cut.
   (a) How long would it take to machine the surface of the work if the feed used is 0.008 in./stroke?
   (b) How long will it take if a hydraulic shaper is used?
   (c) What are the return speeds on both shapers?

6.46. Assume that a company is to manufacture 250 parts per setup and that the job calls for 8 setups. Calculate the cost of production if the following data are used.

Direct labor cost	$12.00 per hour
Machining time	20 min/piece
Nonmachining time	14 min/piece
Setup time	1 hr 30 min/setup
Tool sharpening	2 min/piece
Checking and gaging	20 sec and 8 checks per piece
Fatigue	18%
Personal needs	3%
Tool change	12 min
Tool life	16 hr
Performance factor	1.5

6.47. When determining material costs of machined items, which items are included?

6.48. Which items are included as material costs for tools, jigs, and fixtures?

6.49. Figure 6.5 is a pattern to be used to cast a cast iron part. Calculate the weight of the casting from the volume of the pattern.

6.50. Figure 6.8 shows a pattern. The casting that will be made from this pattern will be brass. Calculate the weight of the casting from the volume of the pattern.

6.51. Calculate the weight of a brass casting made from a mahogany pattern that weighs 12 lb.

6.52. Assume a sand core in Prob. 6.51 weighing 4 lb. What is the weight of the cored casting?

6.53. Confirm the answer that you obtained in Prob. 6.50 by calculating the weight of the pattern. Assume the pattern to be made from white pine.

6.54. **(a)** Calculate the weight of an aluminum casting made from a cherry-wood pattern that weighs 28 lb.
   (b) If a sand core weighs 8 lb, what is the weight of the casting?

6.55. Select any casting from any engineering drawing book and calculate its weight.

6.56. Check your library for a pattern drawing. Calculate the weight of the material from the pattern.

6.57. **(a)** Using the single-row single-pass scrap strip in Fig. 6.9a, calculate the number of parts that can be punched from an 8-ft-long strip of stock.

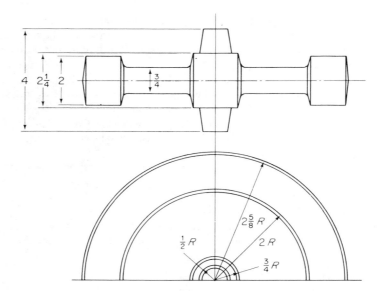

Figure 6.8

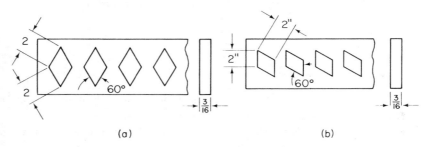

(a)   (b)

Figure 6.9

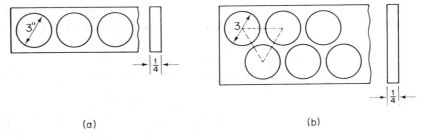

(a)   (b)

Figure 6.10

(b) Repeat for Fig. 6.9b and compare the results.

(c) Calculate the waste end for parts (a) and (b).

(d) Calculate the area per piece for parts (a) and (b).

6.58. Repeat Prob. 6.57 for Fig. 6.10.

6.59. Repeat Prob. 6.57 for Fig. 6.11.

6.60. Repeat Prob. 6.57 for Fig. 6.12 (a), (b) and (c) and compare the area of the blank.

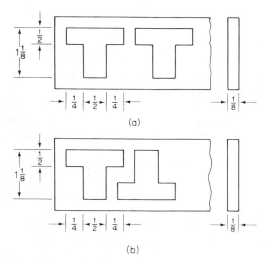

Figure 6.11

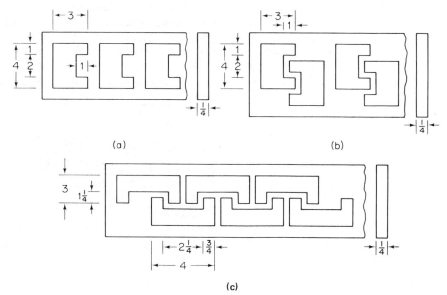

Figure 6.12

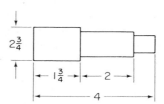

Figure 6.13

6.61. Calculate the weight of the material in Fig. 6.13 if 500 parts are to be made. Assume that the nearest diameter for raw material is 3 in. The material is brass.

6.62. The bushing shown in Fig. 6.14 is to be finished all over. The material is cast iron and can be purchased as tubing; outside diameter 3 in., inside diameter 1 in. Calculate the weight of 150 bushings.

6.63. If $\frac{1}{8}$ in. material is added to the diameter for finishing the part in Fig. 6.15, how much will it cost to buy the material necessary to make 200 parts if the cost of the steel is \$1.40/lb? The parts are made from steel. The finished part is $2\frac{1}{8}$ in. long.

6.64. Make a list of those items that are charged to overhead. Explain how they affect the cost of production.

6.65. List the various methods of allocating overhead cost. Briefly describe how each is allocated.

6.66. **(a)** How is maintenance applied to overhead?
  (b) How does maintenance differ from improvement?

6.67. How does obsolescence differ from deterioration?

6.68. Explain the statement: ''The use of a fixture in production has as its main purpose the making of a profit by paying for itself from savings that result because of its use.''

6.69. A fixture has been judged to have a scrap value of \$75. Its original cost was \$300. The cost of the new fixture is \$500. The percentage of the overhead charged to this fixture is 25%. The expected saving per piece is \$0.60. The fixed costs are: interest 6%, upkeep 2%, taxes and insurance 15%, amortization 2 years. Calculate the number of pieces that should be produced to break even so that it may be paid off in 1 year.

6.70. Use the data from Prob. 6.69. How much may be spent for fixtures and gages if the number of pieces to be manufactured is 2000?

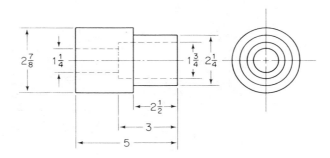

Figure 6.14

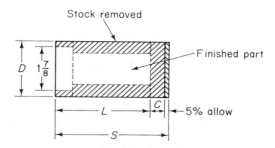

Figure 6.15

6.71. Assume the cost of a new fixture to be $1500 for a yearly production of 1800 pieces. Using the same data as in Prob. 6.69, determine the length of time needed to amortize the cost.

6.72. If the fixture in Prob. 6.70 results in a production of 3000 pieces per year, what is the saving?

6.73. What is the saving for the fixture in Prob. 6.71 if the production is 2500 pieces per year?

Chapter 7

Quality Control

7.1 MEASURING INSTRUMENTS

It is assumed that measuring instruments such as scales, calipers, and micrometers have been studied in other courses. In this section we deal with them only as they relate to fixtures used in design. Thus instead of studying the micrometer and the vernier caliper directly, we will deal with the construction of vernier scales so that the tool designer may use them if the opportunity presents itself.

Most measuring instruments used in the shop are constructed to read in thousandths of an inch (0.001 in.). It is, of course, not practical to engrave a scale to 0.001-in. increments. The vernier scale "amplifies" these small increments so that they may be read conveniently.

The process is to develop a *main scale*, Fig. 7.1, which can easily be read. A movable scale is graduated to one more (or less) division than the length of the matching main-scale division. Thus if the smallest main-scale division is 0.025 in., the *vernier scale* may be divided into increments of 0.024 in. Each division on the vernier scale is therefore 0.001 in. less than each division of the main scale. It should be noted that each vernier division could have been made 0.001 in. longer than each main-scale division.

If the zero on the vernier scale coincides with the zero on the main scale, the next line on the main scale will be out of alignment with the next line on the vernier scale by exactly 0.001 in. This is shown in Fig. 7.1. Thus, when the two zeros coincide, the jaws of the calipers, or anvils of the micrometer, will be closed. When the vernier is moved so that the next two lines coincide, the zeros will separate by 0.001 in. and the jaws will also separate by 0.001 in.

The following example shows how a vernier may be constructed.

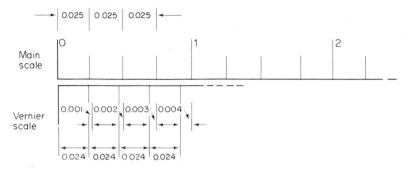

Figure 7.1

Example 7.1

Construct a vernier that is to have 40 divisions per inch on the main scale and is to read in increments of 0.001 in. (a) What is the length of each division on the main scale? (b) What is the length of each division on the vernier scale? (c) Calculate the minimum length of the vernier that matches the main scale when the zeros coincide. (d) when the zeros match, how many divisions on each scale are needed to match?

Solution

(a) The length of each division on the main scale will be

$$D_m = \frac{1}{40} = 0.25 \text{ in./div} \qquad \begin{aligned} D_m &= \text{main-scale division} \\ D_v &= \text{vernier-scale division} \end{aligned}$$

(b) The length of each vernier division is

$$D_v = 0.025 - 0.001 = 0.024 \text{ in./div}$$

(c) The minimum length of the vernier scale is

$$L_v = 0.024 \text{ in./div} \times 25 \text{ div} \qquad L_m = \text{length of main scale}$$
$$L_v = \text{length of vernier scale}$$

$$= 0.600 \text{ in.}$$

The matching length of the main scale is

$$L_m = 0.025 \text{ in./div} \times 24 \text{ div} = 0.600 \text{ in.}$$

(d) When the main and vernier scales match at zero, 24 divisions on the main scale match 25 divisions on the vernier.

(e) The partial construction is shown in Fig. 7.1.

Example 7.2

Calculate and construct a vernier that has a 50-divisions-per-inch main scale and is to read in increments of 0.005 in.

Solution   The length of each main-scale division is

$$D_m = \frac{1.000}{50} = 0.020 \text{ in./div (see Fig. 7.2)}$$

The length of each vernier division may be 0.005 in. smaller or longer than each main-scale division. We will use 0.005 in. smaller. Thus

$$D_v = 0.020 - 0.005 = 0.015 \text{ in./div}$$

The number of divisions on the main scale that matches the number of vernier scale divisions is found as follows: The main scale is constructed with 50 div/in., each division 0.020 in. long. The vernier is constructed so that each division is 0.015 in. long. Thus when the zeros coincide, the numbers 60 on the main scale and the vernier scale will coincide. The minimum length of the vernier scale is 0.060 in. long. This may be arrived at in the following manner: The length of the main scale equals the length of the vernier scale

$$L_m = L_v$$
$$0.020 \text{ in.} \times \overset{3}{\cancel{15}} = 0.015 \times \overset{4}{\cancel{20}}$$
$$0.060 \text{ in.} = 0.060 \text{ in.}$$

Since the minimum length of the vernier scale should be 0.060 in./div, 3 divisions of the main scale matches 4 divisions on the vernier scale, as shown in Fig. 7.2. It should be evident that matching also takes place at 120, 180, 240, and so on, inches. However, it is not necessary to repeat matching lengths, since the smaller vernier (the first matching points) is less expensive to manufacture and can be moved to additional desired positions.

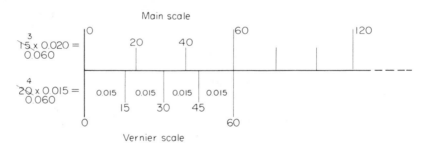

Figure 7.2

Verniers are used in the construction of micrometers, Fig. 7.3a; vernier calipers and depth verniers, Fig. 7.3b; height gage, Fig. 7.3c; and bevel protractors, Fig. 7.3d.

There are other methods that may be used to mechanically amplify small distances. One such device is a two-dimensional scale which may produce amplification of up to 100 or more parts. One such array, Fig. 7.4, shows 10 rows and 10 columns. The 0.5 line fits between the 0.3 and 0.4. Thus Figure 7.4 reads 0.536.

Gage blocks, Fig. 7.5, are precision blocks that are made to high accuracy in size, parallelism, and flatness. Two blocks can be rubbed (wrung) together and, because of their extreme flatness, will adhere. Many blocks can be wrung together to

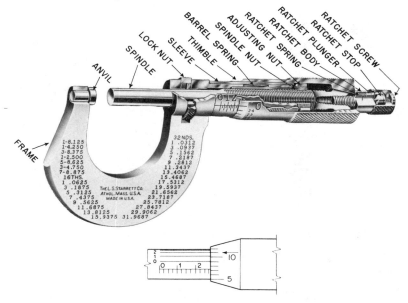

Figure 7.3(a)

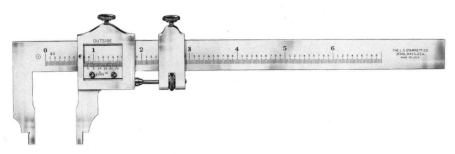

Figure 7.3(b)

construct many sizes. They can also be stacked and inserted into various holders to form a height gage, scribers, half-round jaws, and so on.

Their accuracy is designated with letters such as *A, A+, AA, master.* Table 7.1 shows the accuracy of each class of blocks.

The set of 88 blocks in Fig. 7.5 is made up of the blocks in Table 7.2.

When stacking a dimension, the procedure is to eliminate the right digit first.

Example 7.3

Using Table 7.2, the stack dimension = 5.69345 in. The stack is shown in Fig. 7.6.

Solution

Figure 7.3(d)

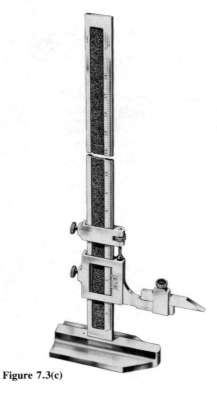

Figure 7.3(c)

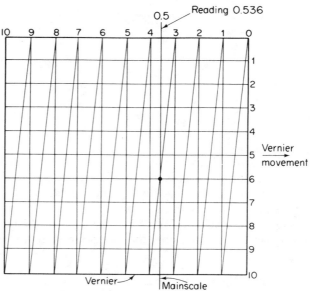

Figure 7.4

Figure 7.5

TABLE 7.1   CLASSES OF GAGE BLOCKS (IN.)

	A	A+	AA	Master
Length	+0.000006	+0.000004	+0.000002	+0.000001
	−0.000002	−0.000002	−0.000002	−0.000001
Flatness	0.000004	0.000003	0.000003	0.000002
Parallel	0.000004	0.000003	0.000002	0.000001

TABLE 7.2   GAGE BLOCK SIZES

Number of Blocks	Range (in.)	Increments (in.)
9	0.1001–0.1009	0.0001
49	0.101–0.149	0.001
19	0.050–0.950	0.050
4	1.000–4.000	1.000
3	1.00025–1.00075	0.00025
4	$\frac{1}{16}, \frac{5}{64}, \frac{3}{32}, \frac{7}{64}$	

Dimension = 5.69345

Block 1 $\dfrac{-1.00025}{4.69320}$

Block 2 $\dfrac{-0.1002}{4.5930}$

Block 3 $\dfrac{-0.1430}{4.450}$

Block 4 $\dfrac{-0.450}{4.000}$

Block 5 $\dfrac{-4.000}{0.000}$

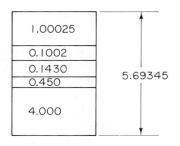

Figure 7.6

Stacked gage blocks may be used to transfer dimensions from the top of the blocks to the workpiece using a height gage as shown in Fig. 7.7a; or with a sine bar to set up or check angles, Fig. 7.7b.

The height gage, Fig. 7.7a, is set so that the needle reads zero when the ball is in contact with the gage blocks. The height gage is moved so that the ball contacts the step on the workpiece. If the dial gage reads plus, the step is higher than the gage block stack by the amount of the plus reading. If the dial reads minus, that quantity should be subtracted from the stack reading.

The sine bar, Fig. 7.7b, also is used with precision gage blocks. The sine bar becomes the hypotenuse of the triangle as shown in Fig. 7.7b. The diameter of the buttons and the distance between them may be accurate to 50 millionths of an inch over 5 in.

This setup may be used to determine height A or the angle α in Fig. 7.7b. The equation for a 5-in. sine bar is

$$\sin \alpha = \frac{A}{5}$$

where

$$A = h_2 - h_1$$

It should be noted that a 10-in. sine bar simply requires that the decimal point be moved in the sine function to determine height A. These instruments may be useful in fixture designs where angles need to be machined on workpieces.

Gage blocks may also be used with optical flats. Optical flats are very accurately ground for parallelism. They are accurate from 0.000001 to 0.000004 in., and are available in sizes $1 \times 1\frac{1}{2}$ in. to $12 \times 2\frac{3}{4}$ in.

Bands created by wave interference are shown in Fig. 7.8a. The use of monochromatic (single-wavelength) light with an optical box is shown in Fig. 7.8b. In Fig. 7.8c the incoming light divides at the lower surface of the optical flat and reflects

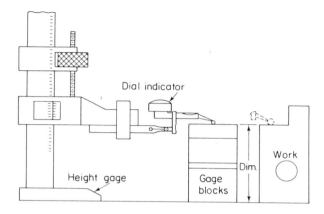

Figure 7.7(a)

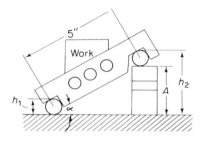

Figure 7.7(b)

from the upper surface of the work. These waves will reflect from these two surfaces and either reinforce or destroy each other. Dark bands indicate destruction when the wave path is $\frac{1}{2}$, or some half-wave multiple, wavelength long. There is a half-wave shift at the lower face of the optical flat which causes the dark band.

Therefore, in Fig. 7.8c when helium light with a wavelength of 23.2 millionths of an inch is used, four dark bands appear. Considering the contact point of optical flat and the gage block as the zero point, the first dark band will indicate an air wedge of one-half wavelength of helium light, or 11.6×10^{-6} in. (11.6 millionths in.); the second dark band will be 23.2×10^{-6} in.; the third dark band will be 34.8×10^{-6} in.; and the fourth dark band will be 46.4×10^{-6} in. The process for measuring the height of workpieces is therefore a matter of counting dark bands and multiplying by 11.6×10^{-6}. Thus

$$y = N(11.6 \times 10^{-6})$$

y = height of air wedge
N = number of bands

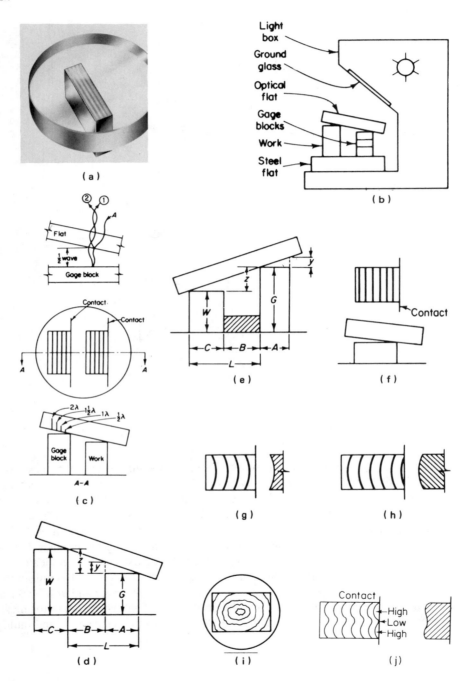

Figure 7.8

Figure 7.8d shows the relationship of the air gap y and the difference in height z between the gage block G and the work W.

The equation for finding z is

$$z = N(11.6 \times 10^{-6})\frac{L}{A}$$

y = air gap at gage block
z = correction factor
A = width of gage block

In Fig. 7.8d, W is greater than G. The factor z is added to the height of the gage block stock G (Fig. 7.8d).

$$W = G + z$$

B = length of spacer
C = width of work
L = distance between contact points

In Fig. 7.8e, W is shorter than G. In this case z is subtracted from G (Fig. 7.8e). Thus

$$W = G - z$$

N = number of bands
W = height of work
G = height of gage blocks

Example 7.4

A workpiece is measured with a micrometer to be 3.500 in. Assume that 12 bands appear when the setup is the same as in Fig. 7.8e. The width of the gate block is $\frac{7}{8}$ in., the width of the spacer is 1.000 in., and the width of the work is $\frac{1}{2}$ in. Calculate (a) the length of the work when helium light is used and (b) the number of bands if G is taken 0.0002 in. less than the height of the work and it measures 3.4998 in.

Solution

(a) In Fig. 7.8e the gage block stack is taken 0.0002 in. longer than 3.5000 in. Thus

$$z = N(11.6 \times 10^{-6})\frac{L}{A} = N(11.6 \times 10^{-6})\left(\frac{C + B}{A}\right)$$
$$= 12(11.6 \times 10^{-6})\left(\frac{0.500 + 1.000}{0.875}\right)$$
$$= 0.000239 \text{ in.}$$

G = 3.500 + 0.0002
    = 3.5002 in.
N = 12 bands
A = $\frac{7}{8}$ in.

and

$$W = G - z = 3.5002 - 0.000239$$
$$= 3.499961 \text{ in.}$$

B = 1.000 in.
C = $\frac{1}{2}$ in.

(b) If G is taken 0.0002 in. less than nominal 3.4998 in., as shown in Fig. 7.8d, the number of bands that appear for this workpiece is

$$N = \frac{A(W - G)}{11.6 \times 10^{-6}(A + B)} = \frac{0.875[3.499961 - (3.4998 - 0.0002)]}{11.6 \times 10^{-6}(0.875 + 1.000)}$$

$$= 14 \text{ bands}$$

$$W = 3.499961 \text{ in.}$$
$$G = 3.4996 \text{ in.}$$

The flatness of a surface may be checked with an optical flat. The flat is placed on a work surface to create a wedge. If the bands appear regularly spaced as shown in Fig. 7.8f, the surface is flat. If they curve toward the contact edge as shown in Fig. 7.8g, the surface of the work is concave, or low in the center. If the bands curve as shown in Fig. 7.8h, the work is convex, or high in the center. Figure 7.8i shows a high spot in the center. This happens when the center of the flat is pressed against the surface. Figure 7.8j shows two ridges and a concave surface between them.

Interferometry is being used more widely than ever before for measuring or checking surfaces, and for aligning work in machine tools. In the Fizeau interferometer the interference rays are collimated, Fig. 7.9, so that a correct reproduction of the work surface may be viewed. If the instrument is used to scan a large surface, a topographical map may be made of that surface.

Optical flats use interference patterns from narrow band wavelengths of light. Mercury, cadmium, krypton, thallium, sodium, helium, and neon, as well as gas lasers, are elements that will release discrete wavelengths of energy when their atoms are excited. The most common method for exciting these atoms is with an electric current while the element is in the gaseous state.

All elements mentioned above have characteristic wavelengths. Most of them are not pure, but have small amounts of other wavelength mixed in with the predominant characteristic wavelength for a particular material. The gas laser, however, is highly monochromatic and much more intense than any of the monochromatic sources mentioned above.

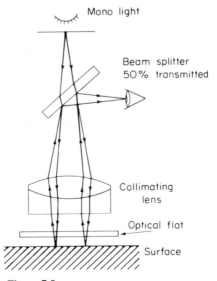

Figure 7.9

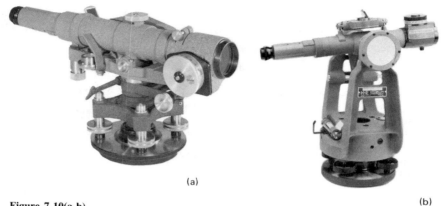

(a)

Figure 7.10(a,b)   (b)

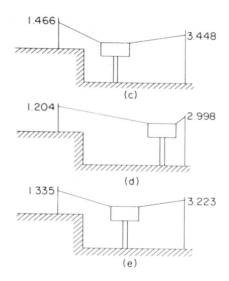

Figure 7.10(c,d,e)

Another system used for aligning fixtures, workpieces, and so on, uses the *optical tooling level* shown in Fig. 7.10a, or the jig transit, Fig. 7.10b. Both use the earth's gravitational field as a vert reference. Leveling the telescope perpendicular to this vertical line requires that a circular bubble remain centered in one position of the telescope and also when it is rotated through 180°.

After it has been determined that the instrument is level (circular level), the instrument is focused on a reference scale, and the tubular level is centered with the micrometer adjusting head. If the instrument is set properly, two ends of the bubble may be brought together and adjusted to indicate that the line of sight is horizontal.

Two objects are set a fixed distance apart. With the bubbles coincided and the instrument set as shown in Fig. 7.10c, two readings are taken. When subtracted, the difference is found to be

$$3.448 - 1.466 = 1.982 \text{ in.}$$

The instrument is then moved to the position shown in Fig. 7.10d. and the difference in the two readings is found to be

$$2.998 - 1.204 = 1.794 \text{ in.}$$

If the instrument is set in the center of the two reference poles (Fig. 7.10e), and two readings are taken, their difference would be

$$3.223 - 1.335 = 1.888 \text{ in.}$$

When the average of the two readings in Fig. 7.10c and d are taken, the result should be the same as the center reading, Fig. 7.10e.

$$\frac{1.982 + 1.794}{2} = 1.888 \text{ in.}$$

This indicates that the instrument is level.

Scales may be marked with pairs of lines, so that the line of sight observed is such that the cross hairs split the space between two lines on the scale, Fig. 7.11. The reading in Fig. 7.11 is

$$3.200 + 0.031 = 3.231 \text{ in.}$$

The alignment telescope, Fig. 7.12a, is mounted on a cup, and can be aligned slightly with adjustment screws. A target, Fig. 7.12b, is also mounted on a cup. Once set they are used as reference targets for checking and rechecking alignment.

Figure 7.12c shows a penta prism and Fig. 7.12d a leveling mirror. The prism is used to redirect the line of sight through 90°. The leveling mirror is used to redirect the line of sight in any direction in the horizontal plane.

The alignment of parallel shafts, machine tool beds, large fixtures, and so on, may be accomplished using these instruments. See the examples in Fig. 7.12e.

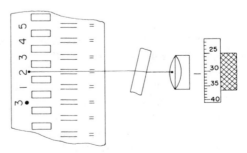

Figure 7.11

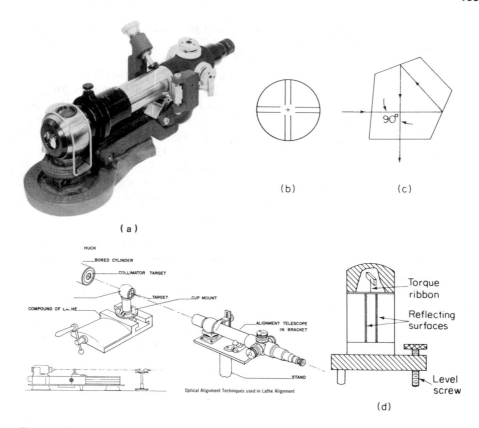

(b)   (c)

(a)

(d)

Figure 7.12

7.2 GAGING DEVICES

Limit gages.   These gages determine whether or not the workpiece is within acceptable dimensional limits. They generally do not attempt to determine the nature of a dimension.

Basically, there are three general classifications of gages: working, inspection, and master. The main distinction between the three is the tolerances to which they are manufactured. The working gage tolerances are greater than the tolerances for inspection gages, which in turn are greater than the tolerances applied to inspection gages.

Working gages.   These gages are used during production on the machine or at the machine station. Because they are used so much, they should have limits so that the workpiece will always be inside the limits of the inspection gage.

Inspection gages.   These gages are used by the manufacturing company or customer when checking workpieces received from the production department.

Master gages. These gages are very accurately made. They are used to set or check inspection and working gages.

Plug gages. These gages may be either single ended, Fig. 7.13a and b, or progressive, as shown in Fig. 7.13c. The GO gage is generally longer than the NO GO gage so that it may be distinguished from the other visually. As the size gets larger, the design changes. Figure 7.13a has a straight cylindrical gaging pin inserted into a handle. Figure 7.13b has plugs which have tapered ends to match the taper in the handle. The GO gage is longer than the NO GO plug. The handle is hollow and may be used to accommodate a range of plugs. Figure 7.13c is a progressive plug gage. The NO GO gage is part of the same cylinder as shown. Figure 7.13d, e, and f are large enough so that it is not practical to have the GO and NO GO gages as part of one gage.

Ring gages. Figure 7.14a and b show the two designs generally used for ring gages to check diameters. Both plug and ring gages may be made in the shop. Depending on the job requirements, the carbon content of the steel used is machine steel, plain carbon, or alloy steels. Machine steel may range from 0.20 to 0.50%

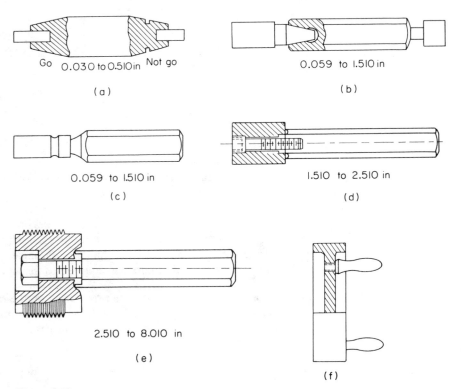

Go   0.030 to 0.510 in   Not go

(a)

0.059 to 1.510 in

(b)

0.059 to 1.510 in

(c)

1.510 to 2.510 in

(d)

2.510 to 8.010 in

(e)

(f)

Figure 7.13

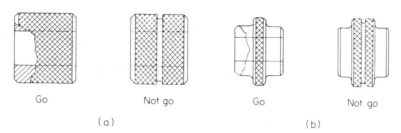

     Go   Not go   Go   Not go

         (a)   (b)

Figure 7.14

carbon. Carbon steel may have as much as 1% manganese and small quantities of sulfur and phosphorus. It should not contain silicon, as this element causes warping during heat treatment. The preference is to use high carbon or tool steels. The carbon content of the high carbon steels used is about 0.90% carbon, 0.35% manganese, small amounts of phosphorus, sulfur, and silicon. Special alloy steels have also been developed for high-precision gages.

    Commercial gages are made from materials such as carbide, glass, and sapphire. Carbide materials have high wear and corrosion resistance. Glass is relatively low in cost compared to steel, is corrosion resistant, has a low coefficient of thermal conductivity, and high resistance to wear. If dropped, the breakage factor is high. Sapphire has high resistance to wear and abrasion. It has greater resistance than glass to shock.

    Snap gages.   These gages have a C-frame construction. They may be set or adjusted to within the limits of their construction. They are used to check diameters, thicknesses, or lengths. Figure 7.15a is a commercial snap gage that uses four precision pins. The first two, which engage the workpiece, are set to the largest dimension and labeled the "GO" gage. The second set, which does not allow the workpiece to pass, is set to the smallest dimension or low limit. It is labeled the "NO GO" gage.

    Figure 7.15b shows four buttons instead of pins, and Fig. 7.15c shows two buttons and one anvil. It is to be noted that all gages are set to the same limits. Figure 7.15d shows a *progressive snap gage* which has been designed and manufactured at the home plant. The upper limit, 1.002, will reject all workpieces larger than 1.002 in. The lower limit is 0.998 in. and workpieces smaller than 0.998 will be rejected. Workpieces that have dimensions between 1.002 and 0.998 in. will be accepted by the GO (1.002) portion of the gage, and stopped at the NO GO pad.

    Figure 7.15e shows a commercial dial snap gage fitted with a dial indicator which reads in increments of 0.0001 in. and has ranges from 0–1 in. to 3–4 in. The anvils are carbide and adjustable.

    If manufacturing working gages do not indicate the tolerances, the tolerance applied to the gage is approximately 10% of the workpiece tolerances, except where greater precision is needed. Rounding off should take place in the direction of greater

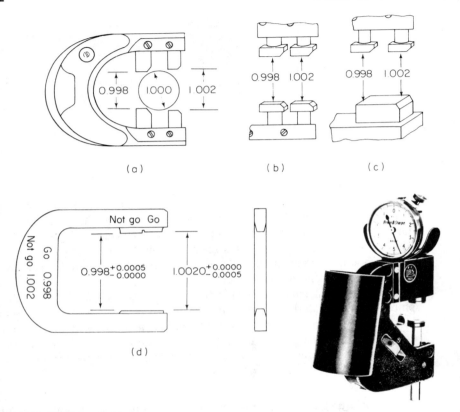

(a)   (b)   (c)

(d)

Figure 7.15(a,b,c,d,e)

precision. Allowance for wear should be added. This is usually 5% of the workpiece tolerance. There are five classes of plug and ring gages:

XX   Precision lapped (masters)
X   Precision lapped plugs and rings
Y   Lapped plugs and rings (commercial)
Z   Ground and polished
ZZ   Ground only (rings)

The tolerances applied to these gages, by size, are shown in Table 7.3. They are usually applied unilaterally to *GO* and *NO GO* gages. The tolerances are applied to *GO plugs* and *NO GO rings* in the *plus* direction. They are applied in the *minus* direction to *NO GO plugs* and *GO ring* gages. Bilateral application is used for master gages.

A GO plug will wear so that the dimension of the plug will be outside the permissible tolerance zone. Therefore, any allowance for wear should provide mate-

TABLE 7.3   GAGE TOLERANCE BY SIZE

Size (From–Including)	Class × 10^{-5}				
	XX	X	Y	Z	ZZ
0.029–0.825	2.0	4.0	7.0	10.0	20.0
0.825–1.510	3.0	6.0	9.0	12.0	24.0
1.510–2.510	4.0	8.0	12.0	16.0	32.0
2.510–4.510	5.0	10.0	15.0	20.0	40.0
4.510–6.510	6.5	13.0	19.0	25.0	50.0
6.510–9.010	8.0	16.0	24.0	32.0	64.0
9.010–12.010	10.0	20.0	30.0	40.0	80.0

Source: Federal Standards, U.S. Department of Commerce, *Handbook H-28.*

rial, which when worn off, will leave the diameter of the plug within the tolerance zone. This will result in the need for a slightly larger plug diameter. Thus wear allowance is added to the GO plug. Since the same argument can be applied to the GO snap gage, the material added to take care of wear will result in a smaller gage opening. Since wear reduces the NO GO plug diameter and increases the NO GO snap gage size, and since these dimensions after wear are within the tolerance zones, no allowances need be provided for wear. Figure 7.16 shows the application of the gage tolerances and the wear allowances.

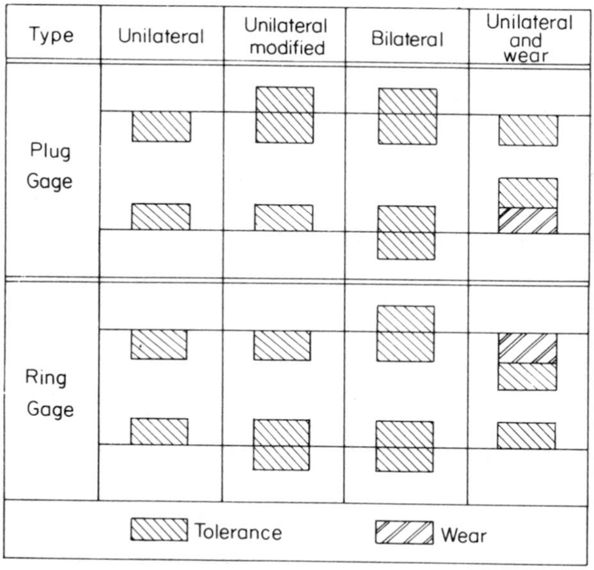

Figure 7.16

Example 7.5

Assume a diameter dimension of 1.875 ± 0.0015. Apply the tolerances (a) to the GO and NO GO plug gages and (b) to the snap (ring) gages. Draw the plugs and rings, and dimension them.

Solution   The tolerance of the gages is

$$10\% \times 2(0.0015) = 0.0003 \text{ in.}$$

The class is ZZ in Table 7.3.

(a) For the GO and NO GO plug gage dimensions:

1. The GO plug is

$$1.8750 - 0.0015 = 1.8735 \text{ in.}$$

The tolerance is applied in the *plus* direction:

$$1.8735 \, ^{+0.0003}_{-0.0000}$$

2. The NO GO plug is

$$1.8750 + 0.0015 = 1.8765 \text{ in.}$$

The tolerance is applied in the *minus* direction:

$$1.8765 \, ^{+0.0003}_{-0.0000}$$

3. The dimensioned plugs are shown in Fig. 7.17a.

(b) For the GO and NO GO snap (ring) gage dimensions:

1. The GO snap gage is

$$1.8750 + 0.0015 = 1.8765 \text{ in.}$$

The tolerance is applied in the *minus* direction:

$$1.8765 \, ^{+0.0000}_{-0.0003}$$

2. The NO GO snap gage is

$$1.8750 - 0.0015 = 1.8735 \text{ in.}$$

The tolerance is applied in the *plus* direction:

$$1.8735 \, ^{+0.0003}_{-0.0000}$$

3. The dimensioned snap gage is shown in Fig. 7.17b.

Example 7.6

In Ex. 7.5 add the allowance for wear, redraw the plug and snap gages, and dimension them.

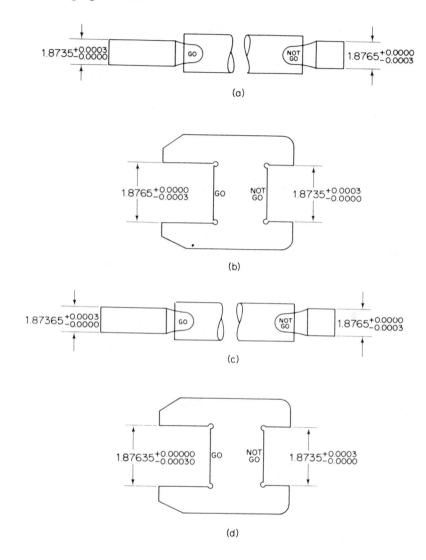

Figure 7.17

Solution   The allowance for wear is

$$5\% \times 0.003 = 0.00015$$

The GO plug dimension would be

$$1.8735 + 0.00015 = 1.87365$$

The tolerance applied would be

$$1.87365 \, {}^{+\,0.0003}_{-\,0.0000}$$

The NO GO plug gage will wear in the direction of the tolerance range. Therefore, no allowances need be made for wear.

The GO snap gage dimension is

$$1.87650 - 0.00015 = 1.87635 \text{ in.}$$

The tolerance applied would be

$$1.87365 \begin{smallmatrix} +0.0003 \\ -0.0000 \end{smallmatrix}$$

The NO GO snap gage will wear in the direction of the tolerance range. Therefore, no allowance need be made for wear.

The new dimensions for the plug are shown in Fig. 7.17c. The new dimensions for the snap gage are shown in Fig. 7.17d.

Flush pin gages.   A simple flush pin gage is shown in Fig. 7.18a. The plunger has a step machined at one end so that the low step represents the minimum dimension $0.998 + 0.500 = 1.498$ in. The high step represents the maximum dimension $1.002 + 0.500 = 1.502$ in. The pin is pushed down until it registers in the

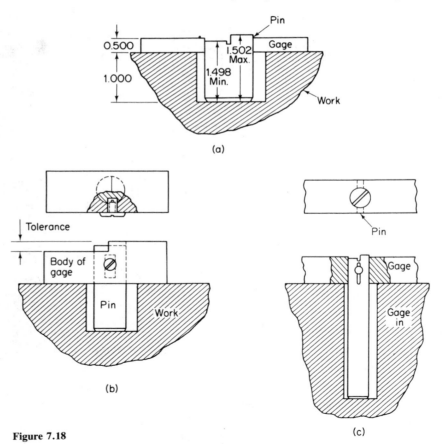

Figure 7.18

bottom of the hole. Since the hole is 1.000 in. deep, the flush pin will register as shown in Fig. 7.18a.

The limit step may also be applied to the body of the gage as shown in Fig. 7.18b. If the hole is deep, a long rod is used. Again one end is stepped to give the maximum and minimum limits. Figure 7.18c shows such a flush pin gage.

Flush pin gages may be used to check holes, shoulders, and hole or boss positions. They should be used for tolerances or limits of 0.005 in. and greater. If the tolerances or limits are less than 0.005 in., a dial indicator should be used.

Templates.   Template gages are used to check profiles, or the relationship of surfaces and shapes to one another, or to some reference plane. They are especially useful where the precision of the dimensions is such that it will permit visual inspection.

Radius gages, Fig. 7.19a, are templates. These gages may be purchased, or manufactured from $\frac{1}{16}$-in. sheet steel. When production requires special shapes, a profile of these shapes is machined into the surface of a flat piece of easily machined steel. If the demands of the job warrant such action, these gages may be hardened. Figure 7.19b shows such a template. Another type of template gage is a screw pitch gage shown in Fig. 7.19c. It is used to check the number of threads per inch of thread.

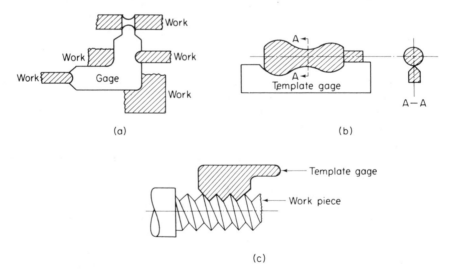

Figure 7.19

7.3 INDICATOR GAGES AS AMPLIFIERS

At times direct measurements or comparisons of surfaces to be checked with standards are not reliable. This results when the gaging system does not have enough precision built into it, or because the person making the check does not have the physical sensitivities to detect small variations in dimensions. It should be remembered that it is impossible to machine a surface precisely to a base dimension. The method of checking a dimension is therefore controlled to a great degree by the tolerances. As the tolerances get smaller and the precision increases, the method of checking the dimension becomes more selective. It soon becomes necessary to amplify the dimension, which also amplifies the tolerances. This may be done mechanically, electrically, pneumatically, and optically.

7.4 MECHANICAL AMPLIFICATION

Mechanical amplification is done with levers, gear trains, or reeds. Figure 7.20a shows the construction of a simple lever that can be used in fixtures. The amplification ratio can be chosen to suit the purpose at hand. These levers, if machined to a 10:1 ratio, will be sufficient for most purposes. One such commercial indicator is shown in Fig. 7.20a.

Dial indicators use gears to amplify readings. Amplifications with indicators can range from 40:1 to as much as 1500:1. In this case a pointer or ball probe is attached to the end of a rack. The linear motion of the rack is transferred to a gear train. A needle is attached to the output gear. The ratio of input to output of the gears in the train determines the amplification. The dial is graduated accordingly. A hairspring prevents backlash and a pullback spring holds the probe in a positive contact with the work at all times.

Since indicators are widely used with fixtures—especially checking fixtures; several types are shown in Fig. 7.20. It should be noted that standards for dial indicators (American Gage Designers—AGD) have been established. They have been standardized to sensitivities of 0.001 to 0.00005 in./div. They may also be graduated continuously from zero to maximum in one revolution, Fig. 7.20b, or balanced as shown in Fig. 7.20c.

Another type of widely used indicator is the dial test indicator shown in Fig. 7.20d. The dial is generally graduated to read 0–15–0 in 0.001- or 0.0005-in. increments with a total range of 0 to 0.030 in.

These dial gages may be used in a variety of ways in fixture design. Figure 7.21a shows a dial indicator mounted on an overarm so that the assembly may be used as a comparator. The overarm is set to a desired height and the fine adjustment made with the fine-adjustment screw. The system is used in the following manner. The lever is depressed to raise the probe. A gage block or some other standard is placed on the table. The lever is released slowly so that the probe registers on the standard. The needle is set at zero by rotating the bezel and is locked. The lever is depressed and the workpiece substituted for the standard. The needle will immediately indicate the plus or minus variations from zero. Attachments may be purchased or manufactured which quickly convert the unit into a checking fixture. Several possibilities are shown in Figs. 7.7a and 7.21b.

The dial bore gage, Fig. 7.21c, relies on three point contacts to center the gage so that holes may be quickly checked for irregular shapes, bell-mouth, taper, or out-

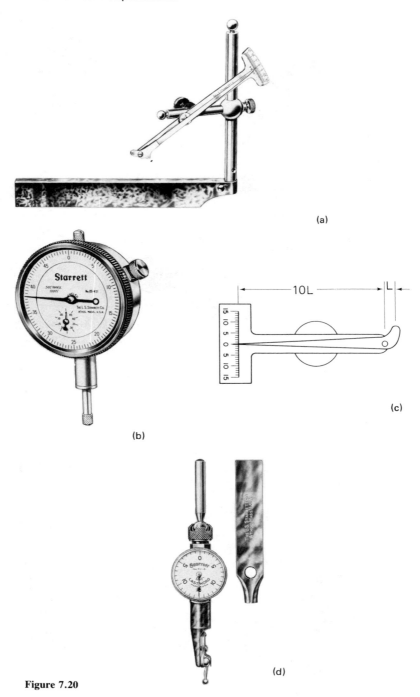

(a)

(b)

(c)

(d)

Figure 7.20

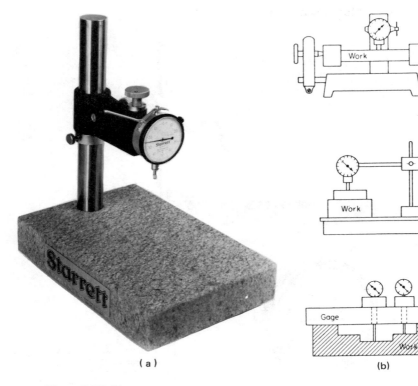

(a)

(b)

Figure 7.21(a,b)

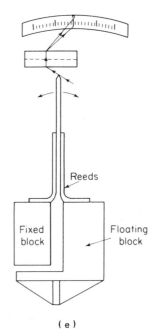

(d)

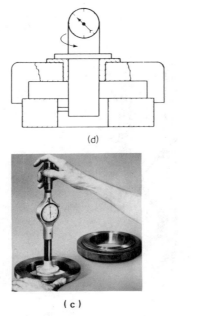

(c)

Figure 7.21(c,d,e)

(e)

of-roundness. Other designs rely on the fact that this gage may be built into a fixture and rotated in the workpiece hole as shown in Fig. 7.21d.

The reed method of amplification can achieve ratios of from 500:1 to as much as 20,000:1. The method is to mount a floating block on reeds made from spring steel. One reed is attached to the floating block and the second reed is attached to a fixed block. The fixed block is attached to the mounting. A slight motion of the contact point flexes one reed and causes the pointer attached to the far end to move. The high magnifications are achieved when a system of lenses projects light on a graduated scale. The pointer casts a shadow on the graduated scale. The process is shown in Fig. 7.21e.

7.5 ELECTRONIC AMPLIFICATION

Electronic amplification may be accomplished through five gaging systems: limit-type electric switches; induction bridges; electronic amplification; spark gap; and radiation. They are generally used when the tolerance is below 0.002 in. In some instances combinations of the five systems are used. These gates produce instantaneous responses which have or may be compared with a standard, or may be used to perform such jobs as sorting or actuating machine tools. They can also achieve magnifications as much as 100,000:1.

Electrical gaging, Fig. 7.22a, although not electronic in nature, is used in bench comparators. The indicator probe, which is similar to the standard dial gage, contacts the work and actuates one of two on–off switches. If the switches are both open, the dimension indicated with a light should be within the permissible limits. If the dimension is too low (undersize), an amber light is flashed on. If the size is too high (oversize), a red light may flash on. Adjusting screws permit the setting of limits of accuracies of 25 millionths of an inch. The gages can be read from 0.001 to 0.0001 in. This depends on the dial graduations.

Figure 7.22b shows a type of amplifier and a gage head which can be used to read divisions from 0.0001 to 0.000001 in. with amplifications of 650 to 65,000. They may be used for first-piece, in-process checking; inspection at the machine; and so on. They are also used to check gages and other critical measurements. Figure 7.22b shows a setup that permits the showing of the sum and difference in the readings of two heads for comparing one piece with a standard, checking parallelism, concentricity, or diameters.

In contrast to the electrical gages, which are essentially mechanical, a second type of system, the electronic gages, have continuous outputs. Again a probe is moved by contacting the workpiece. This movement acts on the alternating current brought to the gaging head, through the movement of an armature. In Fig. 7.22c the meter needle is caused to move off the zero position when a bridge is unbalanced.

In the third system, electronic amplification, the movement of the contact probe moves a coil in an electromagnetic field. This sets up an induced current flow in two oppositely wound coils. This induced current is amplified and fed into a voltmeter. The voltmeter dial is graduated to read maximum or minimum variations from the

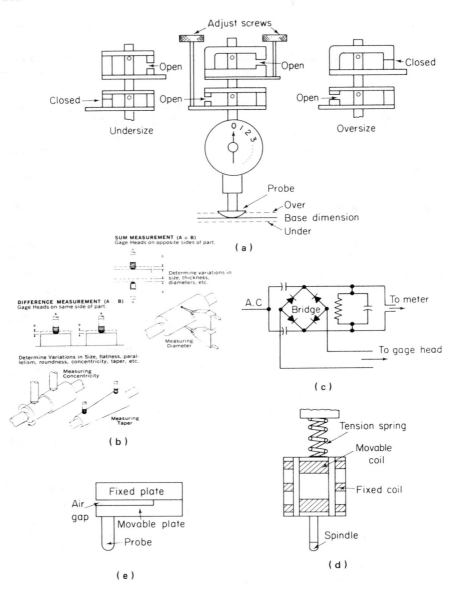

Figure 7.22

base dimension. This is shown in Fig. 7.22d. Another method is to construct the gaging head, as shown in Fig. 7.22e, so that when the probe is moved, it changes the volume of an air gap. This change in the air gap causes a change in the capacitance between the two plates, which in turn changes the capacitance of a bridge. This produces high magnifications.

A fourth method of gaging is that which creates a spark gap between the probe (probes) and the workpiece. This method permits continuous comparison of the shape of the workpiece with a master. With this method contours may be traced out on a tape, or read out on a suitable meter.

Radiation may also be used to check thicknesses of materials. A beam from a radioactive source is directed at the work surface. The thicker the workpiece, the less radioactivity will pass through the material. The radiation that does pass through the workpiece is collected in a scintillation detector which converts the radiation into a current. Therefore, the thicker the cross section of the workpiece, the less radiation, the smaller the current generated. This current is amplified and directed to a suitable indicator.

7.6 PNEUMATIC INDICATORS

Pneumatic gages operate according to two basic methods. There are those that register size, or variation in size from a standard. A differential pressure is created, Fig. 7.23a, when the workpiece actuates and regulates the airflow in one part of the gage. This is compared with the present flow in another part of the gage. The size of the workpiece determines the amount of air that will escape from the orifice in the gaging head.

The other gaging process operates according to a constant pressure but uses a varying airflow per unit time, as shown in Fig. 7.23b. The gage head meters the quantity of air permitted to escape from the orifice between the work and the walls of the gage head. Assuming that the air pressure is constant, the space between the head and the work is changed for different workpieces.

The flow of air per unit time must change. The bob goes up or down according to the change in *airflow* per unit time.

A back-pressure gage, Fig. 7.23c, is a flow gage that operates when the change in back pressure at the gage head is registered on a calibrated gage. The probe is inserted into the master. This creates a back pressure for a given air escapement. The indicator is set to read zero. If more air escapes between the workpiece and the gaging head, the back pressure drops; if less air escapes, the back pressure increases. The indicator is calibrated accordingly.

Figure 7.23d shows various types of pneumatic tooling designs that may be used in fixtures. When these types of gaging orifices are incorporated into fixtures, they may be used to check depths, outside or inside diameters, heights, out-of-roundness, thickness, warpage, tapers, or bell-mouths.

The advantage of using air is that metal from the gage does not need to touch the metal from the workpiece, and therefore the possibility of damaging the workpiece or the gage is greatly reduced. Figure 7.23e shows a noncontact gaging head where the quantity of air escaping is controlled by the diameter of the orifice and the clearance between the work and the gage. In Fig. 7.23f is shown a gaging head that houses a probe. This probe is depressed when it contacts the work, thus controlling the amount of air released. The cartridge heads used with tooling may be purchased to gage tolerances from 0.0001 in. to 0.100 in., and from 50 to 2000 magnifications.

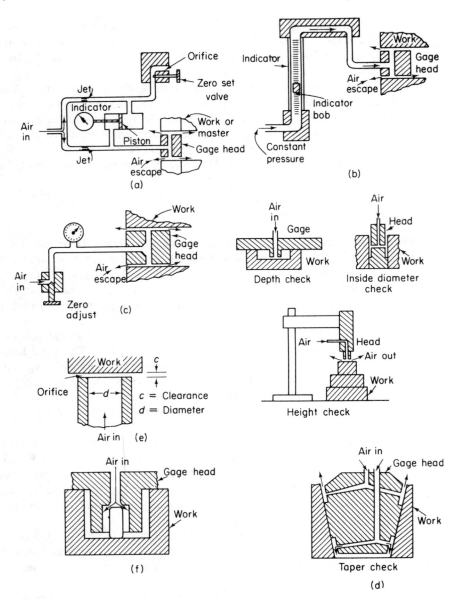

Figure 7.23

7.7 OPTICAL INDICATORS

Gage blocks, Sec. 7.1, are precision reference standards that are used with optical flats to measure size and work surface conditions. Another of the optical tooling devices discussed in Sec. 7.1 is the level and jig transit.

The optical comparator, Fig. 7.24a, is an instrument that sends a light beam, obstructed by the workpiece, into a magnifying glass, which then reflects an image to the rear of a viewing screen. With the use of several mirrors, the image may be projected vertically up or down so that it may be viewed from above.

The techniques used with these comparators are GO and NO GO comparisons with a master drawing, measuring by moving the part, or measuring by tracing the contour of the part.

Measuring by comparison.   This method depends on an accurately made master layout or gaging chart. The part is held in an appropriate fixture or standard holding device on the staging table. This positions the work so that the light beam will project a shadow of the desired contour of the workpiece on the screen. The master drawing is mounted on the screen, and the shadow allowed to fall so that a match can

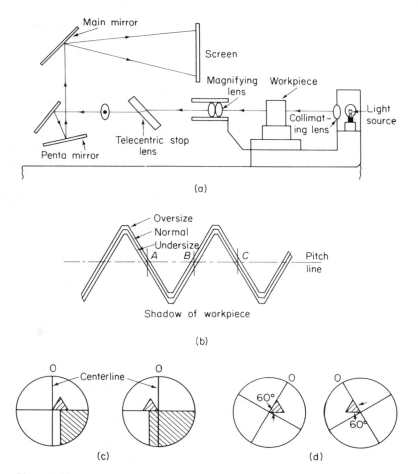

Figure 7.24

be made. Figure 7.24b shows a thread contour. The shadow is compared with the master and must fall within the limits of the drawing. In some cases the workpiece is caused to register against a reference line, plane, or hole. The rest of the workpiece contours must fall within the limits of the lines on the master chart, or be rejected.

Measurement by movement.   This method relies on the accurate longitudinal movement of the machine table. The vertical movement of the table produces vertical movement of the shadow on the screen on a *horizontal beam* machine. The movement of the table in and out produces a vertical movement of the shadow on a *vertical beam* machine. This type of inspection will not warp or distort the workpiece because it produces no forces that are capable of producing these variations in the work. Generally, the horizontal and vertical cross lines are used as reference lines. In addition, a ring, graduated in parts of a degree, may be rotated to check angles. The cross lines rotate with this ring.

The procedure is to place the workpiece in the fixture so that the light beam creates a shadow of the contour of the workpiece on the screen. The table is moved so that one edge of the shadow coincides with the centerline on the screen. The micrometer dial is zeroed. Then the table is moved until the opposite side of the shadow corresponds with the centerline. A reading is taken from the micrometer dial. This is shown in Fig. 7.24c. By rotating the screen, Fig. 7.24d, so that the vertical centerline coincides with the edge of the shadow in two positions, the angle at the corners of the workpiece may be measured.

It should be noted that the crossed centerlines *A* at the pitch line, Fig. 7.24b, may be used as a reference point. After the micrometer dial is zeroed, the table is moved so that point *B* coincides with the crossed centerlines. This will give the width of the space at the pitch line. If the movement is to point *C*, the micrometer reading will produce the pitch of the thread. If point *B* is zeroed and the table moved to point *C*, the thickness of the thread at the pitch line may be measured. Using the horizontal centerline and the pitch line of the thread, the vertical movement of the table may be used to measure its pitch diameter.

Starting at *A*, Fig. 7.24b, as a reference, the distance from point *A*, along the pitch line, to the corresponding point several threads away is calculated. The table is moved this distance. The vertical centerline should correspond to the same point on the pitch line of the shadow. This is a good check of the lead accuracy. The angle of the thread may also be checked in the same manner, as shown in Fig. 7.24d.

Measurement by translation.   This method uses a 1:1 tracer mechanism. This tracer may be a ball, an opaque dot on a glass reticle, a circle, or two circles, on a chart. The projected reticle contour (gage) moves past the fixed circle as the workpiece is being traced. These three methods are shown in Fig. 7.25.

The method of measurement by translation is used to measure or trace contours which are not directly accessible with a light beam. Undercut sections, V-cuts, recessed odd shapes, internal gears, and so on, may be traced with a pantograph arrangement. The stylus of the tracer follows the surface to be checked as the other

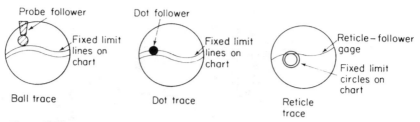

Figure 7.25

arm of the pantograph carries a follower that intersects the light beam and is projected as a shadow on the screen. The ratio of input to output is usually 1:1.

Many types of checks, comparison, and measurements may be made with comparators of this type. Standard charts and accessories permit a wide variety of uses.

7.8 SURFACE INSPECTION

Surface conditions of a workpiece affect the service life, appearance, and function of the workpiece. The surface characteristics with which this section deals are roughness, waviness, and lay. Surface hardness, wear resistance, luster, corrosion, and so on, are not dealt with in this section.

The simplest method for measuring surface quality is the visual comparison of the workpiece with an acceptable sample. This sample surface may be a machined surface set aside by someone in authority. The comparison of the sample with the machined workpiece may be visual as well as by running a fingernail first across the sample and then across the workpiece. The workpiece may also be compared with standard samples which may be purchased in sets. One such set is shown in Fig.

Figure 7.26(a)

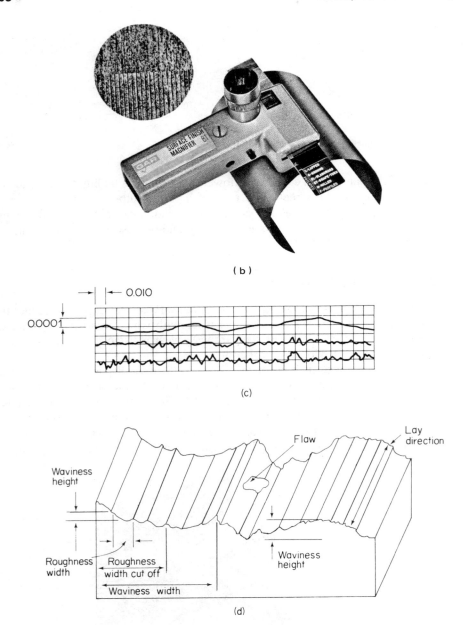

(b)

(c)

(d)

Figure 7.26(b,c,d)

7.26a. A magnifier and comparator are shown in Fig. 7.26b. These systems allow comparisons of from 2 to 500 microinches.

Surface may also be visually examined after being recorded on a transparent

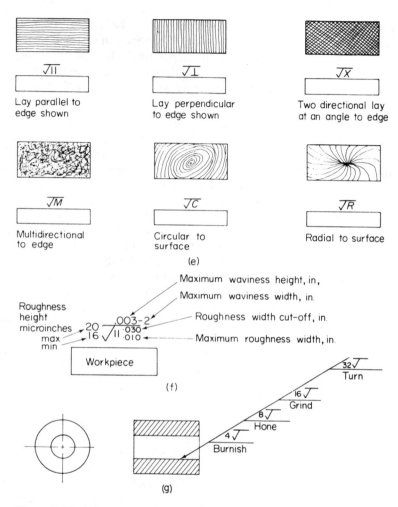

Figure 7.26(e,f,g)

film which is placed on the surface to be inspected. This film, when partially hardened, may be stripped off the workpiece. The hardened film records the surface of the workpiece, which then may be studied.

Profilometers, which use a very finely pointed diamond tracer, are used to read surface irregularities. They produce readouts as arithmetic or rms (root mean square) averages. With a diamond tracer which has a tip radius of 0.0005 in. and a 90° cone angle, very finely finished surfaces may be averaged. The tracer may be operated manually or mechanically.

Figure 7.26c shows a trace where the variation in depth is read to 0.0001 in. and the length of trace is 0.010 in. The lower trace is a trace of the surface of the workpiece in terms of roughness, waviness, and flaws such as scratches. The *middle trace* shows fine irregularities of roughness in the work surface. The averages of the irregularities are shown by this trace. When the roughness is spaced out it is called a *wave*. Waves are recorded by the upper trace. Recent experiments with electrical condenser plates, very fine air jets, and light beams seem to hold some hope for their use as tracers. Commonly used terms are shown in Fig. 7.26d.

Roughness.   These are the surface irregularities that result from the various fabricating processes. They are the finely spaced irregularities which combine to form the surface texture. The height, width, and general direction of the pattern are included. The *height* is averaged above and below, perpendicular to a mean line, and stated in microinches. The *width* is stated in inches and is the maximum spacing allowed between successive peaks, measured parallel to the mean line.

This establishes the *roughness-width cutoff,* which must always be greater than the roughness width to establish the rating values for roughness height. These cutoffs may be 0.003, 0.010, 0.030, 0.100, 0.300, and 1.000. Unless another value is used, the preferred value is 0.030. It should be noted that the roughness-width cutoff is a matter of instrument setting rather than surface quality. If the surface of the work has very fine irregularities, the machine may be set at 0.003 roughness-width cutoff. If the roughness is wide, such as would be the case in machining chatter, the roughness-width cutoff should be chosen to include the major variations in roughness height of the work surface. Thus the higher the number of the roughness-width cutoff, the more inclusive but discriminating are the evaluations. A 0.030 number includes most of the coarse and some of the medium and all the fine irregularities. The 0.003 includes only the fine irregularities.

Waviness.   This refers to the irregularities that are outside the roughness-width cutoff values. As seen in Fig. 7.26d, the *waviness width* has the roughness superimposed on it. It results from bending, strain, work deflection, and so on. It is measured from one peak to the next peak. It is given in inches. *Waviness height* is the peak-to-valley distance. *Waviness-width cutoff* is also selected as the maximum permitted variation in waviness height. Anything outside this cutoff is said to be an *imperfection*. It is given in inches.

Cracks, scratches and ridges are called *flaws*. They are not regularly recurring and are imperfections outside the regular pattern of surface texture.

Lay.   Lay is directional and reflects the machining or fabricating method. The symbol $\sqrt{}$ is used to indicate surface conditions. The notations $\parallel$, $\perp$, X, M, C, and R are used with the symbol to indicate lay. This is shown in Fig. 7.26e.

The terms above are shown in Fig. 7.26f. The symbol is drawn so that it touches the surface line, extension line, or leader line which indicates the surface described. Figure 7.26g shows the finish requirements for four machining operations.

Surface finishes that can be expected from the various fabricating processes are shown in Table 7.4. These vary with the type of manufacturing company considered.

TABLE 7.4 SURFACE FINISHES PRODUCED BY VARIOUS PROCESSES

Operation	Rms (10^{-6} in.)	Operation	Rms (10^{-6} in.)
Lapping	2–8	Turning	20–300
Honing	2–10	Shaping	20–300
Polishing	2–10	Sand casting	500–1000
Reaming	8–50	Extrusion	10–250
Grinding	5–150	Sawing	250–1000
Broaching	15–60	Blanking	30–100
Drilling	75–200	Forging	100–400
Milling	20–300	Die casting	15–100
Burnishing	2–4		

The rms is obtained by taking the square root of the sum of the squares of the y-measurements divided by the number of y-measurements. The mathematics is

$$\Sigma\, y_n = y_1 + y_2 + y_3 + \cdots + y_n$$

The arithmetic average is

$$\frac{\Sigma\, y_n}{n}$$

The root mean square is

$$\text{rms} = \sqrt{\frac{\Sigma\, y_n^2}{n}}$$

Example 7.7

Calculate (a) the arithmetic average and (b) the rms for Fig. 7.27.

Solution The y and y^2 values are:

n	y	y^2
1	6	36
2	8	64
3	15	225
4	10	100
5	0	0
6	8	64
7	0	0
8	7	49
9	0	0
	$\Sigma\, y = 54$	$\Sigma\, y^2 = 538$

(a) The arithemetic average is

$$\Sigma\frac{y_n}{n} = \frac{54}{9} = 6.00 \ \mu\text{in.}$$

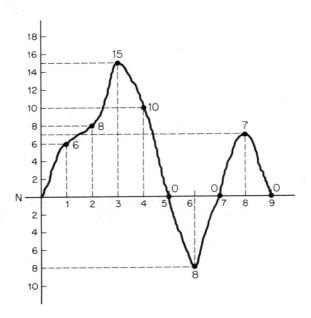

Figure 7.27

(b) The root mean square is

$$\text{rms} = \sqrt{\frac{\Sigma\, y_n^2}{n}} = \sqrt{\frac{538}{9}} = 7.73 \ \mu\text{in.}$$

7.9 THREAD INSPECTION

Threads are usually checked to determine whether the pitch diameters of the machined thread meet the tolerance limits established by the tool designer. Pitch diameters may be checked with special thread micrometers, Fig. 7.28a. This check is not very accurate because the anvils are made for a *range* of pitches and is accurate only at the center of the range.

Measuring threads with three precision wires is much more accurate. The measurement over the wires may be made using a micrometer or a supermicrometer. Wires should be accurate to 0.000020 in. (20 millionths of an inch) to be able to check thread pitch diameters to 0.0001-in. accuracies. The wire diameters are selected so that they are tangent to the sides of the thread at the pitch line as shown in

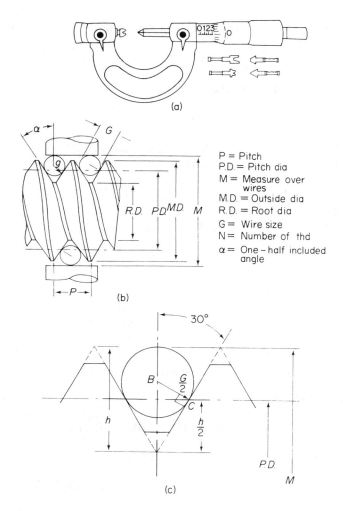

P = Pitch
P.D. = Pitch dia
M = Measure over wires
M.D. = Outside dia
R.D. = Root dia
G = Wire size
N = Number of thd
α = One – half included angle

Figure 7.28(a,b,c)

Fig. 7.28b. If this is the case, the wire is said to be the *best wire size*. The *general* equation for the best wire size is

$$G = \frac{p}{2}\sec\alpha \qquad \begin{array}{l} p = \text{pitch} \\ \alpha = \tfrac{1}{2} \text{ included angle of thread} \\ G = \text{best wire size} \end{array}$$

For threads that have a 60° included angle, the best wire is

$$G = 0.57735p = \frac{0.57735}{N} \qquad N = \text{number of threads per in.}$$

To calculate the pitch diameter using three wires, a simplified general equation is

$$\text{P.D.} = M + \frac{\cot \alpha}{2N} - G(1 + \csc \alpha)$$

M = measurement over wires
$P.D.$ = pitch diameter

For 60° threads this equation reduces to

$$\text{P.D.} = M + \frac{0.86603}{N} - 3G$$

The derivation of this equation is as follows: From Fig. 7.28c the depth of the 60° thread over the sharp corners is

$$h = \frac{0.86603}{N}$$

The distance from the sharp V to the pitch line is

$$\frac{h}{2} = \frac{0.86603}{2N}$$

In the right triangle ABC,

$$\csc 30° = \frac{AB}{BC} = 2.000$$

Solving for AB yields

$$AB = 2BC = G$$

The pitch diameter is

$$\text{P.D.} = M - 2(BC) - 2(AB) + 2\left(\frac{h}{2}\right)$$

Substituting the values from above gives us

$$\text{P.D.} = M - G - 2G + \frac{0.86603}{N}$$

Collecting terms yields

$$\text{P.D.} = M + \frac{0.86603}{N} - 3G$$

Example 7.8

Assume a 1–12UNF thread is measured with three wires. The measurement over wires is 1.018037 in. Calculate (a) the best wire size, (b) the pitch diameter of this thread, and (c) the measurement over wires if the pitch diameter should be 0.94455 in.

Solution

(a) The best wire size for this thread is

$$G = \frac{0.57735}{N} = \frac{0.57735}{12}$$

$$= 0.048112 \text{ in.}$$

$N = 12$ threads/in.

(b) The pitch diameter is

$$\text{P.D.} = M + \frac{0.86603}{N} - 3G$$

$M = 1.018037$ in.

$$= 1.018037 + \frac{0.86603}{12} - 3(0.048112)$$

$$= 0.94587 \text{ in.}$$

(c) The measurement over wires when the pitch diameter is 0.94455 in. is

$$M = \text{P.D.} - \frac{0.86603}{N} + 3G$$

$$= 0.94455 - \frac{0.86603}{12} + 3(0.048112)$$

$\text{P.D.} = 0.94455$ in.

$$= 1.016717 \text{ in.}$$

Figure 7.28d shows a measurement being taken with a thread-measuring machine. The measurement of thread angles, pitch, and lead, as indicated earlier, may be accomplished with an optical comparator and a suitable chart. High magnifications make visual inspection of thread gages possible with the comparator. Leads may also be checked using accurate test points and appropriate gage blocks, Fig. 7.28e.

Another very common method for checking threads is plugs and rings. The thread plug gages, used for checking internal threads, are marked GO or HI and LO (formerly NO GO) gages. Thread ring gages, used for checking external threads, are also marked GO or HI and LO. Thus a *GO thread plug gage* would pair with a *HI*

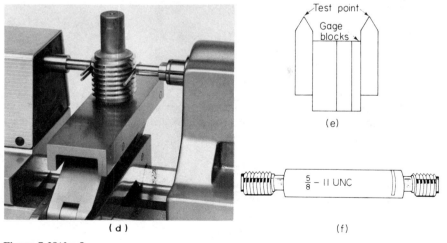

Figure 7.28(d,e,f)

thread plug gage. The GO gage must enter freely and the HI gage may enter only *three* threads and be stopped by the drag from the contact. A *GO thread ring gage* would be paired with a *LO thread ring gage*. Thus the GO ring must accept the entire external thread being checked. The LO ring may accept only *one* thread.

The *GO plug* checks all thread elements except the minor diameter of the workpiece. The *GO* ring checks all thread elements except the major diameter. It should be noted that the HI/LO gages are truncated to ensure engagement at the pitch diameter. Thread plug and thread ring gages are shown in Fig. 7.28f.

PROBLEMS

7.1. Describe the construction of a vernier and how it operates as an amplifier. Choose units to suit your purpose.

7.2. Construct a vernier that is to have 40 divisions per inch of the main scale and is to read in increments of 0.002 in. What is the length of each division on
  (a) the main scale and
  (b) the vernier scale?
  (c) Calculate the minimum length of the vernier that matches the main scale when the zeros coincide.
  (d) How many divisions on each scale are needed to match, when the zeros match?
  (e) Construct the vernier.

7.3. Repeat Prob. 7.2 when the increment is 0.005 in.

7.4. Repeat Prob. 7.2 when the vernier is to have 25 divisions per inch and is to read in increments of 0.004 in.

7.5. List the measuring instruments that use vernier as length amplifiers.

7.6. What is the reading on the row-column vernier in Fig. 7.29 when the main-scale line is as shown at
  (a) *A* and
  (b) *B?*

7.7. Repeat Prob. 7.6 for
  (a) *C* and
  (b) *D.*

7.8. **(a)** State the accuracies of the various sets of gage blocks.
  (b) Indicate the attachments and adapters that may be used with gage blocks.

7.9. Using Table 7.2, stack the dimension
  (a) 6.33455 in. and
  (b) 1.74335 in.

7.10. Using Table 7.2 stack a dimension 3.28865 in.

7.11. Explain the use of stacked gage blocks and a height gage when the transfer method is used to check workpieces.

7.12. Repeat Prob. 7.11 when a sine bar is used.

7.13. Indicate the blocks needed to set up the angle shown in Fig. 7.30 when points *A* and *B* are 0-0 on the height gage.

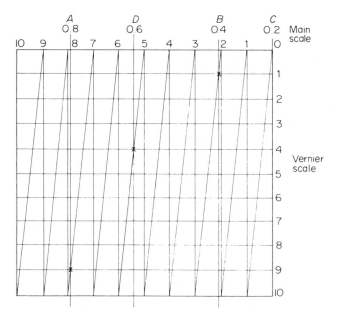

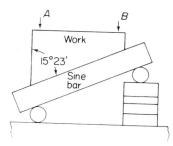

Figure 7.30

Figure 7.29

7.14. (a) Explain the principle underlying the use of optical flats with monochromatic light to check dimensions of workpieces very accurately.

(b) What else can be checked with these flats? Explain.

7.15. A workpiece is measured, using a micrometer, to be 2.8756 in. Assume 18 bands when the setup is as shown in Fig. 7.8e. The width of the gage block is $1\frac{1}{8}$ in., the width of the spacer is $1\frac{7}{8}$ in., and the width of the work is $2\frac{1}{2}$ in. Calculate

(a) the length of the work when helium light is used and

(b) the number of bands if G is taken 0.0002 in. less than the height of the work, which measures 2.8748 in.

7.16. Repeat Prob. 7.15 if 24 bands are counted.

7.17. A workpiece is measured with a micrometer to be 1.4475 in. Assume 10 bands when the setup is as shown in Fig. 7.8d. The width of the gage block is $\frac{3}{4}$ in., the width of the spacer is $1\frac{5}{8}$ in. and the width of the work is $1\frac{3}{8}$ in. Calculate the length of the work when helium light is used.

7.18. Repeat Prob. 7.17 for 16 bands.

7.19. A workpiece is measured with a micrometer to be 3.000 in. Assume that bands appear when the setup is the same as in Fig. 7.8e. The width of the gage block is $\frac{7}{8}$ in., the width of the spacer is $1\frac{1}{2}$ in., and the width of the work is $\frac{3}{4}$ in. Calculate

(a) the length of work when helium light is used and

(b) the number of bands if G is taken 0.0002 in. less than the height of the work and it measures 2.9998 in.

7.20. Draw the interference patterns and explain the contour of the surfaces for Fig. 7.8f through i.

7.21. (a) Name at least six other elements that may be used as monochromatic light for producing interference patterns.

(b) Explain the use of a gas laser beam for producing patterns.

7.22. **(a)** How does an optical level or jig transit use the earth's gravitational field for leveling?
(b) What is the purpose of the tubular level?

7.23. Assume the readings in Fig. 7.10c to be 2.332 and 3.555 in., and those in Fig. 7.10d to be 2.076 and 3.177 in. Calculate the instrument reading at the center Fig. 7.10e if the instrument is properly set.

7.24. Repeat for readings of 1.888 and 2.766 in Fig. 7.10c and 2.008 and 2.624 in Fig. 7.10d.

7.25. What is the reading on the scale in Fig. 7.31a?

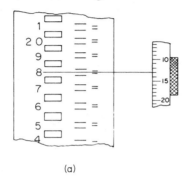 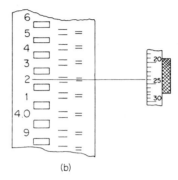

(a)   (b)

Figure 7.31

7.26. What is the reading on the scale in Fig. 7.31b?

7.27. What are penta prisms and leveling mirrors?

7.28. Define and illustrate limit gages.

7.29. State the three general classifications of gages. Illustrate by explanation and example the purpose of each class.

7.30. Explain the various plug gage designs shown in Fig. 7.13.

7.31. Explain the design of the ring gages shown in Fig. 7.14.

7.32. Discuss the various types of steels used when gages are
(a) made in the "home" shop and
(b) purchased commercially.

7.33. Draw the snap gages in Fig. 7.15a through d. Explain their uses.

7.34. **(a)** If working gages do not indicate the tolerance applied to the gage, what tolerance should the toolmaker apply?
(b) How much should the toolmaker allow for wear?
(c) Explain the direction in which these tolerances and allowances should be applied to plug and ring gages.

7.35. List the five classes of plug and ring gages.

7.36. Assume a diameter of 2.625 ± 0.001. Apply the tolerances
(a) to the GO and NO GO plug gages and
(b) to the snap gages.
(c) Draw the plugs and snap gages and dimension them.
(d) Calculate the wear allowances.
(e) Draw the plugs and snap gages and dimension them according to part (d).

7.37. Plug and snap gages are to be designed to check a dimension of $1.125^{\pm 0.005}$.
   (a) Calculate the plug gage dimensions and tolerances, and insert the dimensions in a drawing of the gages.
   (b) Repeat part (a) for snap gages. Add 5% for wear.

7.38. Make a sketch of a flush pin gage for checking a step block which has a dimension of $0.875^{\pm 0.005}$.

7.39. Select two objects at random and make sketches of flush pin gages for checking steps, holes, or shoulders.

7.40. Select two objects at random and make sketches of templates that may be used to check the shapes of the objects.

7.41. Name three templates that may be purchased. Make a sketch of each and indicate their purpose.

7.42. Discuss the need for sophisticated methods of checking high-precision dimensions.

7.43. Discuss the use of levers, gear trains, and reeds as mechanical amplifiers.

7.44. Make a sketch of a lever amplifier that you would design to check the concentricity of a center punch layout hole in a lathe.

7.45. Describe several types of dial indicators that may be used in fixture design, and show how they may be used.

7.46. **(a)** What is a dial bore gage?
   (b) How may it be used in fixtures?
   (c) List the conditions of a hole that may be checked with this gage.

7.47. How does reed amplification work in checking instruments?

7.48. List the five methods of electronic amplification. Briefly indicate how each is distinct from the other.

7.49. Make a sketch of an electrical switching system used in gaging. Explain how it may be used to indicate acceptable, oversize, or undersize dimensions.

7.50. Describe the electronic gage in Fig. 7.22b. Show how it may be used to check sizes, parallelism, and concentricity.

7.51. Explain the use of an electrical bridge in gaging dimensions
   (a) by induction and
   (b) by capacitance.

7.52. How is the spark gap method used to check work contours?

7.53. How may radioactivity be used to measure the thickness of materials?

7.54. Discuss the differential pressure and the constant airflow methods of gaging workpieces.

7.55. Explain the operation of the back-pressure gage.

7.56. Air orifices may be used to check many conditions. List them.

7.57. What is the major advantage to using air in gaging precision surfaces? Explain.

7.58. What two factors control the process of gaging with air? Explain.

7.59. **(a)** Make a sketch of an optical comparator used to check workpieces.
   (b) List the three techniques that are used when checking with a comparator.

7.60. Explain the use of meters with a comparator when checking by comparison. Give at least one example that is not in this book.

7.61. Using the reference lines, the graduated bezel, and the controlled movement of the table, explain and illustrate how the following parts of V-thread may be checked:
   (a) the 60° included angle;

   (b) the thickness of the thread at the pitch line;

   (c) the pitch;

   (d) the lead;

   (e) the depth of the thread; and

   (f) the pitch diameter.

   Use a $\frac{3}{4}$–10 UNC thread in your explanation.

7.62. Explain the operation of measurement by translation.

7.63. Discuss surface inspection of workpieces by comparison. There are at least three methods discussed in this book.

7.64. Discuss the use of a profilometer and its operations in conjunction with its readout.

7.65. Roughness is defined in terms of height, width, and pattern direction. Explain each term.

7.66. **(a)** What is roughness-width cutoff?

   (b) What is its significance?

   (c) Explain its effect on roughness height.

7.67. **(a)** Describe and illustrate waviness.

   (b) How does it differ from roughness?

   (c) What is the significance of waviness-width cutoff?

7.68. Describe a flaw in terms of

   (a) roughness limits and

   (b) waviness limits.

7.69. **(a)** Define the term "lay."

   (b) Using the surface quality symbol and the notations for lay, describe the surface which they represent.

7.70. Write the surface quality symbol when the following are the conditions for a cylindrically ground rod.

   Roughness height   8–16
   Roughness width   0.008
   Roughness cutoff   0.030
   Waviness width   3
   Waviness height   0.002

7.71. List the rms value ranges for

   (a) lapping,

   (b) reaming,

   (c) grinding,

   (d) drilling,

   (e) milling,

   (f) turning, and

   (g) blanking.

7.72. Calculate

   (a) the arithmetic average and

   (b) the rms for Fig. 7.32a.

7.73. Find

   (a) The arithmetic average and

   (b) the rms for Fig. 7.32b.

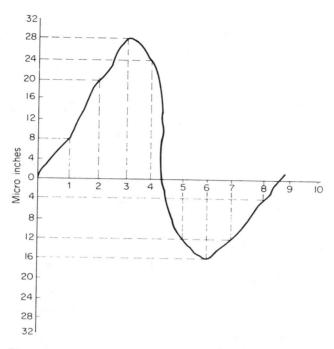

Figure 7.32(a)

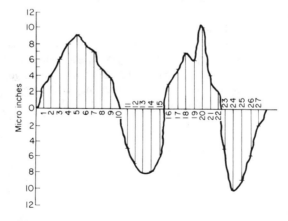

Figure 7.32(b)

7.74. Calculate
   (a) the arithmetic average and
   (b) the rms for Fig. 7.32c.

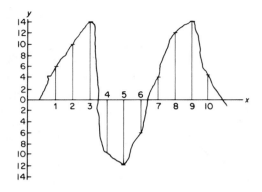

Figure 7.32(c)

7.75. Define the term "best wire size." Illustrate.

7.76. Given a $\frac{13}{16}$–20 UNEF thread. Calculate
(a) the best wire size and
(b) the measurement over wires if the P.D. is 0.7785 in.

7.77. Assume a $\frac{3}{4}$–10 UNC thread is measured with three wires. The measurement over wires is 0.7669 in. Calculate
(a) the best wire size,
(b) the pitch diameter, and
(c) the measurement over the wires if the pitch diameter should be 0.6832 in.

7.78. Given a $1\frac{1}{8}$–12 UNF thread, calculate
(a) the best wire size and
(b) the pitch diameter if the measurement over the wires is 1.1431 in.

7.79. (a) Explain why a GO thread plug gage is coupled with a HI thread plug gage and a GO thread ring gage is coupled with a LO thread ring gage.
(b) How many threads on the workpiece may each accept?

7.80. Why are the HI/LO thread gages truncated?

Chapter 8

Geometric Control

8.1 GEOMETRIC DIMENSIONING

Geometric dimensioning and tolerancing provide a method whereby drawing and design requirements relate to the *functional relationships* of components. It is a method that allows engineering, design, quality control, and production to interpret and understand what is needed to meet the functional requirements of a part as it relates to assembly.

To accomplish this goal, the objectives of the standards developed by the American National Standards Institute Committee (Y14), a subcommittee of industrial people and liaison representatives were to provide a uniform single standard for the United States, so that the functional aspect of design and manufacturing of components would be considered. In essence it is a "language" applied to a drawing that communicates its *functional* requirements. The purpose is to reduce significantly the need for notes on a drawing. With this system the question is always "How does it work?" The system enhances the concept of interchangeability.

To accomplish the above, symbols are used together with a *feature control symbol*, Fig. 8.1a. The FCS, Fig. 8.1b, says that the 1.500-in. hole diameter $\varnothing$, must be perpendicular $\perp$ to the datum face $\boxed{-A-}$ to within 0.001 in. when the hole is at $\textcircled{M}$ maximum material condition. We will see how an analysis of the feature control symbol applies to the dimensioning of this part and in turn how the part is to function with its mating part.

In the FCS are other symbols. Table 8.1 shows the symbols that define geometric characteristics. Table 8.2 shows the modifying symbols.

The *maximum material condition*, MMC $\textcircled{M}$, is that condition where the maximum amount of material is permitted. Thus, in Fig. 8.2a, the MMC $\textcircled{M}$ for the shaft

223

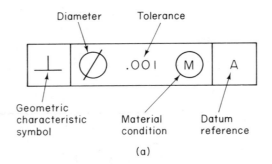

(a)

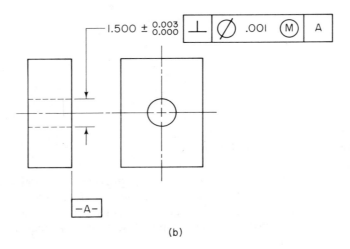

(b)

Figure 8.1

TABLE 8.1 GEOMETRIC CHARACTERISTIC SYMBOLS

Feature	Tolerance Type	Characteristic	Symbol
Individual feature	Form	Straightness	—
		Flatness	▱
		Circularity	○
		Cylindricity	⌀
Individual or related feature	Profile	Of a line	⌒
		Of a surface	⌓
Related feature	Orientation	Angularity	∠
		Perpendicularity	⊥
		Parallelism	//
	Location	Position	⊕
		Concentricity	◎
	Runout	Circular runout	↗
		Total runout	↗↗

Source: Dimensioning and Tolerancing, ANSI Y14.5M–1982, Fig. 68, p. 30.

TABLE 8.2 MODIFYING SYMBOLS

Term	Symbol
Maximum material condition MMC	M
Regardless of feature size RFS	S
Least material condition LMC	L
Projected tolerance zone	P
Diameter	⌀
Spherical diameter	S⌀
Radius	R
Spherical radius	SR
Reference	( )
Arc length	⌒
Datum feature	-A-
Datum target	(AT)

Source: Dimensioning and Tolerancing, ANSI, Y14.5M–1982, Fig. 72, p. 31.

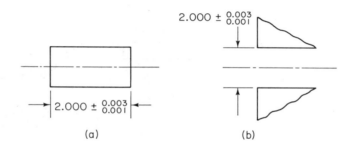

Figure 8.2

is 2.003 in. The *least material condition,* LMC(L), means the smallest amount of material permitted. For the shaft, the LMC is 1.999 in. For the hole, the reverse is true. The smallest hole has the most material and the largest hole has the least material. Thus in Fig. 8.2b the MMC dimension is 1.999 in. and the LMC dimension is 2.003 in.

Example 8.1

In Fig. 8.3, what are (a) the MMC dimensions and (b) the LMC dimensions?

Solution

$$
\begin{aligned}
\text{(a) MMC: Outside dimension} &= 3.502 \text{ in.} \\
\text{Hole} &= 1.497 \text{ in.} \\
\text{(b) LMC: Outside dimension} &= 3.498 \text{ in.} \\
\text{Hole} &= 1.503 \text{ in.}
\end{aligned}
$$

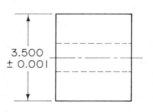
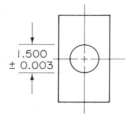

Figure 8.3

Virtual condition. A virtual condition is the envelope within which the part must lie after all permissible conditions have been applied. It is a collection of size and form that permits the part to fit its mating part. Figure 8.4c shows a round pin at its maximum material size and straightness variation. The virtual condition is 1.010 in.

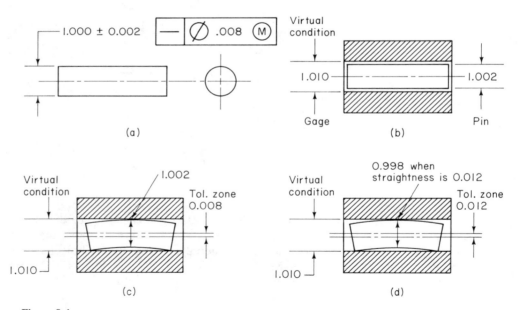

Figure 8.4

Example 8.2

Figure 8.4a shows a 1.000-in.-diameter pin that can vary in straightness by 0.008 in. under maximum material conditions. Design the gage for this pin.

Solution   Under maximum material conditions the pin is

$$1.000 + 0.002 = 1.002 \text{ in.} \qquad \text{(perfectly straight, Fig. 8.4b)}$$

When the pin is 1.002, the gage may be

$$1.002 \text{ (MMC)} + 0.008 \text{ (straightness tolerance)} = 1.010 \text{ in.}$$

The pin may be bent a maximum of 0.008 in. The envelope (virtual conditions) that is possible, Fig. 8.4c, is 1.010 in. The maximum diameter of the pin allowed is 1.002 in., but the straightness can vary by 0.008 in. Thus the virtual condition permitted is 1.010 in.

TABLE 8.3

If Feature Size $\varnothing$	$\varnothing$ Tolerance Zone
1.002 MMC	0.008
1.001	0.009
1.000	0.010
0.999	0.011
0.998 LMC	0.012

If the pin is at least material condition (LMC), the straightness tolerance may increase to 0.012 in. The two parts should still go together. Table 8.3 shows the analysis. At maximum material condition, (M) , the cylindrical part is 1.002 and the straightness allowed is 0.008 in. If the size of the pin is 0.001 in. smaller than MMC, or 1.001, the straightness tolerance may be increased from 0.008 to 0.009 in. This pin will still go into a 1.010-diameter hole because now

$$1.001 \text{ (pin)} + 0.009 \text{ (straightness)} = 1.010 \text{ (virtual)}$$

If the analysis is continued, when the pin is at least material condition,   L   , it can be out of straightness by 0.012 in. and still assemble with the hole. Thus

$$0.998 \text{ (pin)} + 0.012 \text{ (straightness)} = 1.010 \text{ (virtual)}$$

In summary, the pin in Fig. 8.4a may have a maximum diameter of 1.002 in. and be straight within 0.008. When the pin is at maximum material condition, it must not exceed 0.008 in. in straightness. This means that the hole must accept a pin diameter of 1.010 in. As the diameter of the pin gets smaller, the hole will accept a greater straightness tolerance. The increase must take place in the straightness. The diameter of the pin is locked into ± 0.002 in., so that when the pin size is 0.998 in., the straightness allowance can be 0.012 in., Fig. 8.4d. The *pin* will still function in its mating part.

8.2 TOLERANCE OF FORM (FIG. 8.5)

A few definitions are in order before discussing form. The concept of datum needs to be defined at this point.

Single surface
forms
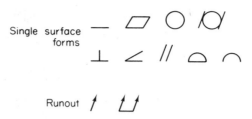

Runout

Figure 8.5

A *datum* may be a point, line, plane, or axis of a configuration (cylinder, cone, etc.) that is considered theoretically exact. It is used as a reference for other features of a part or assembly.

A *datum feature* is the actual configuration of the surface of the part (see Fig. 8.6).

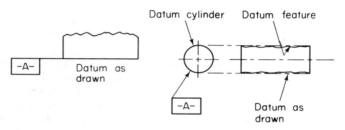

Figure 8.6

Figure 8.7 shows a *feature control frame* with three datums. Note that the primary datum letter is in the first box, followed by the secondary and tertiary datums. The letters need not be in alphabetical order. Figure 8.7 shows the three datums *P, L,* and *R.*

Flatness applies to the entire surface of a *tolerance zone.* Figure 8.8a shows that the entire upper surface of the block must be flat, or lie within a 0.005-in. tolerance zone even though the plane itself may be at an angle to the base of the part.

Straightness, Fig. 8.8b, applies to only one element of the block at MMC. Where cylinders, or cones, are considered, the element may be the centerline (Fig. 8.9a). Each element must not be out of straightness by more than 0.002 in. Fig. 8.9b The indicator is moved from one end of the piece to the other. Then the piece is rotated and checked again, and so on. Note that elements are checked at MMC. If the Ⓜ is omitted in Fig. 8.9a, the checks must be made regardless of feature size, RFS, Figure 8.9a. The axis of the cylinder must be straight—within the cylindical tolerance zone Ø of 0.002 in. at maximum material condition   Ⓜ

Figure 8.9c shows three possibilities of straightness. Note that all are at MMC, or 1.005 in.

As the feature size (1.005 in.) decreases, the straightness tolerance zone may increase. With the feature size and tolerance zone combinations shown in Table 8.4,

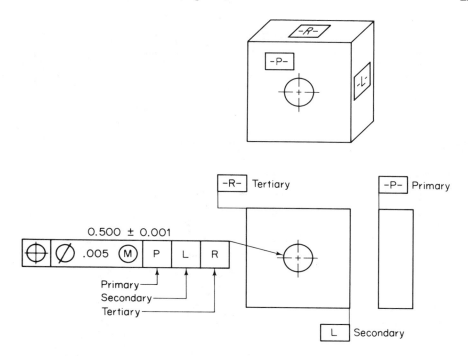

Figure 8.7

TABLE 8.4

Feature Size $\varnothing$	$\varnothing$ Tolerance Zone
1.005	0.002
1.004	0.003
1.003	0.004
1.002	0.005
1.001	0.006
1.000	0.007
0.999	0.008
0.998	0.009
0.997	0.010
0.996	0.011
0.995	0.012

this pin will assemble with its mating part. So that if the pin is at 1.005 in., the straightness must be within 0.002 in. If the pin diameter is at 1.000 in., the straightness may be 0.007. If the pin diameter is 0.995 in., the straightness must be 0.012 in. or less.

When the pin is at MMC and perfectly straight, and the mating part is at virtual

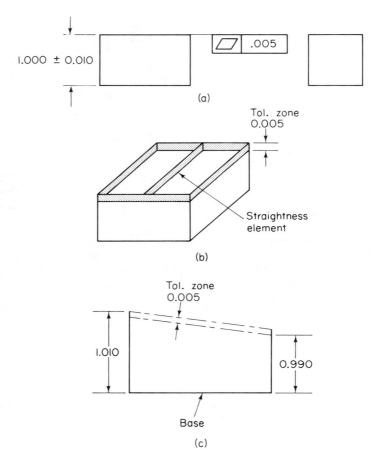

Figure 8.8

condition (1.007 in.), the parts will go together, Fig. 8.9d. When the pin is at MMC (1.005 in.), Fig. 8.9e, and the straightness is at 0.002 in., the sum of

$$1.005 + 0.002 = 1.007 \text{ in.}$$

the parts will assemble. When the pin is at 0.995 in. at the LMC, the pin and hole will still assemble (see Fig. 8.9f). The straightness tolerance zone may be 0.012 in., so that

$$\text{tol.} = 1.005 \text{ (MMC)} - 0.995 \text{ (LMC)}$$
$$= 0.010 \text{ in.}$$

$$\text{total tol.} = 0.010 \text{ (tol.)} + 0.002 \text{ (straightness)}$$
$$= 0.012 \text{ in.}$$

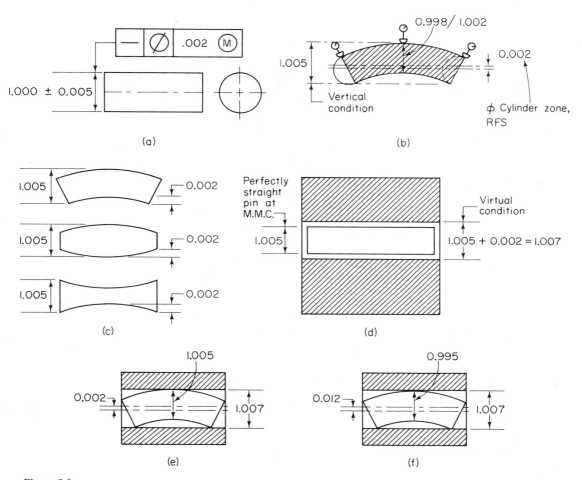

Figure 8.9

8.3 ROUNDNESS ⌭

Roundness is defined as the surface of revolution where all points at the surface in a plane perpendicular to a common axis are equidistant from that axis. In the case of a sphere the plane passes through the center.

Tolerance, when applied to roundness, defines the zone of two concentric circles in the plane of the roundness. Figure 8.10b shows the tolerance zone for Fig. 8.10a. If the largest diameter is at MMC, the smallest diameter is

$$0.755 - 2(0.004) = 0.747 \text{ in.}$$

The tolerance zone is round ⌭ to 0.004 in. in the plane A-A perpendicular to the axis of the cylinder.

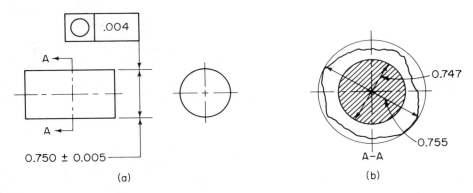

Figure 8.10

Example 8.3

Consider Fig. 8.11. Assume that the MMC measurement is 1.630 in. (a) What is the width of the roundness tolerance zone? (b) What is the smallest diameter of that tolerance zone?

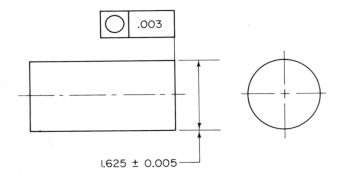

Figure 8.11

Solution

(a) The width of the tolerance zone is 0.003 in.

(b) At a particular cross section of the cylinder, the smallest diameter is

$$\text{MMC} = 1.630 - 2(0.003) = 1.624 \text{ in.}$$

(c) Assume that the MMC decreases by 0.001 in. Develop a table showing the possible large and small dimension combinations. Table 8.5 shows the maximum large measurement coupled with the minimum small measurement in the plane perpendicular to the centerline of the cylinder. Note that the maximum measurement cannot be smaller than 1.626 in. Should this dimension measure 1.625 in. and should the out-of-round small measurement use the entire 0.003 in., the small measurement would violate the −0.005 tolerance.

TABLE 8.5

Max. Large Measurement	Min. Small Measurement
1.630	1.624
1.629	1.623
1.628	1.622
1.627	1.621
1.626	1.620
1.625	1.619 (violates 0.005)

8.4 CYLINDRICITY

Cylindricity deals with all points on a surface that must be equidistant from a common axis. Thus, in Fig. 8.12(a), the tolerance zone lies between two concentric circles 0.003 in. apart. The diameter of the large cylinder is 0.006 in. larger than the small cylinder. Cylindricity controls roundness.

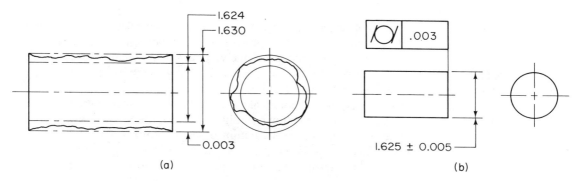

(a)   (b)

Figure 8.12

Example 8.4

In Fig. 8.12(b) use the MMC dimension. (a) What is the smallest dimension? (b) How does it apply to the cylinder? (c) Assume that the large cylinder is 1.622 in. How wide is the tolerance zone? (d) What is the small measurement? (e) Assume a tolerance zone of 0.002 in. What is the small measurement?

Solution

(a) If the largest diameter is 1.630 in., the smallest diameter is

$$1.630 - 2(0.003) = 1.624 \text{ in.}$$

The entire surface must lie inside these two cylinders, Fig. 8.12. (a)

(b) It applies to the entire surface.

(c) The base dimension 1.625 in. can vary by 0.005 in. The cylindricity can vary by 0.003 in. Thus, if the largest measurement is 1.622 in., the tolerance zone must be 0.001 in.

(d) The small measurement is

$$1.622 - 2(0.001) = 1.620 \text{ in.}$$

This is so even though the feature control symbol says that the cylindricity can vary by 0.003 in.

(e) In this case, should the tolerance zone vary by 0.002 in. (cylindricity by 0.004 in.), the small measurement would be

$$1.622 - 2(0.002) = 1.618 \text{ in.}$$

Although cylindricity includes roundness, cylindricity and roundness are not the same. The reverse is not true. *Roundness* refers to measurements taken in one place. *Cylindricity* applies to roundness over the entire surface. It applies only to cylindrical objects, such as cylinders and cones.

8.5 DATUMS

Thus far we have been considering configurations that may not relate to datums, but rather refer to idealized features. As an example, perpendicularity may use the base as a datum for a vertical side of an object; or use the vertical side as a datum for the base. A datum may be a point, line, axis, plane, and so on, that is taken to be theoretically exact and that is used as a reference feature.

The datum surface may be a theoretical plane that contacts three high points on the primary feature datum surface as shown in Fig. 8.13a. Figure 8.13b shows a boss that is required to be perpendicular to the datum base -A- to 0.004 in. Figure 8.7 shows three datum surfaces. -P- is the primary datum surface, -L- the secondary datum surface, and -R- the tertiary datum surface, referenced to the center hole.

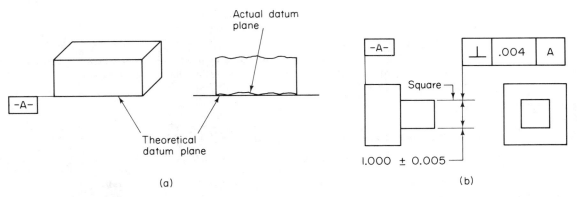

(a)   (b)

Figure 8.13

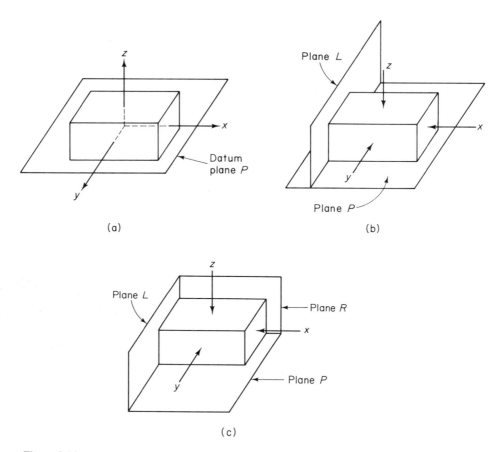

Figure 8.14

To establish these three datums, the concept of "three degrees of freedom" needs to be considered. The manner in which the three degrees of freedom are confined determines the establishment of the three datum planes. The three datum planes may be established by the manner in which the workpiece is to function. They are theoretically perfect planes at 90° to each other. They are related to the conventional x-, y-, and z-axes. Usually, the x-plane is established by the functionally most important feature of the part. For stability, the plane must pick up at least three points of contact.

Figure 8.15a shows a workpiece resting on plane P. At least three points of contact are needed for stability. This plane is the primary datum surface. It confines the workpiece along the z-axis. In Fig. 8.15b, the theoretical vertical plane L restricts movement along the x-axis. Two points of contact are needed to stabilize this movement. In Fig. 8.15c, the movement along the y-axis is stabilized with plane R. Only one point of contact is needed to stabilize the workpiece.

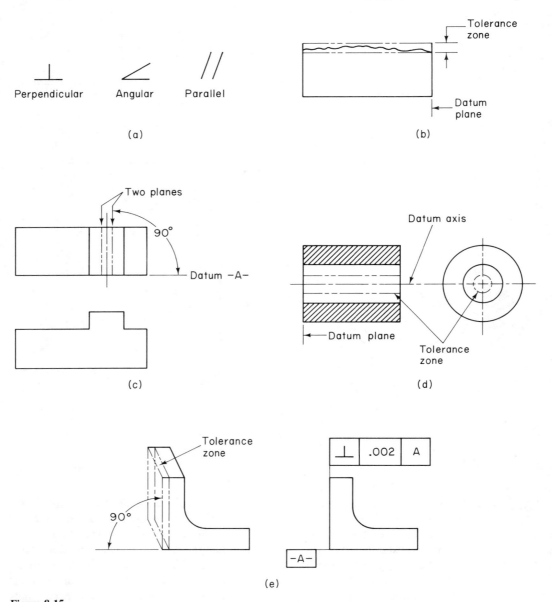

Figure 8.15

8.6 PERPENDICULARITY $\perp$

Tolerance of form, Fig. 8.16a, relates to single surfaces referenced to a datum. Thus, in Fig. 8.16b the tolerance zone is generated by two parallel planes perpendicular to the datum plane. A medium plane may be the center of two planes that generate a

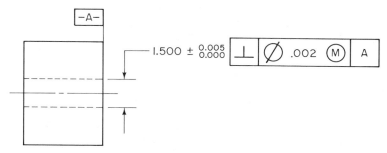

Figure 8.16

tolerance zone. This tolerance zone is perpendicular to a datum plane, Fig. 8.16c. Figure 8.16d shows a cylindrical tolerance zone perpendicular to the face of the cylinder. In Fig. 8.16e the feature control symbol, $\boxed{\perp\ .002\ A}$ calls for a tolerance zone 0.002 in. wide to be perpendicular to the datum base $\boxed{\text{-A-}}$ to within 0.002 in. This is shown in Fig. 8.16e.

Example 8.5

Given the block, Fig. 8.16f show the effect of the tolerance on the perpendicularity as it deviates from exact perpendicularity.

Solution The feature control symbol means that the hole must be perpendicular to the datum face $\boxed{\text{-A-}}$ to within a 0.002-in. tolerance zone at MMC. When the hole is at MMC, the dimension is 1.500 in. Note that the maximum material condition for a hole is the smallest allowable dimension. The analysis is shown in Table 8.6. Also note that in all cases the change of the actual size of the hole allows a change in the perpendicularity tolerance. In all cases in Table 8.6 the difference between the actual hole size and the perpendicularity is 0.002 in. So that when the hole is at 1.505 in. LMC, the largest hole size permitted, the perpendicularity is 0.007 in. This exceeds the 0.002 in., but does not affect the functional assembly with the mating part.

TABLE 8.6

Actual Hole Size Diameter (in.)	Tolerance on Perpendicularity Diameter (in.)
1.500 MMC	0.002
1.501	0.003
1.502	0.004
1.503	0.005
1.504	0.006
1.505 LMC	0.007

8.7 ANGULARITY $\boxed{\angle}$

In Fig. 8.17a the angle may be in a tolerance zone of 0.005 in. when the dimension is at MMC. The 0.250-in. dimension may vary by as much as 0.008 in. The angle may vary but is not permitted to fall outside the tolerance zone, as shown in Fig. 8.17b.

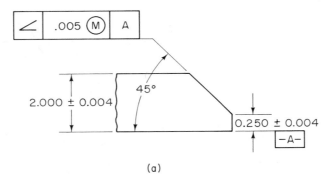

(a)

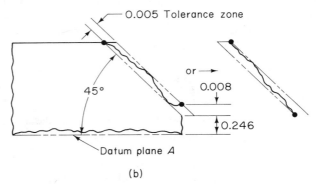

(b)

Figure 8.17

8.8 PARALLELISM $\boxed{/\!/}$

Parallelism applies to a surface, or axis. In either case all points must be equidistant from a datum plane, or axis. For flat surfaces the tolerance zone is bounded by two planes. These planes are parallel to a datum plane (see Fig. 8.18a). The tolerance zone planes may define the parallelism of the centerline of a hole with reference to a flat surface. In Fig. 8.18b the parallel planes border the tolerance zone, which in turn is parallel to the base datum plane.

Example 8.6

In Fig. 8.19 calculate the allowable deviation from parallelism of the ½-in. hole.

Solution   The hole, Fig. 8.19a, at MMC must be within two parallel planes that create a tolerance zone 0.004 in. wide. These planes also must be parallel to plane $\boxed{\text{-A-}}$. As the hole increases in size from MMC, the parallel tolerance may increase. Note that the parallel planes touch the bottom left and the top right of the hole, Fig. 8.19b. As the hole size increases, these planes can be farther apart. Table 8.7 shows the combination of hole size and tolerance zone size.

Assume that the planes are the sides of a pin. When the hole is 0.498 in. MMC, the

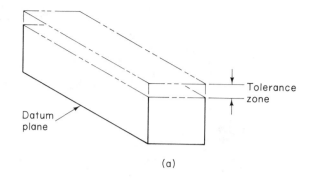

(a)

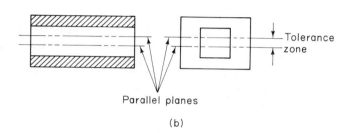

Parallel planes

(b)

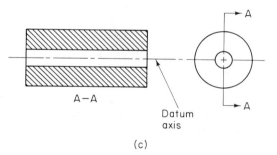

(c)

Figure 8.18

0.500 ± 0.002

| // | .004 Ⓜ | A |

-A-

(a)

Tolerance zone

Datum plane

Centerline (hole)

Parallel planes

(b)

Figure 8.19

TABLE 8.7

Feature Size (in.)	Parallel Tolerance (in.)
0.498 MMC	0.004
0.499	0.005
0.500	0.006
0.501	0.007
0.502 LMC	0.008
0.503 (not acceptable)	0.009

hole can be out of parallelism by 0.004 in. and still assemble with its mating part. When the hole is at 0.502 in. LMC, the parallelism can be out as much as 0.008 in. and still assemble with the pin.

If the tolerance zone is out of parallel by 0.009 in., the feature size of the hole would need to be 0.503 in. to assemble with the pin. This would not be acceptable, because the 0.503-in. dimension would exceed the +0.002 in. allowed.

8.9 PROFILE OF A LINE* $\frown$

The *profile of a line* defines the profile of a *plane* cut through a particular profile. An example would be a plane cut through a propeller blade perpendicular to the centerline of the blade. It defines the contour of the blade only at the point where it cuts through the blade. *Profile of a surface* applies to the entire surface of the blade. The latter deals with many planes cut through the blade perpendicular to the centerline.

Figure 8.20 says that the tolerance zone from 270 to 330° is 0.001 in. wide. These elements must be perpendicular for the entire surface to the face of the cam -A-.

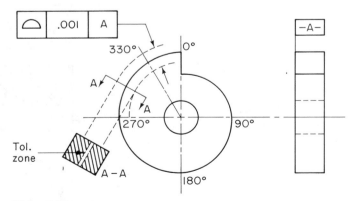

Figure 8.20

*For the profile of a surface and of a line, consult ANSI Y14.5M–1982.

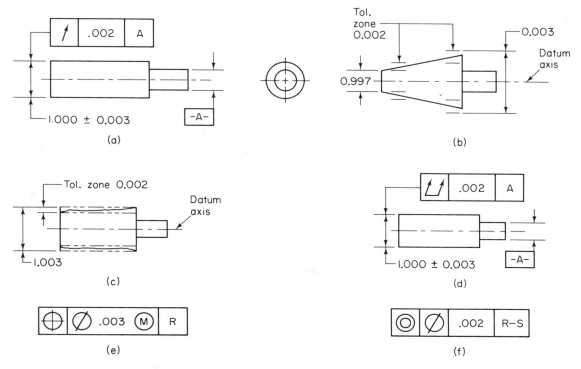

Figure 8.21

8.10 RUNOUT ↗

Circular runout ↗. Figure 8.21a shows that each *circular* element of the 1.000-in. diameter must be within 0.002 in. of full indicator movement (FIM). The diameter must be between 1.003 in. and 0.997 in. and each element is to be perpendicular to the centerline by not more than 0.002 in. The centerline is the datum axis for both diameters. Note that concentricity is not the same as runout. Concentricity is a matter of two separate axes. It is possible to have two concentric diameters that are out of round.

In Fig. 8.21b each circular element must be inside the tolerance zone of 0.002 in. In Fig. 8.21c the entire surface must be within 0.002 in. of total runout.

Total runout ↗↗. Total runout applies to the *entire* surface about a common axis, Fig. 8.21d. Both circular and total runout area applies "regardless of feature size," RFS, and not on an MMC basis.

An MMC situation implies assembly, or interchangeability. If it is necessary to control the runout on mating parts so that they will assemble, both axes and the surfaces need to be controlled. Under these conditions, MMC be considered and position ∅ is used, Fig. 8.21e.

   If a part has two or more diameters and the axes need to be controlled, without concern for assembly, RFS is implied and concentricity ⊕ is used, Fig. 8.21f.

8.11 POSITION ⊕

Tolerance of location relates to geometric characteristics such as symmetry, concentricity, and position. In order to discuss tolerance of location, the features of size, the relationships that exist between mating parts and MMC, are considered.

   Position by definition describes the theoretically exact location of a feature (line, point, or plane) with reference to a datum. The tolerance of position is the permissible variation from this theoretical location. As an example, assume that there are two holes in a plate that are to be assembled at MMC with another plate with two pins. Ideally, if the *distance* between the holes is theoretically correct and the diameters of the pins is at MMC, the centerline of the pins and the holes will be the same.

   Assume that the diameter of the holes get larger; then the pin location can increase, or decrease, by the same amount. In Fig. 8.22, the MMC of the hole is 0.498 in. while the position tolerance is 0.005 in. When the LMC exists, or 0.502 in., the position tolerance of the hole is 0.009 in.

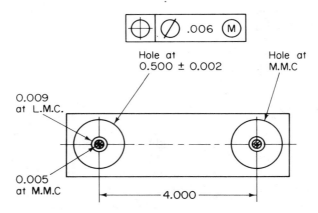

Figure 8.22

   Location tolerance is related to the location of the position of the centerline, axis, or datum plane. When considering location, the size feature of a pin, or hole, is important to the feature of location. When considering position, one of the useful applications of MMC is interchangeability. In Fig. 8.23a when the hole is at MMC, the location of the pin with reference to another pin may vary by the position tolerance. As the hole size increases toward LMC, the location tolerance may increase by the same amount.

   There are decided advantages to the theory of position over the coordinate system, Fig. 8.23. In the coordinate system the hole and pin are round, but the position of the pin relative to the theoretical position, generates a square tolerance

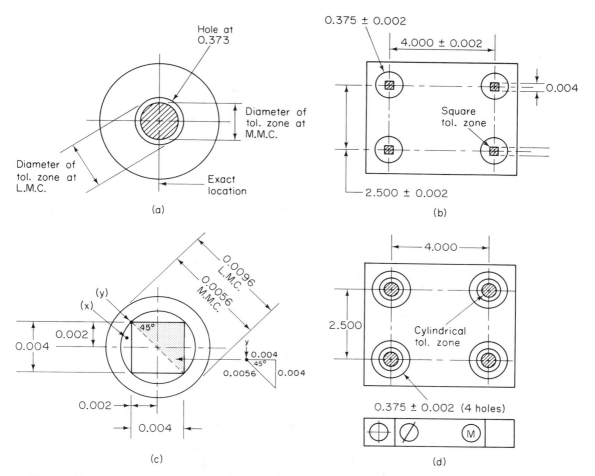

Figure 8.23(a,b,c,d)

zone. In Fig. 8.23b the diagonal of the tolerance zone only permits positioning at the four corners of the square. With the coordinate system, locations outside this square are rejected. In Fig. 8.23c a dimension x outside the square tolerance zone but closer to the theoretically exact centerline than a measurement y at the extremities of the diagonals would be rejected. Only measurements at the borderline, or inside the square tolerance zone, are acceptable. With the application of positioning theory, both x and y, Fig. 8.23c would be acceptable.

In the coordinate system, the tolerance zones for the centers of the holes are square, Fig. 8.23b. In the positioning system, the tolerance zones are round, Fig. 8.23d. Table 8.8 shows all the acceptable hole sizes and position tolerances.

Also, the dimensions of 4.000 and 2.500 in. are used as exact values. The location tolerances relate to the hole size, as shown in the control symbol, not with the location dimensions 4.000 and 2.500 in.

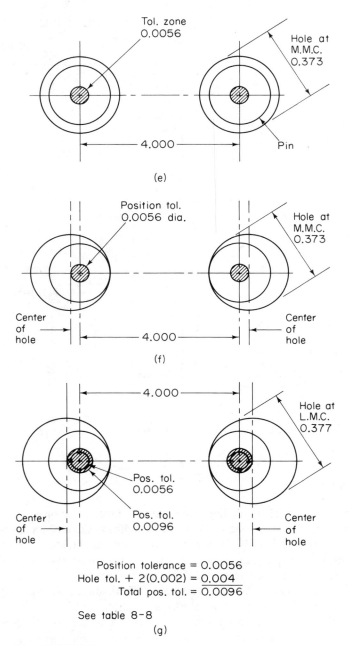

Position tolerance = 0.0056
Hole tol. + 2(0.002) = 0.004
Total pos. tol. = 0.0096

See table 8-8

(g)

Figure 8.23(e,f,g)

TABLE 8.8

Hole Size (in.)	Position Tol. (in.)
0.498 MMC	0.005
0.499	0.006
0.500	0.007
0.501	0.008
0.502 LMC	0.009

When the hole is at MMC (0.373 in.), the position tolerance is 0.0056 in. When the hole is at LMC (0.377 in.), the position tolerance is 0.0096 in. This is true for the entire diameter of the hole; whereas in the coordinate system, it is only true at the extremities of the diagonals of the tolerance zone.

Figure 8.23e shows the ideal condition. The centerlines of the pins and the holes are both at the *exact* 4.000 in. dimension when the holes are at MMC (0.373 in). Figure 8.23f shows the holes at MMC (0.373 in.) but offset at 0.0056 in. on location. In Fig. 8.23g the holes are at LMC and the possible pin locations vary by 0.0096 in. In all cases in Table 8.8, the pins will assemble with the holes.

8.12 GEOMETRIC CHARACTERISTICS: A REVIEW

The following is a capsule review of the geometric characteristics discussed in this chapter.

Feature control frame.   This frame describes the tolerance zone as it applies to the shape, size, and type of zone. It also addresses the datum (*A, B, C*) features as they apply to the setup contact points necessary to confine a workpiece; or the confinement of the degrees of freedom, see Sec. 9.1.

The most important fixturing direction, or datum, is *A*. The second most important locator, or datum, is *B*. The least important datum surface is *C*. There are three contact points on datum *A*, two at *B*, and one at *C*.

Straightness.   There are two straightness tolerance zones: surface and axis. These zones define the limits within the tolerance zones.

In surface straightness the control frame is attached to the surface and axis. These zones define the limits within the tolerance zones.

In surface straightness the control frame is attached to the surface being considered. If attached to, or in the vicinity of, the size dimension, the axis straightness is being considered.

When the control frame is attached to either front view, the straightness applies to the length of the workpiece. It may also be attached to the side view. In the latter case, straightness should apply to the width of the work. Axis straightness in a longitudinal direction addresses the deviation from the axis along the length of the workpiece.

Flatness.   All elements must lie within the boundaries of two parallel flat planes. The control panel is attached to the surface considered.

Cylindricity and circularity.   Cylindricity describes a surface of revolution where all parts of that surface are equidistant from a common axis. Circularity (roundness) describes a surface of revolution equidistant from a common axis at a specific point on the surface (cone), or through a common center (sphere).

Profile tolerance.   These tolerance zones control the boundaries within the limits of the profile of a workpiece. This zone may be described in unilateral or bilateral terms. Thus the profile as defined may be confined totally above the basic dimension(s), below the basic dimension(s), or partially above and below the basic dimension.

Orientation tolerances.   Orientation tolerances have three elements: parallelism, perpendicularity, and angularity.

Parallelism.   Parallelism defines a surface, or plane, as equidistance from a datum plane. The surface in the datum plane rests on at least three points (surface plate). This concept describes the condition of two planes relative to each other. Flatness means the condition of the surface relative to itself.

Angularity and Perpendicularity.   Angularity and perpendicularity refer to the condition of the surface at some specific angle to the datum plane.

Location tolerances.   Location tolerances may deal with position, concentricity, or symmetry.

Positional Zones.   Positional zones define the variation of a zone from true position. This is shown in a control frame by a positional symbol, a tolerance, and a datum reference. The tolerance zone may vary from the true position of a datum axis, or plane.

Concentricity.   Concentricity refers to the variation of two cylinders from a common, or datum, axis.

Runout.   Runout refers to control of the datum axis and a surface perpendicular to the datum axis. Runout always applies to RFS, never to MMC. There are two types of runout: circular and total.

Circular Runout.   Circular runout relates to two datum cylinders. It deals with the cumulative effect of circularity and coaxality. Thus all points may be equidistant from a central axis of one diameter, and another set of points may be equidistant from the central axis of another diameter. However, the axes of both cylinders need not be common, or coaxial. If they are not, there is runout.

Total Runout.   Total runout refers to the control of all the elements discussed in ths section as they refer to a datum surface, or axis.

PROBLEMS

8.1. Why is geometric dimensioning and tolerancing important to the designer?

8.2. What does the phrase "functional requirements" mean?

8.3. The following table lists the geometric terms used in the geometric tolerancing and dimensioning system. Insert the exact symbol to the right of each term.

Term	Symbol
Diameter	
Spherical diameter	
Regardless of feature size	
Datum feature symbol	
Spherical radius	
Arc length	
Least material condition	
Datum target symbol	
Projected tolerance zone	
Maximum material condition	
Radius	

8.4. Repeat Prob. 8.3 for the following terms:

Term	Symbol
Straightness	
Profile of a line	
Position	
Total runout	
Cylindricity	
Perpendicularity	
Concentricity	
Flatness	
Angularity	
Circular runout	
Profile of a surface	
Circularity	
Parallelism	

8.5. In your own words, explain eight of the characteristic terms in Prob. 8.4.

8.6. How does cylindricity differ from circularity? Give examples.

8.7. How does concentricity differ from circular runout? Give examples of circular runout.

8.8. State the difference between the profile of a line and the profile of a surface. Give examples.

8.9. Discuss position. What is position tolerance?

8.10. State the difference between straightness and flatness. Give examples.

8.11. Explain each of the symbols in the Feature Control Frame, Fig. 8.24(a).

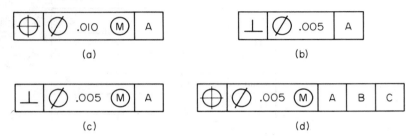

(a)   (b)

(c)   (d)

Figure 8.24

8.12. Repeat Problem 8.11 for Fig. 8.24(b).

8.13. Repeat Problem 8.11 for Fig. 8.24(c).

8.14. Repeat Problem 8.11 for Fig. 8.24(d).

8.15. Given a hole size of 3.875 ± 0.010 in.:
  (a) What is the MMC and the hole?
  (b) What is the LMC of the hole?

8.16. Given a pin size of 3.875 ± 0.010 in.:
  (a) What is the MMC of the pin?
  (b) What is the LMC of the pin?

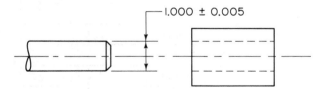

Figure 8.25

8.17. In Fig. 8.25 state
  (a) the MMC of the pin, and
  (b) the LMC of the hole.
  (c) State the MMC of the hole and
  (d) the LMC of the pin.

8.18. **(a)** Define and give an illustration of a virtual condition.
  (b) What is the virtual condition of Fig. 8.26?

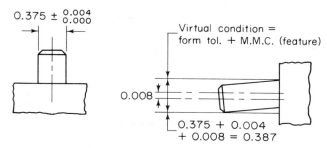

Figure 8.26

8.19. Define the term "datum".

8.20. What is a "datum feature"?

8.21. Using Fig. 8.27, develop a table of feature sizes $\varnothing$ and zone tolerances. Explain the chart in terms of straightness of a mating part.

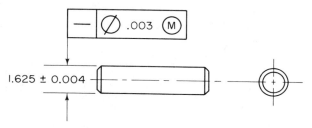

Figure 8.27

8.22. Calculate:
  (a) The tolerance zone in Fig. 8.28.
  (b) What is the smallest cross sectional diameter permitted?
  (c) What are the combinations of feature sizes and zone tolerances? Develop a table of the latter.

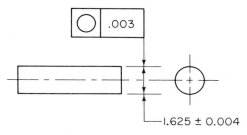

Figure 8.28

8.23. Give an illustration of how datums are used.

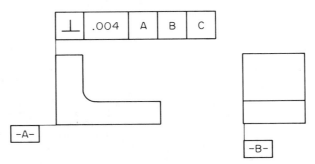

Figure 8.29

8.24. Given Fig. 8.29, which datum is
   (a) primary,
   (b) secondary,
   (c) tertiary.
8.25. Design a simple fixture using the datum surfaces from Fig. 8.29.
8.26. Show the effect of the tolerance on perpendicularity in Fig. 8.30 as it deviates from exact perpendicularity.

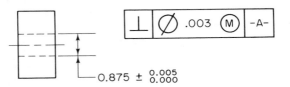

Figure 8.30

8.27. Describe the tolerance zone associated with parallelism. Make a drawing that shows the tolerance zone
   (a) of two parallel planes,
   (b) of a hole and a plane.
8.28. Calculate the allowable deviation from parallelism of the $\frac{3}{4}$ in. hole in Fig. 8.31.

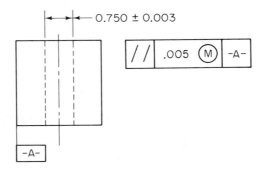

Figure 8.31

8.29. State the difference between circular runout, total runout and concentricity. Give an illustration of each.

8.30. **(a)** Explain Figure 8.23.

   (b) What is the difference between a "coordinate" and a "positioning" system as they relate to tolerance zones?

Chapter 9

Fixture Design

9.1 CONFINEMENT OF THE SIX DEGREES OF FREEDOM

A *fixture* has, as its purposes, to hold and accurately position a workpiece. A *jig* is also used to hold and position a workpiece. In addition, it provides some means for guiding the cutting tool. According to these definitions a device which holds many small workpieces so that they may be tack welded, according to some location requirement, is classified as a *fixture*, unless the device provides some method for guiding the welding torch. A device that locates and clamps a workpiece, and uses drill bushings to guide the drill during the drilling operation, is called a drill *jig*.

One of the most important requirements for successful jig or fixture design is that a workpiece may be replaced with successive workpieces in such a manner that they may be machined within the limits of the print tolerances. That is, after a workpiece is machined and removed from the fixture or jig, the operator should be able quickly to put another workpiece into that fixture or jig, clamp it, and machine it to the print dimensions with the tolerances permitted.

In order to be able to repeatedly *locate* a workpiece in a holding device in the same place, the device must provide confinement or control through *six degrees of freedom* in space. These six degrees of freedom are made up of three linear and three rotational motions in space. That is, a body is free to move in a *straight line* along the three linear axes, or to *rotate* about these three axes. In jig and fixture design the conditions of good design are fulfilled if a workpiece is confined or controlled in one each of the six motions. The opposite movement is confined by some type of clamp or holding device, or is utilized during the machining operation.

Figure 9.1a and b show each of these six degrees of freedom in space. Figure 9.1a shows the linear freedoms; Fig. 9.1b shows the rotational freedoms in space.

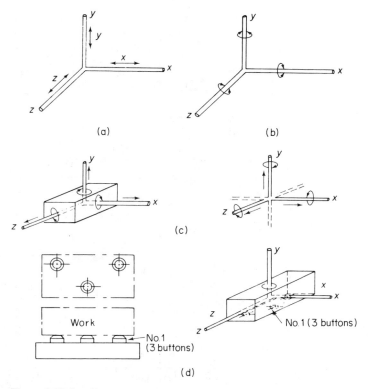

Figure 9.1(a,b,c,d)

Obstructions placed in the path of these motions eliminate the degrees of freedom. Figure 9.1c shows a block of steel floating in space. This block is free to move or rotate in any direction along or about the three axes.

Figure 9.1d shows the same block resting on three support pins marked (1). The simple act of placing the block on three pins means that the block is no longer free to move along the y-axis. It is enough that it cannot move in a downward direction. A clamp can be used to keep it from moving in an upward direction. However, it is still capable of being moved along the x- and z-axes. Therefore, it has lost only *one linear degree of freedom*. The three pins also prevent it from rotating around the x- and z-axes. However, it is still free to rotate around the y-axis. It has lost two rotational degrees of freedom. It has two linear and one rotational degree of freedom remaining with which we must deal.

In Fig. 9.1e two buttons, marked (2), have been added to prevent movement of the block along the x-axis. Thus the block has lost its freedom to move linearly along the x-axis. It now has only one linear degree of freedom remaining. It can still move linearly along the z-axis. It has also at this point lost its third rotational degree of freedom. It can no longer rotate about the y-axis without losing contact with the number 2 pins.

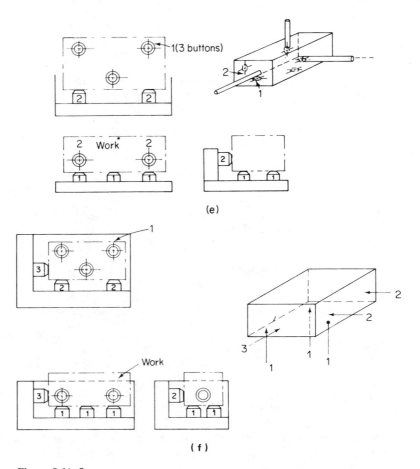

Figure 9.1(e,f)

Pin number (3), Fig. 9.1f, completes the confinement of the block through all six degrees of freedom. It has prevented the possibility of any linear movement along the z-axis. To complete the confining process, clamps may be used to hold the workpiece against the banking pins.

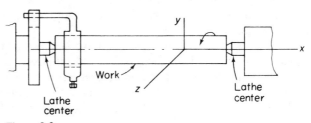

Figure 9.2

In some instances, one or more of the freedoms in space are controlled and may be used during the machining process. For example, Fig. 9.2 shows a workpiece mounted between centers on a lathe. The lathe centers confine *linear* motions along the x-, y-, and z-axes. They also prevent rotation about the y- and z-axes. The workpiece may rotate about the x-axis, but only at the rpm for which the machine has been set.

9.2 THE 3-2-1 PRINCIPLE

Figure 9.1f illustrates a basic principle referred to as the 3-2-1 principle. The principle says that a workpiece will be completely confined when banked against three points in one plane, two points in another plane, and one point in a third plane if the planes are perpendicular to each other.

The support of a workpiece on three buttons is minimum for any holding device. Care must be exercised when three buttons are used because a workpiece surface, whether finished or rough, will always "seat" on all three points, even if there are extreme variations in the flatness or roughness of the surface. For finished surfaces three locating points in one plane are usually enough. For rough surfaces three locating points are ideal, provided that the roughness allows the workpiece to be machined within the print, or operational tolerances. The buttons should be set as far apart as possible. The greater the spread between the buttons, the less the alignment error.

For large or heavy workpieces it is best to use four or more support points. If the workpiece surface is flat and smooth, the four points may be fixed. If there is some question about whether or not the workpiece will rest on all four supports, as is the case with such surfaces as rough-cut surfaces or castings, the fourth pin should be adjustable. Later in this chapter, the various types of support pins are discussed.

Thus in Fig. 9.3a is shown a workpiece which has its lower surface finish machined. All four buttons are fixed and all contact the lower surface of the workpiece. In Fig. 9.3b the conditions are different. If four buttons are to be used, and if the surface is rough, the probability is remote that the workpiece will rest on all four

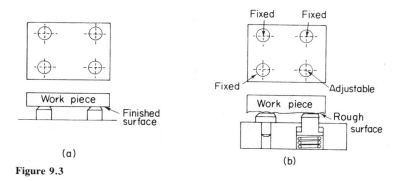

Figure 9.3

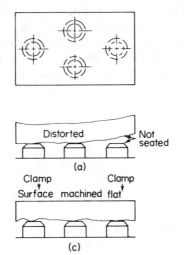
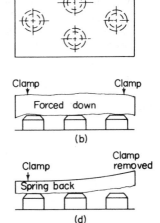

Figure 9.4

buttons, Fig. 9.4a. It may rock on two buttons and be pulled down against the third and fourth buttons with clamps. Drawing the workpiece down against all four buttons will distort the workpiece. If a machine cut is taken while the workpiece is distorted, the cut will remain flat as long as the clamps are tight, Fig. 9.4c. Releasing the clamps will allow the workpiece to spring back to its normal shape, Fig. 9.4d. However, the machined surface will no longer be flat. If the fourth button is a spring load and a button-locking mechanism is provided (not shown in Fig. 9.3b), once the button seats it will be supported adequately on all four buttons without distorting the workpiece.

As a good general rule, it takes six support points to confine a workpiece through six degrees of freedom. Three *fixed* points should be adequate for the first surface, two *fixed* points for the second surface, and one *fixed* point for the third surface. If additional supports are needed, they should probably be *adjustable* so that they will register against the work surface without distorting it. They should also be provided with some type of locking device once they have registered against the work surface. However, in terms of the six degrees of freedom, they have no effect except to provide stability to the workpiece.

9.3 CHOOSING A LOCATING SURFACE

It is very important to choose the proper locating points or surfaces. It is not uncommon to take a machine cut of a surface that would not otherwise need to be machined; or to provide bosses which have no other function except to act as banking surfaces. Once the locating surfaces are chosen, they should be used until all operations on the workpiece are complete. Every time a change is made from one locating surface to another, the error between machined surfaces accumulates.

Locating points may contact surfaces that are flat, or curved. They may be machined surfaces, holes, or points. As indicated, it is good practice to machine surfaces, or holes, which may then be used as locating points for operations which are to follow. If these locating surfaces are part of the finished workpiece, the error for subsequent operations will be reduced. This is especially true if the subsequent operations are dimensioned from this locating surface.

9.4 LOCATORS

The most elementary type of locator is a *machined pad on the fixture*. Its purpose is to support the workpiece. One such pad is shown in Fig. 9.5a. To save the time and

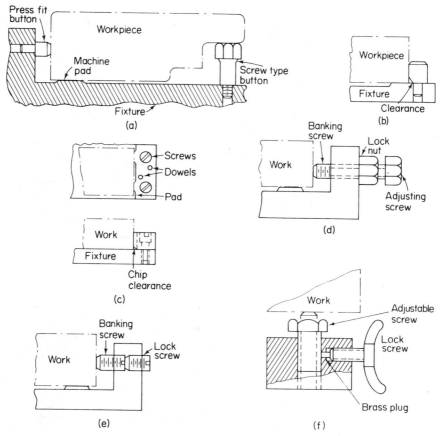

Figure 9.5

expense of machining such a pad, *rest buttons* are used. The heads are usually hardened and ground. The shanks are heat treated for toughness. Hardened shanks break too easily and should be avoided. These buttons may be pressed or screwed into the body of the fixture. Both types are shown in Fig. 9.5a. Buttons have several advantages when compared with machined pads. They are less expensive to machine, easier to replace, and easier to clean.

A variation of a rest button is shown in Fig. 9.5b. The button is pressed into the base of the fixture. The work registers against the side of the button. Chip, dirt, and so on, clearance must be provided, as shown.

Sometimes it is more advantageous to machine the type of plate shown in Fig. 9.5c. This plate provides two banking pads. The plate is usually hardened and ground.

The locators discussed to this point are of the *fixed* type. There is at times a need for locators that can be adjusted, called *jackscrews,* and then locked in position. This sometimes happens if a new mold or forging die is used to produce workpieces. The contour of the workpiece may change slightly and require a fine adjustment. Of course, there are many ways in which to do this. Figure 9.5d, e, and f show three possibilities.

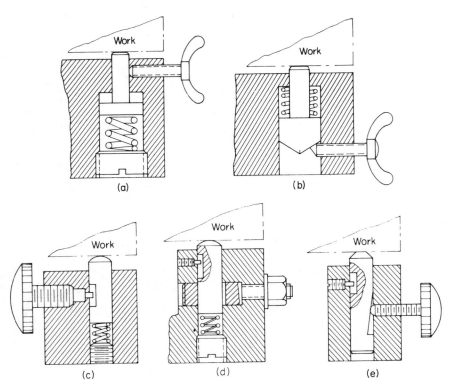

Figure 9.6

If it becomes necessary to adjust the support points for each workpiece, the use of a jackscrew would be too cumbersome and time consuming. In this case pins are used. These *jackpins* must be designed so that they register on the workpiece without applying too much force. There is usually some kind of locking device provided so that once the workpiece is firmly set in the fixture, the pin may be locked. Sometimes the mechanism that is used to cause the pin to register on the workpiece is also the locking device. Figure 9.6 shows several of these jackpins.

9.5 LOCATOR PINS AND JAMMING*

Sometimes locator pins are used to pick up a finished hole, Fig. 9.7a, or to pick up two previously machined holes, as shown in Fig. 9.7b. One of the problems that arises when using pin locators in holes is jamming. This happens when the pin tilts so that the distance across the jamming points is greater than the diameter of the hole from which the pin is being withdrawn. This is shown exaggerated in Fig. 9.7c. Thus if the plug is tilted as it is being withdrawn, it will jam. If it is tipped in the opposite direction to try to free it, the correcting force will probably jam it in the opposite

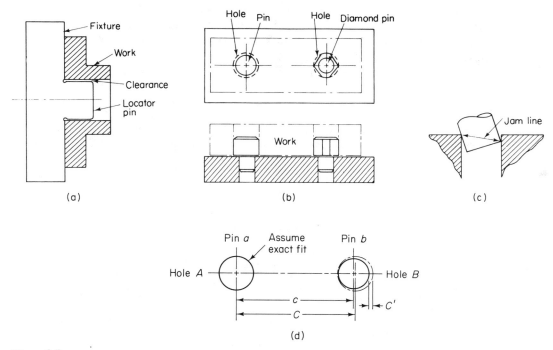

Figure 9.7

*The student should compare this section to Chapter 8 (Section 8.11).

direction. Repeating the force in the other direction will cause it to jam again. This cycle of jamming could make it difficult to take workpieces off pins, or put workpieces on pins. There are several ways in which to handle this problem.

First, the amount of *clearance* allowed between the pin and hole diameters is a factor in controlling the amount of jamming that will take place. The larger the hole diameter, the greater the clearance needed to prevent jamming. Thus 0.005-in. clearance for a 1-in.-diameter pin and hole may be too loose a fit. The same 0.005-in. clearance for a 12-in.-diameter pin and hole may be subject to jamming.

In Fig. 9.7d, if the fit between pin *a* and the hole (workpiece) are an exact fit, the allowances between the pin *b* and the hole must equal the sum of the tolerances of *B* and *P*. Thus if the tolerances for *B* are ±0.010 in. and the tolerances for *P* are ±0.006 in., the minimum clearance *c* between the pin and the hole will be zero if the pin is +0.016 in.; and the clearance will be a maximum when the pin is −0.016 in.

Figure 9.8a shows a workpiece located on a two-in fixture. Pin *a* and the base of the fixture eliminate three degrees of linear freedom (*x, y, z*) and two degrees of rotary freedom about the *x*-axis and the *y*-axis. The freedom of the workpiece to rotate about the *z*-axis is not confined. Thus the rotation about the latter axis, hole *b*, should be confined with a pin.

If the clearance between hole *A* and pin *a* is a few ten-thousandths of an inch and if the locator pin is a cylinder, control of the tolerance between hole *B* and pin *b* must be the sum of the tolerances between the two holes. Also, the *x* and *y* movements of the workpiece are confined by the close tolerance of *A* and *a*. Using a diamond locator pin *b*, Fig. 9.8b, makes it easier to control the rotation about the *z*-axis.

In Fig. 9.8b the movement of the *b* pin in the *B* hole will be

$$2\left(\frac{E}{2}\right) = E$$

The contact between a diamond pin *b* and hole *B* should be between $\frac{1}{8}$ and $\frac{1}{16}$ the nominal size of the hole. Using Fig. 9.8b for reference, the values *D* and *E* may be calculated as follows:

$$OF = \frac{3}{4}\left(\frac{b}{2}\right)$$

$$= \frac{3}{8}b$$

$$cos\beta = \frac{\frac{3}{8}b}{\frac{1}{2}b} = \frac{3}{4}$$

$$\beta = 41.41°$$

$$\alpha = 90 - (41.41 + 30.00)$$

$$= 18.59°$$

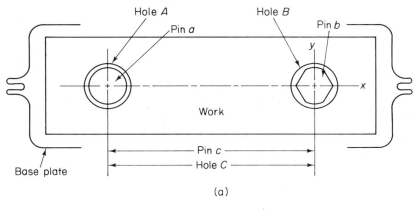

(a)

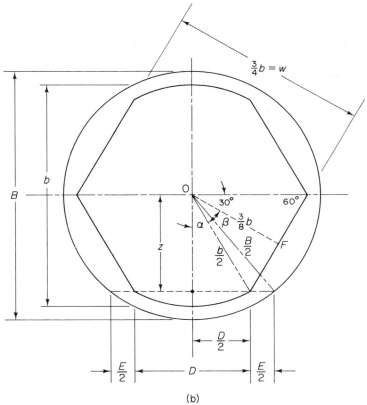

(b)

Figure 9.8(a,b)

To find D:

$$\frac{D}{2} = \frac{b}{2}\sin \alpha$$
$$D = b \sin \alpha$$

To find E:

$$\left(\frac{D}{2} + \frac{E}{2}\right)^2 = \left(\frac{B}{2}\right)^2 - z^2 \quad \text{where } z = \frac{b}{2}\cos \alpha$$
$$\frac{D}{2} + \frac{E}{2} = \sqrt{\left(\frac{B}{2}\right)^2 - z^2}$$
$$E = 2\sqrt{\left(\frac{B}{2}\right)^2 - z^2} - D$$

Example 9.1

Assume that the workpiece in Fig. 9.8b has a center-to-center distance of 10.500 ± 0.005 in. The center-to-center distance between the pins is 10.500 ± 0.0004 in. Hole A is $\frac{3}{4}$ in. in diameter and hole B is $\frac{5}{8}$ in. in diameter. The allowance for the diamond pin is 0.002 in. Calculate (a) the width of the contact surface between the pin b and the hole B and (b) the maximum deviation of the contact pin.

Solution

(a) To find D:

$$\cos \beta = \frac{\frac{3}{8}b}{\frac{1}{2}b} = \frac{3}{4}$$
$$\beta = 41.41°$$
$$\alpha = 60° - 41.41°$$
$$= 18.59°$$
$$\sin \alpha = \frac{D/2}{b/2} = \frac{D}{b}$$

Therefore,

$$D = b \sin \alpha = 0.623 \sin 18.59°$$
$$= 0.1986$$

(b) to find E:

$$z = \frac{b}{2}\cos \alpha = \frac{0.623}{2}\cos 18.59°$$
$$= 0.29456 \text{ in.}$$
$$E = 2\sqrt{\left(\frac{B}{2}\right)^2 - z^2} - D$$
$$= 2\sqrt{\left(\frac{0.625}{2}\right)^2 - 0.29456^2} - 0.1986$$
$$= 0.010 \text{ in.}$$

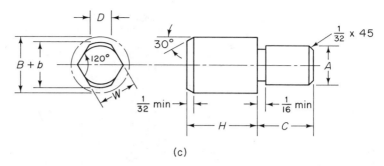

(c)

Figure 9.8(c)

If a diamond pin, d, Fig. 9.8c is used, the width, W, of the pin should be such that the contact surface D is not smaller than $\frac{1}{32}$ in. A good general rule for the dimension D is

$$D = \frac{B}{8}$$

b = diameter of pin, in.
D = contact diameter pin, in.
B = diameter of hole, in.
W = width across flats, in.

A quick calculation for the length of the pin above the fixutre is

$$H = \frac{3b}{2}$$

Several typical dimensions for a diamond pin are shown in Fig. 9.8c and Table 9.1. These pins may be either round or diamond shaped.

The *length of fit* and the shape of the pin are additional conditions that must be considered if jamming is to be avoided.

Length of engagement.   Figure 9.9a shows the limits for designing a plug to avoid sticking in a hole while the workpiece is being removed. If the plug height is

$$H = \sqrt{2(2a + D)(D - d)}$$

H = height or length of plug, in.
a = distance from pivot to hole edge, in.
D = diameter of hole, in.
d = diameter of plug, in.

the plug should not stick. It should be noted that H does not include the chamfered portion of the plug and that the bored hole does not have a rounded or chamfered edge.

TABLE 9.1 FLANGED LOCATING PINS

A	H	C	d (pin) b	D (hole) B
+0.0000				
−0.0002				
0.1878	$\frac{11}{32}$	$\frac{13}{32}$	0.2490–0.2493	$\frac{1}{4}$
0.2503	$\frac{13}{32}$	$\frac{15}{32}$	0.3115–0.3118	$\frac{5}{16}$
0.3128	$\frac{1}{2}$	$\frac{1}{2}$	0.3740–0.3743	$\frac{3}{8}$
0.3753	$\frac{19}{32}$	$\frac{21}{32}$	0.4365–0.4368	$\frac{7}{16}$
0.4378	$\frac{21}{32}$	$\frac{23}{32}$	0.4990–0.4993	$\frac{1}{2}$
0.5003	$\frac{3}{4}$	$\frac{3}{4}$	0.5610–0.5615	$\frac{9}{16}$
0.5628	$\frac{27}{32}$	$\frac{29}{32}$	0.6235–0.6240	$\frac{5}{8}$
0.6253	$\frac{29}{32}$	$\frac{35}{32}$	0.6860–0.6865	$\frac{11}{16}$
+0.0000				
−0.0005				
0.6885	1	$1\frac{1}{8}$	0.7485–0.7490	$\frac{3}{4}$
0.7510	$1\frac{3}{32}$	$1\frac{5}{32}$	0.8110–0.8115	$\frac{13}{16}$
0.8135	$1\frac{5}{32}$	$1\frac{9}{32}$	0.8735–0.8740	$\frac{7}{8}$
0.8760	$1\frac{1}{4}$	$1\frac{1}{2}$	0.9360–0.9365	$\frac{15}{16}$
0.9385	$1\frac{5}{16}$	$1\frac{9}{16}$	0.9972–0.9980	1
1.0010	$1\frac{1}{2}$	$1\frac{3}{4}$	1.1222–1.1230	$1\frac{1}{8}$
1.1260	$1\frac{11}{16}$	$1\frac{15}{16}$	1.2472–1.2480	$1\frac{1}{4}$
1.2510	$1\frac{13}{16}$	$2\frac{3}{16}$	1.3722–1.3730	$1\frac{3}{8}$
1.3760	2	$2\frac{1}{4}$	1.4972–1.4980	$1\frac{1}{2}$
1.5010	$2\frac{3}{16}$	$2\frac{7}{16}$	1.6222–1.6230	$1\frac{5}{8}$
1.6260	$2\frac{5}{16}$	$2\frac{11}{16}$	1.7472–1.7480	$1\frac{3}{4}$
1.7510	$2\frac{1}{2}$	$2\frac{3}{4}$	1.8722–1.8732	$1\frac{7}{8}$
1.8760	$2\frac{11}{16}$	$2\frac{13}{16}$	1.9972–1.9980	2

Source: Vlier Engineering Corp., Burbank, Calif., Catalog, p. 100.

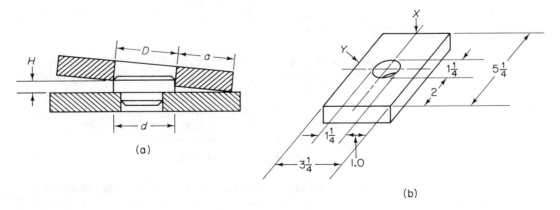

(a)

(b)

Figure 9.9

Example 9.2

A rectangular plate $3\frac{1}{4} \times 5\frac{1}{4}$ in. is in full contact with the base of a fixture. The hole in the workpiece has been machined to $1\frac{1}{4}$ inches at the exact center of the plate. Figure 9.9b shows this plate. The plug diameter is 1.248 in. Calculate (a) the maximum height of the plug if the $3\frac{1}{4}$-in. side X is lifted off the plug and (b) the maximum height of the plug if the $5\frac{1}{4}$-in. side Y is lifted off the plug. (c) Compare the two plug designs. (d) Calculate the maximum diameter of the plug for the $5\frac{1}{4}$-in. lift which will permit the use of a 0.250-in. height plug. (e) Compare the clearance in part (d) with part (b).

Solution

(a) If the $3\frac{1}{4}$-in. (side Y) is raised, the minimum height of the plug should be

$$H = \sqrt{2(2a + D)(D - d)}$$ $D = 1.250$ in.
$$ = \sqrt{2[2(2.000) + 1.250](1.250 - 1.248)}$$ $d = 1.248$ in.
$$ = 0.145 \text{ in.}$$ $a = 2.000$ in.

(b) If the $5\frac{1}{4}$-in. (side X) is raised, the minimum height of the plug should be

$$H = \sqrt{2(2a + D)(D - d)}$$ $D = 1.250$ in.
$$ = \sqrt{2[2(1.000) + 1.250](1.250 - 1.248)}$$ $d = 1.248$ in.
$$ = 0.114 \text{ in.}$$ $a = 1.000$ in.

(c) The greater the distance a, the longer the acceptable interference swing radius permitted.

(d) If the 0.145-in. height is retained, the maximum plug diameter may be calculated from

$$H = \sqrt{2(2a + D)(D - d)}$$

Solve for d:

$$d = D - \frac{H^2}{2(2a + D)}$$ $D = 1.250$ in.
$$ = 1.250 - \frac{0.114^2}{2[2(2) + 1.250]}$$ $a = 2.000$ in.
$$ = 1.2488 \text{ in.}$$ $H = 0.114$ in.

(e) The clearance in part (b) is

$$c = 1.250 - 1.248$$
$$ = 0.002 \text{ in.}$$

The clearance in part (d) is

$$c = 1.250 - 1.2488$$
$$ = 0.0012 \text{ in.}$$

In this problem the clearance is reduced by 0.0008 in., so that the side play between the plug and the hole is reduced by that much. This means the center of the hole-to-machined surface dimension can be located with greater accuracy.

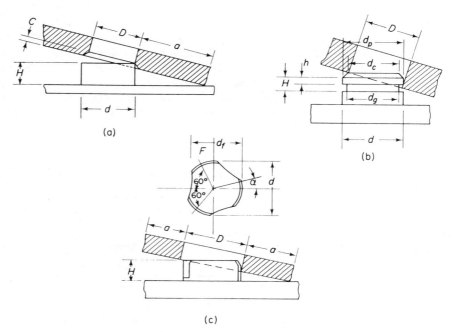

Figure 9.10

Countersunk holes. If a hole is countersunk as shown in Fig. 9.10a the length of the plug should not be greater than

$$H = C + \sqrt{(2a + D)(D - d)}$$

C = height of chamfer, in.
a = distance from pivot to edge of hole, in.
D = diameter of hole, in.
d = diameter of plug, in.
H = height of plug, in.

Example 9.3

Use the data from Ex. 9.2. (a) Calculate the plug height for the $5\frac{1}{4}$-in. workpiece length, if a $\frac{1}{16}$-in. chamfer is machined on the workpiece hole. (b) Compare the plug heights from Exs. 9.2 and 9.3. (c) Assume a plug height of 0.114 in. as in Ex. 2(b). What new plug diameter may be used if a $\frac{1}{16}$-in. chamfer is added to the workpiece hole?

Solution

(a) The height of the plug will be

$$H = C + \sqrt{(2a + D)(D - d)}$$
$$= 0.0625 + \sqrt{[2(2.000) + 1.250](1.250 - 1.248)}$$

$$= 0.165 \text{ in.}$$

$c = \frac{1}{16}$ in.
$a = 2.000$ in.
$D = 1.250$ in.
$d = 1.248$ in.

(b) When a $\frac{1}{8}$-in. chamfer is added, the comparison of the plug heights is

$$0.165 - 0.114 = 0.051 \text{ in. higher}$$

(c) The new plug diameter may be

$$d = D - \frac{(H - C)^2}{2a + D}$$

$$= 1.250 - \frac{(0.114-0.0625)^2}{2(2.000) + 1.250}$$

$$= 1.249 \text{ in.}$$

$C = \frac{1}{16}$ in.
$D = 1.250$ in.
$a = 2.000$ in.
$H = 0.114$ in.
[from Ex. 2(b)]

Alignment groove.   An alignment groove may be machined into the plug as shown in Fig. 9.10b. The depth of the groove, d_g, may be about 95% of the diameter of the body of the plug. That is,

$$d_g = 0.95d$$

The diameter of the pilot end of the plug is given by

$$d_p = \frac{2d^2}{D} - d$$

The clearance between the hole and the pilot is

$$C = D - d_p$$

d = diameter of plug
d_c = diameter of chamfer
d_p = diameter of pilot
d_g = diameter of groove
h = pilot length
H = minimum length
D = diameter of hole
μ = coefficient of friction
C = clearance

The length of the pilot and the groove is a function of the coefficient of friction of the material in the plug and that in the workpiece. For steel against steel it usually is 0.20. The equation for the minimum length of the pilot and the groove is

$$H = \mu d$$

This may be any desired length longer than H.

The equation for the length of the pilot is

$$h = \sqrt{2d_p(D - d_p)}$$

Example 9.4

Use the plug and hole diameters from Ex. 9.2, calculate (a) the groove diameter in the plug, (b) the pilot diameter, (c) the clearance between the hole and the pilot, (d) the minimum length of the groove and the pilot, and (e) the length of the pilot.

Solution

(a) The groove diameter in the plug is

$$d_g = 0.95d = 0.95 \times 1.248$$
$$= 1.1856 \text{ in.}$$

(b) The pilot diameter is

$$d = 1.248 \text{ in.}$$
$$D = 1.250 \text{ in.}$$
$$\mu = 0.15$$

$$d_p = \frac{2d^2}{D} - d = \frac{2(1.248)^2}{1.250} - 1.248$$
$$= 1.244 \text{ in.}$$

(c) The clearance is

$$C = D - d_p = 1.250 - 1.244$$
$$= 0.006 \text{ in.}$$

(d) The minimum length of the pilot and the groove is

$$H = \mu d = 0.15(1.248)$$
$$= 0.1872 \text{ in.}$$

(e) The length of the pilot is

$$H = \sqrt{2d_p(D - d_p)} = \sqrt{2(1.244)(1.250 - 1.244)}$$
$$= 0.122 \text{ in.}$$

Equilateral diamond locator. Another way in which jamming may be prevented is to cut three equal flats from a round plug as shown in Fig. 9.10c. The locating error is increased somewhat. However, if this additional error can be tolerated, this is another way to prevent jamming.

A usable equation that will give the distance from the flat to the centerline of the plug is

$$F = 0.35d \qquad \begin{array}{l} F = \text{distance flat to centerline, in.} \\ d = \text{diameter of solid plug, in.} \end{array}$$

The distance from the flat to the opposite bearing surface is

$$d_f = F + \frac{d}{2} \qquad d_f = \text{distance flat-bearing surface, in.}$$

A plug will not jam with three flats, 120° apart, which are produced using the former equation if the height of the plug to the chamfer is

$$H = 2.4(2a + 0.85d)(D - d) \qquad \begin{array}{l} H = \text{height of plug to chamfer, in.} \\ a = \text{distance pivot to edge of hole} \\ D = \text{diameter of hole, in.} \end{array}$$

The additional error of the equilateral plug when compared to a full round plug is

$$E = 0.207(D - d) \qquad E = \text{additional error}$$

Example 9.5

Use the plug and hole diameters from Ex. 9.2. Calculate (a) the distance from the flat to the center line of the plug, (b) the distance from the flat to the opposite bearing surface, (c) the height of the plug, and (d) the additional error when an equilateral plug is used.

Solution

(a) The flat should be cut so that the distance from the flat to the centerline of the plug is

$$F = 0.35d$$
$$= 0.35(1.248)$$
$$= 0.4368 \text{ in.}$$

(b) The distance from the flat to the opposite bearing surface is

$$d_f = F + \frac{d}{2}$$
$$= 0.4368 + \frac{1.248}{2}$$
$$= 1.0608 \text{ in.}$$

(c) The height of the plug should be

$$H = 2.4(2a + 0.85d)(D - d) \qquad\qquad D = 1.250 \text{ in.}$$
$$= 2.4[2(2.000) + 0.85(1.248)](1.250 - 1.248) \qquad d = 1.248 \text{ in.}$$
$$= 0.0243 \text{ in.} \qquad\qquad\qquad\qquad\qquad a = 2.000 \text{ in.}$$

(d) The additional error when an equilateral plug is used instead of a full round plug:

$$E = 0.207(D - d) = 0.207(1.250 - 1.248)$$
$$= 0.0004 \text{ in.}$$
$$E = \text{additional error}$$

9.6 CONICAL LOCATORS

Conical locators are used to center workpieces. The plug or bore method used to locate finished surfaces, which has been discussed previously in this section, is usually not satisfactory for locating rough surfaces such as cored holes or cast external surfaces. Figure 9.11a shows a rough cast hole where the plug locator fits exactly. Figure 9.11b shows the upper limit of the same series of cored holes. The error thus

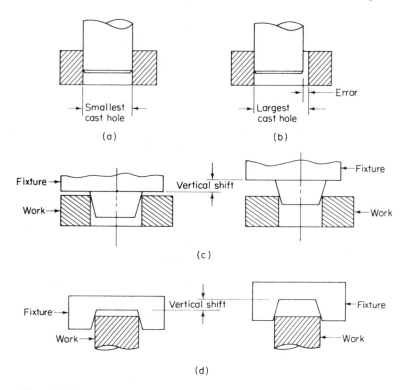

Smallest cast hole

(a)

Largest cast hole

Error

(b)

Fixture→

Work→

Vertical shift

←Fixture

←Work

(c)

Fixture→

Work→

Vertical shift

←Fixture

←Work

(d)

Figure 9.11

introduced may be greater than the print will allow. Because of the much greater size variation of a cast hole when compared to a finish machined hole, this type of locator is not usually acceptable for cast, or cored, holes. Conical locators tend to average out this type of error.

Thus, in Fig. 9.11c it can be seen that the cone locator will shift the workpiece along its vertical axis. However, the relationship of the centerline of the work and the conical locator will remain essentially unchanged. The fixture must make provision for this shift in vertical displacement, and for maintaining the parallelism of the centerlines of the fixture cone and the work.

If the locator has an internal cone, Fig. 9.11d, the same conditions apply. The fixture is a spherical internal cone which shifts up or down, depending on the diameter of the workpiece. There should be a minimum shifting away from the vertical centerline with these cones.

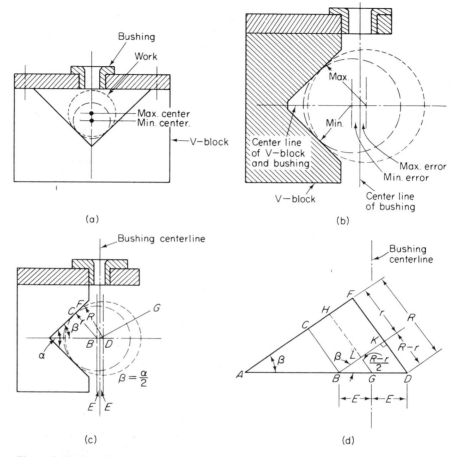

Figure 9.12(a,b,c,d)

9.7 V-LOCATORS

V-locators are widely used because their construction permits automatic location of the workpiece. They are particularly adaptable for locating the centerline of round work along the centerline of the V. Figure 9.12a shows the best method for locating a round workpiece with a V-fixture. Any variation in workpiece size will not affect the centerline accuracy.

Figure 9.12b shows the effect of using the V-block as a locator after it has been rotated 90°. Any variation in the diameter of the work will produce an off-center drilling operation. The variation E between the location of a workpiece that has diameters at the minimum or maximum limits, and the true centerline of the bushing is shown in Fig. 9.12c.

This variation E, Fig. 9.12d, can be obtained from *triangle BLG*, where the angle LBG is β. If α is the included angle of the V-block, then

$$\beta = \frac{\alpha}{2}$$

In triangle BLG, Fig. 9.12d,

$$\beta = \frac{\alpha}{2}; \quad BG = E; \quad LG = \frac{R-r}{2}$$

Since

E = variation, in.
R = maximum radius of work
r = minimum radius of work
α = included angle of V-block
D = maximum diameter of work
d = minimum diameter of work

$$\sin \beta = \frac{LG}{BG}$$

then

$$\sin \beta = \frac{(R-r)/2}{E}$$

Solving for the variation (error) yields

$$E = \frac{(R-r)/2}{\sin \beta} = \csc \beta \left(\frac{R-r}{2} \right)$$

Since most blueprints show dimensions as diameters, this equation becomes

$$E = \frac{D/2 - d/2}{2 \sin \beta} = \csc \beta \left(\frac{D-d}{4} \right)$$

The variation of the workpiece centerline from the centerline of the bushing is seen in Fig. 9.12d to be E. The centerline of the block is the average of the extreme positions of the workpiece centerlines.

As the angle α increases, the function $\csc \alpha$ decreases. The variation is the least where $\alpha/2 = 90$. When this is the case $\alpha = 180°$, which is a straight line. However, a flat surface (straight line) cannot confine a round workpiece. The best angle for a V-block is $\alpha = 90°$, or $\beta = 45°$.

Example 9.6

A hole is to be drilled in the center of a 6-in.-diameter rod which has a tolerance of 0.005 in. Assume conditions such that the V-block drill fixture, Fig. 9.12c, must be used. Determine what variation from the centerline of the drill bushing can be expected if (a) a 90° V-block is used, (b) an 80° V-block is used, and (c) a 100° V-block is used. (d) What observation can be made fron these three answers?

Solution

(a) The variation from the bushing centerline for a 90° V-block is

$$E = \csc \beta \left(\frac{D-d}{4} \right)$$

$$= \csc \frac{90}{2} \left(\frac{6.0025 - 5.9975}{4} \right) \qquad \begin{aligned} D &= 6.0025 \text{ in.} \\ d &= 5.9975 \text{ in.} \end{aligned}$$

$$= 1.414 \times 0.00125$$
$$= 0.00177 \text{ in.}$$

$$\alpha = 90°$$
$$\beta = 45°$$

(b) The variation from the bushing centerline for the 80° V-block is

$$E = \csc \frac{80}{2} \left(\frac{6.0025 - 5.9975}{4} \right)$$

$$\alpha = 80°$$

$$= 0.00194 \text{ in.}$$

(c) The variation from the bushing centerline for a 100° V-block is

$$E = \csc \frac{100}{2} \left(\frac{6.0025 - 5.9975}{4} \right)$$

$$\alpha = 100°$$

$$= 0.00163 \text{ in.}$$

(d) The observation is that an 80° V-block will confine a workpiece more than a 100° V-block. However, a comparable change in the workpiece diameter size will produce a greater variation in the location of the workpiece with reference to the drill bushing centerline when the 80° V-block is used than when the 100°V-block is used.

Figure 9.12e, f, g, and h shows some ways in which V-blocks or variations may be used in fixture design.

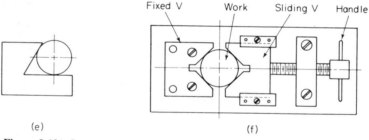

(e)   (f)

Figure 9.12(e,f)

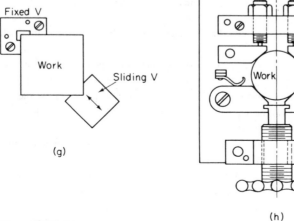

(g)

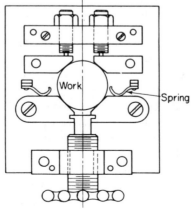

(h)

Figure 9.12(g,h)

PROBLEMS

9.1. Define
  (a) a fixture and
  (b) a jig.
  Give illustrations of each and explain your answers.

9.2. Discuss repetitive nesting of workpieces in a jig or fixture.

9.3. Discuss the six degrees of freedom that must be confined, or controlled, by a jig or fixture. Give examples to support your statements.

9.4. Given the workpiece, Fig. 9.13a, show how it may be confined in space through all six degrees of freedom. Explain your reasons.

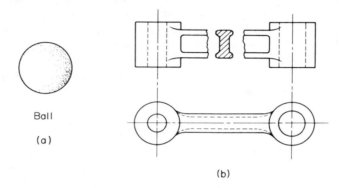

Ball

(a)

(b)

Figure 9.13

9.5. Repeat Prob. 9.4 for Fig. 9.13b.

9.6. Explain the 3-2-1 principle used to determine control of the workpiece in fixture design. Illustrate.

9.7. Why is it necessary to use extreme care when three buttons are used as the principal banking points in a fixture or jig? Illustrate.

9.8. Show by examples that "the greater the spread between the buttons, the less the misalignment error."

9.9. Two surfaces are to be considered. One is smooth, the other is rough. Explain the use of *more than* three support points.

9.10. Discuss the general rule of using six fixed and additional adjustable support points.

9.11. Why is it important that a locating surface be used throughout the various operations when machining a workpiece?

9.12. What are the advantages of using a button over a pad as a banking point?

9.13. List three types of fixed support points.

9.14. What are jackscrews? When are they used as support points? How do jackpins differ in construction and use from jackscrews?

9.15. What are locator pins? How are they used?

9.16. What is jamming? Why is it objectionable?

9.17. Dimensions in Fig. 9.7d are as follows: The hole distance is 8.000 ± 0.015; the plug distance is 8.000 ± 0.003. What are the maximum and minimum clearances if one plug fits exactly?

9.18. Assume that a fixture is equipped with two locator pins as shown in Fig. 9.7d. Assume that hole A is 1.252 in. and pin a is 1.248 in. in diameter, also that hole B is 0.750 and pin b is 0.748 in. in diameter. If the distance between the hole centers is 4 in., calculate the maximum possible angular error when the piece is placed in the fixture.

9.19. Assume that a diamond pin, Fig. 9.8c is used in the small hole, Prob. 9.18. Calculate
    (a) the width of the contact surface between the pin and the hole and
    (b) the length of the pin.

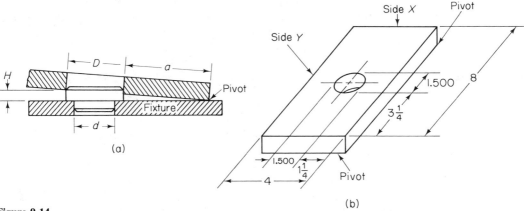

Figure 9.14

9.20. A rectangular plate 4×8 in. is in full contact with the base of a fixture. The hole in the workpiece has been machined to 1.500 in. at the exact center of the plate. Figure 9.14b shows this plate. The plug diameter is 1.495. Calculate
    (a) the maximum height of the plug if the 4-in. side is lifted off the plug and
    (b) the maximum height of the plug if the 8-in. side is lifted off the plug.
    (c) Compare the two plug designs.
    (d) Calculate the maximum diameter of the plug for the 8-in. lift which will permit the use of a 0.200-in. height plug.
    (e) Compare the clearance in part (d) with part (b).

9.21. The plate in Fig. 9.15 is to be located with a plug 1.997 in. in diameter. The plug is to have no chamfer. Assume that the edge marked Z is to remain in contact with the fixture when the work is lifted off the plug. Calculate
    (a) the maximum height of the plug if jamming is to be avoided and
    (b) the maximum plug diameter if a $\frac{1}{2}$-in.-long plug is used.

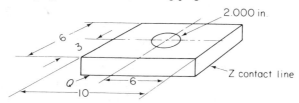

Figure 9.15

9.22. Assume that the plate in Fig. 9.15 is to be lifted off the plug with the edge *opposite* to Z in contact with the fixture. Calculate
   (a) the maximum height of the plug if jamming is to be avoided and
   (b) the maximum plug diameter if a $\frac{1}{2}$-in.-long plug is used.
   (c) If 10-in. edge Q is lifted, what is the maximum plug diameter for a $\frac{1}{8}$-in.-high plug?
   (d) What are the plug-to-hole clearances between parts (b) and (c)?

9.23. Use the data from Prob. 9.20.
   (a) Calculate the plug height for the 8-in. workpiece length if a $\frac{1}{8}$-in. chamfer is machined on the workpiece hole.
   (b) Compare the plug height found in part (a) with that found in Prob. 9.20.
   (c) Assume a plug height of 0.283 in. as in Prob. 9.20b. What new plug diameter may be used if an $\frac{1}{8}$-in. chamfer is added to the workpiece hole?

9.24. Assume that a $\frac{3}{16}$-in. chamfer is added to the hole in Fig. 9.15. Using the data from Prob. 9.21, calculate the maximum height of the plug that may be used.

9.25. Assume that the same $\frac{3}{16}$-in. chamfer is added to Fig. 9.15. Calculate the plug height for Prob. 9.22 when
   (a) $a = 3$ in. and
   (b) $a = 2$ in.

9.26. Assume the conditions of Prob. 9.21(b). Calculate and compare the plug heights if
   (a) a $\frac{3}{16}$-in. chamfer and
   (b) a $\frac{3}{8}$-in. chamfer is added.

9.27. Refer to Prob. 9.21a. Calculate
   (a) the groove diameter in the plug,
   (b) the pilot diameter,
   (c) the clearance between the hole and the pilot,
   (d) the minimum length of the groove and pilot, and
   (e) the minimum length of the pilot.

9.28. Using the plug and the hole diameter from Prob. 9.21a, calculate
   (a) the distance from the centerline of the plug to the flat,
   (b) the distance from the flat to the opposite bearing surface,
   (c) the height of the plug, and
   (d) the additional error used when an equilateral plug is used instead of a solid plug. See Fig. 9.10c.

9.29. Using the plug and hole diameters from Prob. 9.20, calculate
   (a) the groove diameter in the plug,
   (b) the pilot diameter,
   (c) the clearance between the hole and the pilot,
   (d) the minimum length of the groove and pilot, and
   (e) the length of the pilot.

9.30. **(a)** Why are cylindrical locators usually not satisfactory for locating cored holes? Illustrate.
   (b) How is this corrected with conical locators? Illustrate.

9.31. A hole is to be drilled through a round rod perpendicular to the longitudinal axis. The diameter of the rod may vary by as much as 0.030 in. A V-block is to be used in a horizontal position, Fig. 9.12c. The rod is 3 in. in diameter.
   (a) Show that the location error decreases as the included angle of the V-block increases.
   (b) Why is a 90° included angle preferred over a 120° included angle?

9.32. Using the plug and hole diameters from Prob. 9.20, calculate
  (a) the distance from the flat to the centerline of the plug,
  (b) the distance from the flat to the opposite bearing surface,
  (c) the height of the plug and
  (d) the additional error when an equilateral plug is used.

9.33. A hole is to be drilled in the center of a 6 in. diameter rod that has a tolerance of ±0.010 in. Assume conditions are such that the V-block drill fixture, Fig. 9.12c must be used. What variation from the centerline of the drill bushing can be expected if:
  (a) a 90° V-block is used?
  (b) a 70° V-block is used?
  (c) a 110° V-block is used?
  (d) What observation can you make from these three answers?

Chapter 10

Hinges, Components, and Bushings

10.1 HINGES

The *hinge* is one of the most common components used for covers and leaves with jigs and fixtures. The accuracy with which a leaf can be opened and closed repeatedly is of the utmost importance in jigs and fixtures. This is especially true if the leaf is fitted with drill bushings as shown in Fig. 10.1a. In this case the hinge pin is hardened and ground. In Fig. 10.1a the pin may have a press fit in the fixture and a slip fit in the lugs. If it is more convenient to allow the pin to rotate in the fixture, the pin is pressed into the leaf and a slip fit is constructed in the fixture.

In the two instances just discussed, the lugs are part of the fixture. Sometimes the construction is as shown in Fig. 10.1b, where the lugs are integral with the leaf. The construction in Fig. 10.1b is preferred because the leaf is supported at the ends of the pin. This produces less error than the construction shown in Fig. 10.1a.

In either case, if extreme accuracy or severe service is required, hardened washers should be used as shown in Fig. 10.1c. The thickness of the washers should be machined oversize to allow for grinding after hardening. They should be ground very carefully to allow minimum end play between the leaf and the lugs.

In some instances it is desirable to allow for a slip fit in both components. If this becomes necessary, some method must be found to keep the pin from working its way out of the hole. It may be confined with cotter pins, Fig. 10.1d; a C-snap ring, Fig. 10.1e; a clevis pin, Fig. 10.1f; a shoulder screw, Fig. 10.1g; a bushing and socket head screw, Fig. 10.1h; or a cone-point setscrew, Fig. 10.1i.

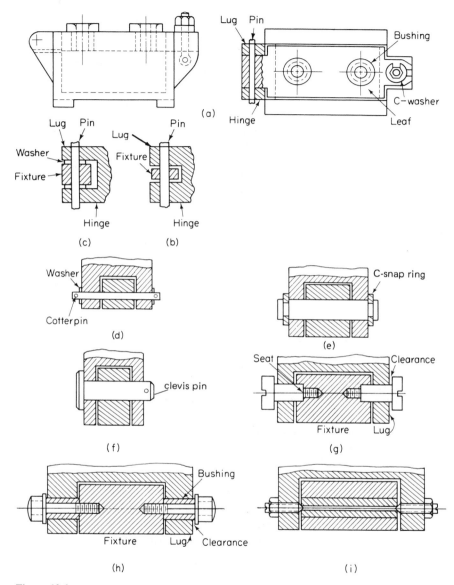

Figure 10.1

10.2 SCREWS AND BOLTS

Screws and bolts may be classified in many ways. Basically, there are eight classifications which may be used: bolts, cap screw, machine screw, shoulder screws, setscrews, studs, self-tapping screws, and wood screws. They are all important to the toolmaker. They are defined as follows:

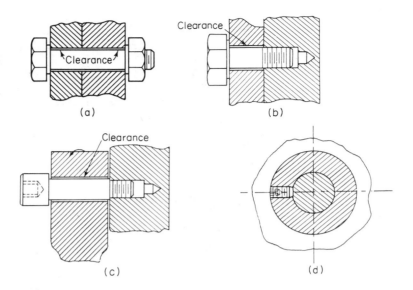

Figure 10.2(a,b,c,d)

A *bolt* is shown in Fig. 10.2a and is generally used with a nut. Therefore, it is passed through a clearance hole in both parts. A nut is tightened at the threaded end which draws the two parts together.

A *cap screw*, Fig. 10.2b, is generally used without a nut. It is passed through a clearance hole in one workpiece and screwed into threads in another workpiece. The threaded workpiece replaces the threaded nut of Fig. 10.2a.

A *machine screw* may be used in the same manner as a bolt or cap screw. They are generally the smaller sizes.

Shoulder screws, Fig. 10.2c, are used to hold workpieces together; with enough clearance so that one part is free to swing or pivot, Fig. 10.2c. Thex may also be used in die sets to act as guides for stripper plates, see Sec. 10.4.

Setscrews, Fig. 10.2d, are used to hold two workpieces together. The outer workpiece is generally threaded. The setscrew is screwed into this thread until it registers against the inner workpiece, locking the two workpieces together.

A *stud*, Fig. 10.2e, is a threaded rod which is screwed into a workpiece so that the threads jam at the top of the hole. This prevents it from turning out when a nut is unscrewed.

Self-tapping screws, Fig. 10.2f, make their own threads. They are hardened and have their tapered starting threads to aid the start of the self-tapping process. The first workpiece through which the screw passes has a clearance hole and should have no threads. The last hole that the screw enters should have a drilled hole equal to the minor diameter of the screw.

Wood screws, several of which are shown in Fig. 10.2g, have 60° interrupted thread forms and a gimlet point.

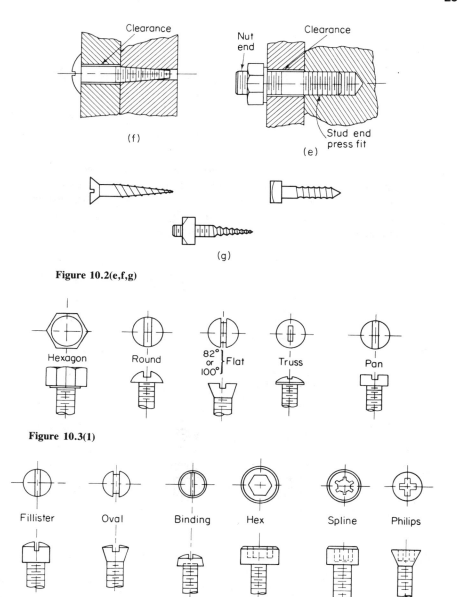

Figure 10.2(e,f,g)

Figure 10.3(1)

Figure 10.3(2)

The various kinds of *heads* available for the types of threaded fasteners just described are shown in Fig. 10.3. They may be external or internal.

Set screw points are shown in Fig. 10.4. Setscrews may have heads. However, in most instances they are headless. They are usually recessed, fluted, or slotted so that they may be screwed into the thread.

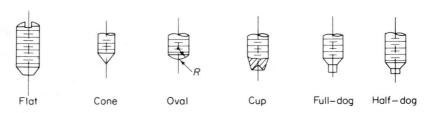

Flat   Cone   Oval   Cup   Full–dog   Half–dog

Figure 10.4

The *nuts* used with bolts are either square or hexagon. They may be plain faced, chamfered, or washer faced; they may be termed unfinished, semifinished, or finished; square, hexagon, jam, slotted, or castle. Figure 10.5a shows plain face nuts. Figure 10.5b shows chamfer face nuts. Figure 10.5c shows washer face nuts. Castle nuts and slotted nuts are used with cotter pins to keep them from working loose.

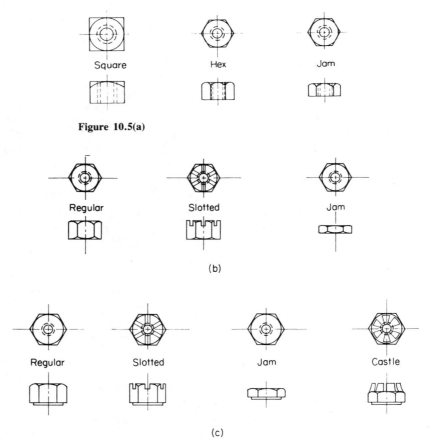

Square   Hex   Jam

Figure 10.5(a)

Regular   Slotted   Jam

(b)

Regular   Slotted   Jam   Castle

(c)

Figure 10.5(b,c)

10.3 LOCKING DEVICES

There are many ingenious ways used to lock bolts or nuts into mechanisms that may vibrate them loose. The simplest way to prevent them from working loose is to use one of the many washer designs available. The plain flat washer is not used as a locking device even though it allows for better action of the nut against the work when it is used.

Washers that are used generally are plain, spring lock, or tooth lock washers. In general, spring lock washers are intended to take up the looseness between parts during their operation. Tooth lock washers are used to lock the nut so that they will not work loose. They are designated by nominal size and standardized as light, medium, heavy, and extra heavy. Figure 10.6 shows the various types of lock washers.

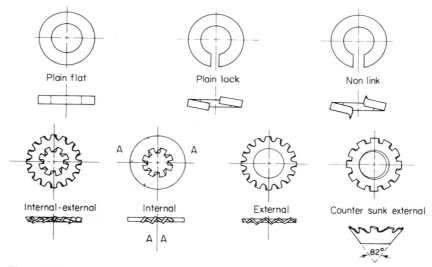

Figure 10.6

As already stated, a common way to lock a nut so that it will not vibrate loose is to use one of the lock washers shown in Fig. 10.6. One such arrangement is shown in Fig. 10.7a. Figure 10.7b shows another type of washer which may be bent once the nut is tightened. This type of washer is sometimes used when a nut is tightened and "backed off" one-quarter of a turn. The bent washer causes the nut to hold its position.

Two nuts may be used in contact with each other as shown in Fig. 10.7c. The lower nut is tightened to exert a desired force, after which the upper nut is tightened to securely lock the first nut. Another method for using a jam nut is shown in Fig. 10.7d.

The use of the *cotter pin* is another of the common ways of locking a nut in position. Several uses of cotter pins are shown in Fig. 10.7e.

There are many additional ways in which to lock nuts so that they will not jar

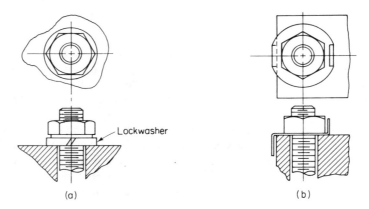

Figure 10.7(a,b)

loose. Figure 10.7f shows an assortment of methods. In addition to those shown, locking may be accomplished with plastic, nylon, or spring steel inserts. Ingenious methods for distorting threads by slotting, expanding or compressing the threads have also been used.

10.4 COMPONENTS FOR JIGS AND FIXTURES

Rest buttons, Fig. 10.8a are used to support work in jigs and fixtures. They are hardened and ground. In most instances they are made by the shop toolmaker. However, they are available as standard parts. Several sizes of rest buttons are shown in Table 10.1. They may be pressed or screwed into the fixture.

Shoulder screws, Fig. 10.2c and 10.8b, are used for swing clamps, washers, and so on. They are caused to seat on the shoulder so that a *part* which has as its dimension C, Fig. 10.8b, and carries a tolerance of $+0.000$ and -0.005 will be able to swivel freely under the head of the screw which has a tolerance of $+0.005$ and -0.000. Some typical dimensions and ranges are shown in Table 10.2.

Thumbscrews, Fig. 10.8c, may be used for quick clamping. Some of the more common sizes are shown in Table 10.3.

The quarter-turn thumbscrew, Fig. 10.8d, is a widely used type screw to lock jig plates. The slot in the plate is made wide enough so that dimension E will fit freely. The chamfers allow for any misalignment of the plate. One-quarter turn is enough to lock the plate. Table 10.4 shows some of the standardized dimensions for quarter-turn thumbscrews.

A variation of the quarter-turn thumbscrew is the *half-turn thumbscrew* shown in Fig. 10.8e and Table 10.5. These thumbscrews are used for half-turn locking of jig plates and covers. Both positions of the screw are shown in Fig. 10.8e.

The *adjusting head screw,* Fig. 10.8f, is used to clamp or hold work when light pressures are required. They are available in sizes of 10–24 to $\frac{1}{2}$–13 and lengths from 1 to $3\frac{1}{2}$ in. The knob is knurled to prevent slipping. Figure 10.8g is the hand knob that

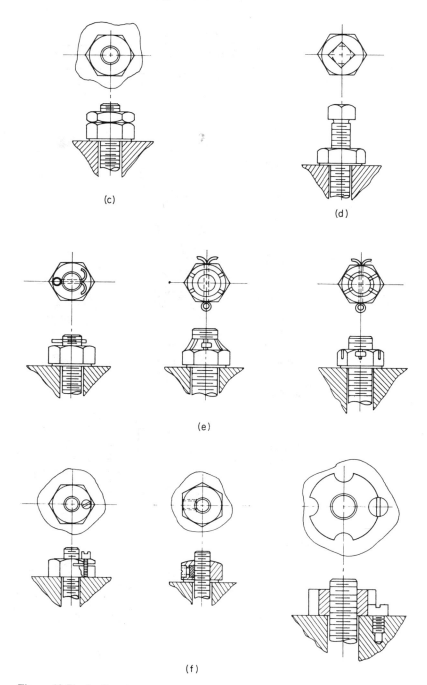

(c)

(d)

(e)

(f)

Figure 10.7(c,d,e,f)

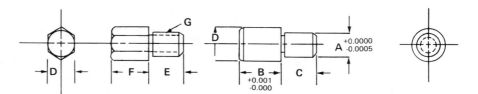

Figure 10.8(a)

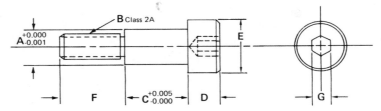

Figure 10.8(b)

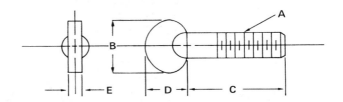

Figure 10.8(c)

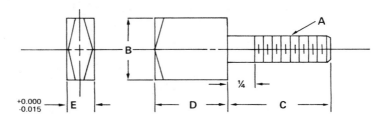

Figure 10.8(d)

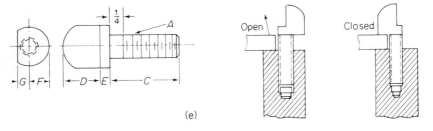

(e)

Figure 10.8(e)

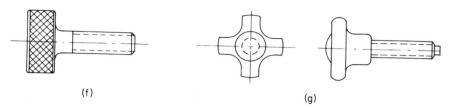

(f)   (g)

Figure 10.8(f,g)

is used where medium clamping pressures are desired. They are available in sizes of $\frac{1}{4}$–20 to $\frac{5}{8}$–11 and in lengths of 1 to $3\frac{1}{2}$ in.

Handles and handwheels are items frequently used with jigs and fixtures. As a design becomes more complex their cost increases rapidly. The simplest of all the handles is the *solid* steel handle shown in Fig. 10.9a. It is generally machined to a slip

TABLE 10.1   REST BUTTONS—FIG. 10.7a

A	B	C	D	E	F	G
0.1885	0.250 0.375 0.500	$\frac{3}{8}$	$\frac{3}{8}$	$\frac{3}{8}$	$\frac{1}{4}, \frac{3}{8}, \frac{1}{2}$ $\frac{3}{4}, 1$	$\frac{1}{4}$–20
0.251	0.250 0.375 0.500	$\frac{1}{2}$	$\frac{1}{2}$	$\frac{1}{2}$	$\frac{1}{4}, \frac{3}{8}, \frac{1}{2}$ $\frac{3}{4}, 1$	$\frac{3}{8}$–16
0.376	0.250 0.375 0.4375 0.500 0.625 0.750	$\frac{5}{8}$	$\frac{5}{8}$	$\frac{5}{8}$	$\frac{1}{4}, \frac{3}{8}, \frac{1}{2}$ $\frac{5}{8}, \frac{3}{4}, 1$	$\frac{1}{2}$–13
0.501	0.375 0.4375 0.500 0.625 0.750	$\frac{5}{8}$	$\frac{7}{8}$			
0.626	0.375 0.4375 0.500 0.625 0.750	$\frac{3}{4}$	1			
0.751	0.375 0.4375 0.500 0.625	$\frac{7}{8}$	$1\frac{1}{4}$			

TABLE 10.2 SHOULDER SCREWS—FIG. 10.8b

A	B	C	D	E	F	G
0.250	10–24	0.250–1.000	$\frac{3}{16}$	$\frac{3}{8}$	$\frac{3}{8}$	$\frac{1}{8}$
0.312	$\frac{1}{4}$–20	0.250–1.000	$\frac{7}{32}$	$\frac{7}{16}$	$\frac{7}{16}$	$\frac{5}{32}$
0.375	$\frac{5}{16}$–18	0.250–1.000	$\frac{1}{4}$	$\frac{9}{16}$	$\frac{1}{2}$	$\frac{3}{16}$
0.500	$\frac{3}{8}$–16	0.375–1.000	$\frac{5}{16}$	$\frac{3}{4}$	$\frac{5}{8}$	$\frac{1}{4}$
0.625	$\frac{1}{2}$–13	0.500–1.000	$\frac{3}{8}$	$\frac{7}{8}$	$\frac{3}{4}$	$\frac{5}{16}$
0.750	$\frac{5}{8}$–11	0.500–1.250	$\frac{1}{2}$	1	$\frac{7}{8}$	$\frac{3}{8}$

Source: Vlier Engineering Corp., Burbank, Calif., Catalog, p. 88.

TABLE 10.3 THUMBSCREWS—FIG. 10.8c

A	B	C	D	E
$\frac{3}{16}$	$\frac{3}{4}$		$\frac{5}{8}$	$\frac{3}{32}$
$\frac{1}{4}$	$\frac{15}{16}$		$\frac{3}{4}$	$\frac{1}{8}$
$\frac{5}{16}$	$1\frac{1}{8}$		$\frac{7}{8}$	$\frac{5}{32}$
$\frac{3}{8}$	$1\frac{1}{4}$		$\frac{15}{16}$	$\frac{5}{32}$
$\frac{7}{16}$	$1\frac{1}{2}$		$1\frac{1}{16}$	$\frac{3}{16}$
$\frac{1}{2}$	$1\frac{5}{8}$		$1\frac{3}{16}$	$\frac{3}{16}$

TABLE 10.4 QUARTER-TURN SCREWS—FIG. 10.8d

A	B	C	D	E
10–24	$\frac{1}{2}$	$1\frac{1}{4}$	$\frac{1}{2}$	$\frac{3}{16}$
10–32	$\frac{1}{2}$	1	$\frac{1}{2}$	$\frac{3}{16}$
$\frac{1}{4}$–20	$\frac{3}{4}$	$1\frac{1}{4}$	$\frac{5}{8}$	$\frac{1}{4}$
$\frac{5}{16}$–18	1	$1\frac{1}{2}$	$\frac{3}{4}$	$\frac{5}{16}$
$\frac{3}{8}$–16	1	$1\frac{1}{2}$	1	$\frac{3}{8}$
$\frac{1}{2}$–13	$1\frac{1}{4}$	$1\frac{3}{4}$	1	$\frac{1}{2}$
$\frac{5}{8}$–11	$1\frac{1}{2}$	$2\frac{1}{4}$	1	$\frac{5}{8}$

Source: Vlier Engineering Corp., Burbank, Calif., Catalog, p. 91.

TABLE 10.5   HALF-TURN SCREWS—FIG. 10.8e

A	B	C	D	E	F	G
10–32	$\frac{3}{4}$	1	$\frac{1}{2}$	$\frac{3}{16}$	$\frac{11}{32}$	$\frac{3}{32}$
$\frac{1}{4}$–20	$\frac{3}{4}$	$1\frac{1}{4}$	$\frac{9}{16}$	$\frac{3}{16}$	$\frac{13}{32}$	$\frac{1}{8}$
$\frac{5}{16}$–18	1	$1\frac{1}{4}$	$\frac{5}{8}$	$\frac{3}{16}$	$\frac{1}{2}$	$\frac{5}{32}$
$\frac{3}{8}$–16	$1\frac{1}{16}$	$1\frac{3}{4}$	$\frac{11}{16}$	$\frac{1}{4}$	$\frac{19}{32}$	$\frac{3}{16}$
$\frac{1}{2}$–13	$1\frac{1}{4}$	$1\frac{3}{4}$	$\frac{3}{4}$	$\frac{1}{4}$	$\frac{21}{32}$	$\frac{1}{4}$

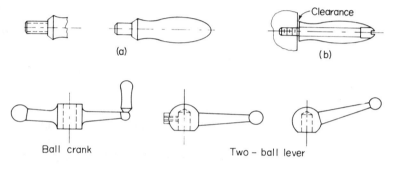

(a)   (b)

Ball crank   Two – ball lever

Figure 10.9(a,b)

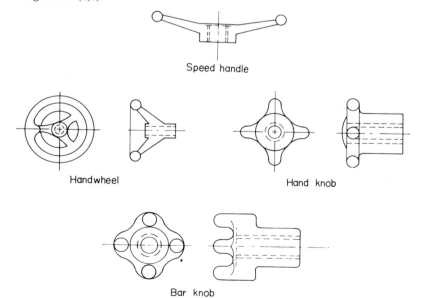

Speed handle

Handwheel   Hand knob

Bar knob

Figure 10.9(c)

fit with a reamed hole. It is locked with a set screw, or pin. It may be threaded at the end as shown in Fig. 10.9a. Another method of using this type of handle is shown in Fig. 10.9b. In this case the handle is free to revolve on the screw. A more expensive series of handles, wheels, and knobs is shown in Fig. 10.9c.

Eye bolts, Fig. 10.10a, and swing bolts, Fig. 10.10b, are used to hold and release cover plates. When used with $\frac{1}{4}$-turn thumbscrews, they make it possible to release, or clamp, a hinged jig plate very quickly. They are manufactured with thread sizes of $\frac{1}{4}$–20 to $\frac{5}{8}$–11 and with lengths from the center of the hole to the end of the thread with a range of $1\frac{1}{2}$ to 6 in.

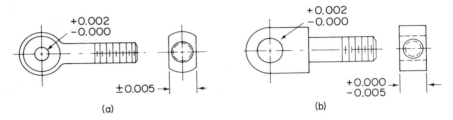

Figure 10.10

Plain and lock washers were discussed in Sec. 10.3. There are three types of washers which are used frequently in jig and fixture design: the C-washer, Fig. 10.11a; the swing C-washer, Fig. 10.11b; and the spherical washer, Fig. 10.11c. Table 10.6 shows some of the important dimensions of the C- and swing washers.

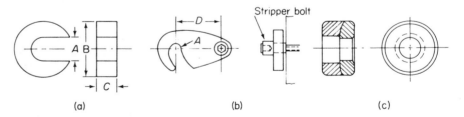

Figure 10.11

The spherical washers are a combination of a concave and a convex washer. They are self-aligning washers which compensate for any misalignment that might exist between a strap clamp and the workpiece. This self-aligning feature helps to reduce the stress on the belt.

10.5 MISCELLANEOUS COMPONENTS

There are various types of spring-loaded plungers. The cross sections of several of these plungers are shown in Fig. 10.12. They may be purchased to apply heavy or light end pressures. The plungers may be hardened steel or nylon noses. The nylon or plastic noses are used with aluminum, brass, or other soft materials. These units may

TABLE 10.6 C- AND SWING WASHERS—
FIG. 10.10

A	B	C	D	E
$\frac{9}{32}$			$\frac{3}{4}$	$\frac{1}{4}$
$\frac{11}{32}$			$\frac{7}{8}$	$\frac{1}{4}$
$\frac{13}{32}$	$1\frac{1}{4}$ $1\frac{1}{2}$ $1\frac{3}{4}$	$\frac{3}{8}$	1	$\frac{1}{4}$
$\frac{17}{32}$	$1\frac{7}{8}$ $2\frac{1}{4}$ $2\frac{3}{4}$	$\frac{3}{8}$	$1\frac{1}{4}$	$\frac{5}{16}$
$\frac{21}{32}$	$2\frac{1}{8}$ $2\frac{1}{2}$ 3	$\frac{3}{8}$	$1\frac{1}{2}$	$\frac{3}{8}$
$\frac{25}{32}$	$2\frac{1}{2}$ $2\frac{7}{8}$ $3\frac{3}{8}$	$\frac{7}{16}$	$1\frac{3}{4}$	$\frac{1}{2}$
$\frac{29}{32}$	$2\frac{1}{8}$ $2\frac{5}{8}$	$\frac{7}{16}$	2	$\frac{1}{2}$
$1\frac{1}{32}$	$2\frac{1}{2}$ 3	$\frac{7}{16}$		

Source: Vlier Engineering Corp., Burbank, Calif., Catalog,
p. 85.

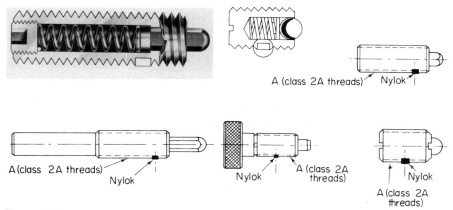

Figure 10.12

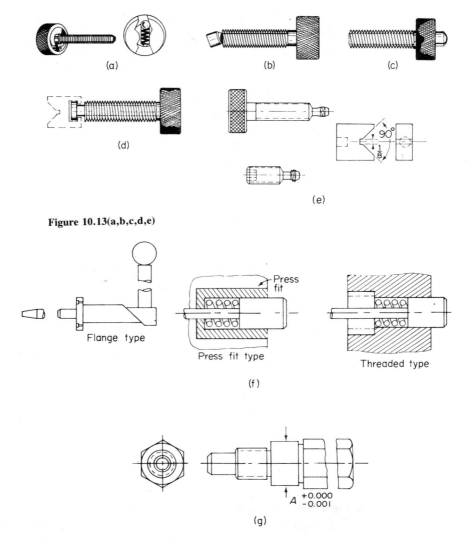

Figure 10.13(a,b,c,d,e)

Figure 10.13(f,g)

be equipped with a nylon plug which compresses when the part is screwed into a fixture. This locks the unit into position.

Special thumbscrews are shown in Fig. 10.13. The *torque thumbscrew*, Fig. 10.13a, prevents the operator from overtightening the screw and damaging the workpiece. When the exact torque is reached, the thumbscrew turns freely.

The *swivel-pad thumbscrew*, Fig. 10.13b, is used to hold off-angle workpieces. The screw is equipped with a torque head.

The *inverted torque thumbscrew*, Fig. 10.13c, allows light positive pressures against the under surface of a workpiece.

TABLE 10.7 V-BLOCKS—FIG. 10.12e

A	B	C	D	E
1.000	0.706	0.875	0.500	0.312
1.000	0.729	0.938	0.562	0.375
1.000	0.739	1.000	0.625	0.438
2.000	1.199	1.500	0.688	0.562
2.000	1.231	1.625	0.750	0.625

The *T-head torque thumbscrew*, Fig. 10.13d, may be used to actuate fixture components. The process is to adapt the component to the T. Figure 10.13d shows the screw adapted to move a V-block.

A series of swivel head screws are shown in Fig. 10.13e. These swivel screws can be made up with small or large pads, or *V-blocks*. Several of these standard V-blocks are listed in Table 10.7. They may be used in fixtures when sliding V-blocks are needed.

At times it becomes necessary to provide fixtures with indexing mechanisms. Figure 10.13f shows a flanged rotary cam *index plunger*. It may be provided with a straight or cylindrical plunger. The plungers are hardened and ground to $+^{0.0003}_{0.0000}$ in. tolerance. Other types of plunger assemblies are shown in Fig. 10.13f. The three types shown are flanged type, press fit, and screw plungers.

Jig legs, Fig. 10.13g, are used to raise the jig off the machine table, thus providing chip clearance. These pins are generally made by the toolmaker, but may be purchased. Diameter A is machined to close tolerances so that it will fit into a counterbored hole, thus ensuring that the leg will be straight and that the counterbore will take the forces rather than the thread.

10.6 BUSHINGS

There are three main types of bushing that may be used in drill jigs: *shoulder, plain,* and *special* bushings. In each of these categories the bushings may have a press or a slip fit into the jig.

Figure 10.14a shows a plain and a shoulder-type press-fit drill bushing. These bushings are pressed directly into the body of the jig and are intended to remain in the jig until they wear out. Replacement is difficult but possible if the design is correct. Their purpose is to guide the tool. Because they are pressed into the jig, they are apt to work through the hole, or be forced back out of the hole by the chip or drill forces. Shoulder press bushings are preferred to plain bushings. However, should it be necessary that the bushing be flush with the fixture, a plain bushing should be used. Sometimes holes must be drilled close together in such a way that shoulder bushings will interfere with each other or with the fixture. In this case a plain bushing should be used.

Renewable bushings are used with *liners*. There are two types of renewable bushings, fixed and slip, Fig. 10.14b. *Fixed renewable bushings* are installed in the

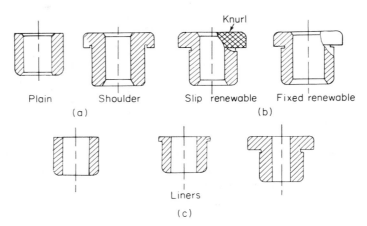

Figure 10.14(a,b,c)

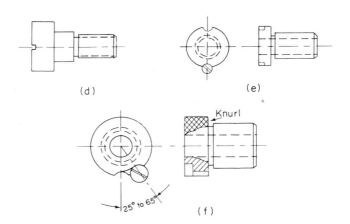

Figure 10.14(d,e,f)

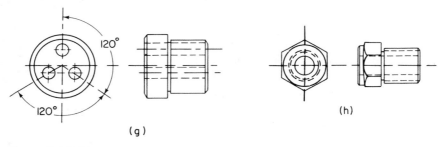

Figure 10.14(g,h)

liner with the intention that they will remain in place until worn out. It is also possible to change the bushing if it becomes necessary to change the hole size in the workpiece.

Slip renewable bushings are available with several inside diameter sizes for each outside diameter size. Therefore, the bushing may be used, removed, and a larger-ID (inside diameter) bushing placed into the same liner. This makes it possible to drill a hole, remove the bushing, and replace it with one that has the reamer size. With slip renewable bushings it is possible to follow the drilling operation with a larger drill, a spot face, counterbore, tap, or reamer. These bushings are usually knurled to aid in their removal. The fixed renewable bushings are not knurled. Liners, Fig. 10.14c, may be shoulder or plain types and are installed permanently into the body of the jig.

Since slip bushings tend to work free, they may be secured with a large head shoulder screw, Fig. 10.14d. They may be secured permanently with these screws, as shown in Fig. 10.14e. They may also be secured so that there is clearance under the head of the screw, Fig. 10.14f, which permits rotating the bushing counterclockwise so that it may be removed. Tables 10.8, 10.9, and 10.10 show permanent and renewable bushings and liners, respectively.

TABLE 10.8 PRESS FIT—PLAIN AND SHOULDER BUSHINGS

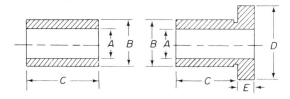

Nominal Hole Size A	Body Diameter B	Body Length C	Head Diameter D	Head Thickness E
	$B^{+0.0000}_{-0.0003}$			
0.0135 to 0.0625	0.1578	$\frac{1}{4}$–1	$\frac{1}{4}$	$\frac{3}{32}$
0.0135 to 0.0995	0.2046	$\frac{1}{4}$–$1\frac{3}{8}$	$\frac{5}{16}$	$\frac{3}{32}$
0.0980 to 0.1406	0.2516	$\frac{1}{4}$–$1\frac{3}{4}$	$\frac{3}{8}$	$\frac{3}{32}$

TABLE 10.8   (*Continued*)

Nominal Hole Size A	Body Diameter B	Body Length C	Head Diameter D	Head Thickness E
0.1250 to 0.1935	0.3141	$\frac{1}{4}-1\frac{3}{4}$	$\frac{7}{16}$	$\frac{1}{8}$
0.1875 to 0.2570	0.4078	$\frac{1}{4}-2\frac{1}{8}$	$\frac{9}{16}$	$\frac{5}{32}$
0.1875 to 0.3160	0.5017	$\frac{1}{4}-2\frac{1}{8}$	$\frac{5}{8}$	$\frac{7}{32}$
0.3125 to 0.4375	0.6267	$\frac{1}{4}-2\frac{1}{2}$	$\frac{13}{16}$	$\frac{7}{32}$
0.3125 to 0.5312	0.7518	$\frac{1}{4}-2\frac{1}{2}$	$\frac{15}{16}$	$\frac{7}{32}$
0.5000 to 0.6562	0.8768	$\frac{1}{2}-3$	$1\frac{1}{16}$	$\frac{1}{4}$
0.5000 to 0.7656	1.0018	$\frac{1}{2}-3$	$1\frac{1}{4}$	$\frac{5}{16}$
	$B\,^{+0.0000}_{-0.0004}$			
0.6250 to 1.0312	1.3772	$\frac{1}{2}-3$	$1\frac{5}{8}$	$\frac{3}{8}$
1.0000 to 1.3906	1.7523	$\frac{3}{4}-3$	2	$\frac{3}{8}$
1.3750 to 1.7656	2.2525	$1-3$	$2\frac{1}{2}$	$\frac{3}{8}$

Source: Jergens, Inc., Cleveland, Ohio, Catalog No. 10.

TABLE 10.9   SLIP FIT—STATIONARY OR RENEWABLE

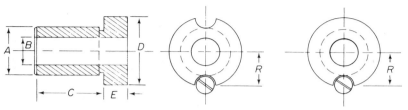

Nominal Hole Size A	Body Diameter B	Body Length C	Head Diameter D	Head Thickness E	Center of Screw R
	$B^{+0.0000}_{-0.0002}$				
0.0135 to 0.0625	0.1875	$\frac{1}{4}-\frac{3}{4}$	$\frac{5}{16}$	$\frac{3}{16}$	$\frac{17}{64}$
0.0135 to 0.1935	0.3125	$\frac{1}{4}-1\frac{3}{8}$	$\frac{9}{16}$	$\frac{3}{8}$	$\frac{1}{2}$
0.1405 to 0.3437	0.5000	$\frac{5}{16}-2\frac{1}{8}$	$\frac{13}{16}$	$\frac{7}{16}$	$\frac{5}{8}$
0.2812 to 0.5625	0.7500	$\frac{1}{2}-3$	$1\frac{1}{16}$	$\frac{7}{16}$	$\frac{3}{4}$
0.4687 to 0.7812	0.1000	$\frac{1}{2}-3$	$1\frac{7}{16}$	$\frac{7}{16}$	$\frac{59}{64}$
	$B^{+0.0000}_{-0.0003}$				
0.7187 to 1.0625	1.3750	$\frac{3}{4}-3$	$1\frac{13}{16}$	$\frac{7}{16}$	$1\frac{7}{64}$
0.9687 to 1.4062	1.7500	$\frac{3}{4}-3$	$2\frac{5}{16}$	$\frac{5}{8}$	$1\frac{25}{64}$
	$B^{+0.0000}_{-0.0004}$				
1.3437 to 1.8750	2.250	$\frac{3}{4}-3$	$2\frac{13}{16}$	$\frac{5}{8}$	$1\frac{41}{64}$

Source: Jergens, Inc., Cleveland, Ohio, Catalog No. 10, p. 8.

TABLE 10.10 LINERS—PLAIN AND SHOULDER

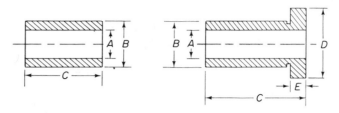

Hole Size A	Body Diameter B	Length C	Head Diameter D	Head Thickness E
+0.0003 −0.0000	+0.0000 −0.0003			
0.1876	0.3142	$\frac{1}{4}-\frac{3}{4}$		
0.3126	0.5017	$\frac{1}{4}-1\frac{3}{8}$	$\frac{5}{8}$	$\frac{3}{32}$
0.5002	0.7518	$\frac{5}{16}-2\frac{1}{8}$	$\frac{7}{8}$	$\frac{3}{32}$
0.7503	1.0018	$\frac{1}{2}-3$	$1\frac{1}{8}$	$\frac{1}{8}$
1.0004	1.3772	$\frac{1}{2}-3$	$1\frac{1}{2}$	$\frac{1}{8}$
+0.0004 −0.0000	+0.0000 −0.0004			
1.3756	1.7523	$\frac{3}{4}-3$	$1\frac{7}{8}$	$\frac{3}{16}$
1.7508	2.2525	$\frac{3}{4}-3$	$2\frac{3}{8}$	$\frac{3}{16}$
+0.0005 −0.0000				
2.2510	2.7526	$\frac{3}{4}-3$	$2\frac{7}{8}$	$\frac{3}{16}$

Source: Jergens, Inc., Cleveland, Ohio, Catalog No. 10, p. 14.

Special bushings are those which are designed to do a specific job. They may have two or three holes in one bushing to ensure the relationship between all holes. A three-hole bushing is shown in Fig. 10.14g. Bushings are sometimes threaded into the jig plates. Such a bushing is shown in Fig. 10.14h.

When installing bushings there are several items with which the designer must be familiar.

   1. The length of the bushing, or liner, should match the standard jig plate thickness as shown in Fig. 10.15a. The standard plate thicknesses are:

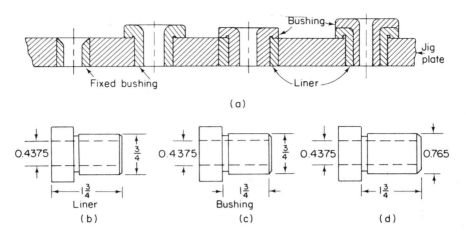

Figure 10.15

$\frac{5}{16}$, $\frac{3}{8}$, $\frac{1}{2}$, $\frac{3}{4}$, 1, $1\frac{3}{8}$, $1\frac{3}{4}$, $2\frac{1}{8}$, $2\frac{1}{2}$, and 3 in.

2. The inside diameter of a bushing may be stated as a decimal, fraction, number, or letter dimension.

3. The various types of bushings have been standardized so that a code number may be interpreted to yield the bushing type, body diameter, body length, and whether or not the body diameter is unfinished. Thus:

(a) The bushing types are specified by a letter or letters.

P	press fit, plain type	F	fixed renewable bushing
H	press fit, shoulder type	L	liner, plain type
S	slip fit, renewable bushing	HL	liner, shoulder type

(b) The body diameter is given in increments of $\frac{1}{64}$ in. A bushing that has a body diameter of

$$\frac{5}{8} = \frac{40}{64} \text{ in.}$$

is designated as having "40 body diameter size."

(c) The body length is given in increments of $\frac{1}{16}$ in. A bushing has a body length of

$$1\frac{3}{8} = \frac{11}{8} = \frac{22}{16}$$

is designated as having a "22 body length size."

Example 10.1

Given a number 0.4375–HL–48–28 bushing. (a) Interpret this number. (b) Make a sketch of this bushing and insert the dimensions. (c) If the bushing were an H-bushing instead of an HL (liner bushing), how would the drawing be dimensioned?

Solution

(a) The hole size in the bushing is

nominal hole size $= \frac{7}{16}$ in.
decimal hole size $= 0.4375$ in.

HL indicates a shoulder-type liner. The 48 gives the body diameter of the bushing. It is

$$\frac{48}{64} = \frac{3}{4} \text{ in.}$$

The 28 gives the length of the body. It is

$$\frac{28}{16} = 1\frac{3}{4} \text{ in.}$$

(b) The dimensioned bushing would be as shown in Fig. 10.15b. (*Note:* $1\frac{3}{4}$ dimension.)
(c) The dimensioned H-bushing would be as shown in Fig. 10.15c. (*Note:* $1\frac{3}{4}$ dimension.)

4. At times it becomes necessary to purchase bushings with extra material on the body so that they may be ground to fit the jig plate. These bushings may be purchased with body dimensions from 0.010 to 0.015 in. oversize. They are designated by a U following the number.

Example 10.2

Assume that the bushing number in Ex. 10.1 reads

0.4375–H–48–28–U

Make a drawing showing all dimensions.

Solution   The dimensioned bushing is shown in Fig. 10.15d. The body is assumed to be 0.015 in. oversize for grinding.

Because drill bushings need to resist the rotation of the drill, all drill bushings except renewable, or slip, bushings should have a slight interference fit into the fixture or liner. For most bushings, the practice is to create an interference fit of about 0.0005 in.

The chip clearance between the bottom of the bushing and the top of the workpiece is a function of the type of chip generated. Thus as the accuracy of the hole in the workpiece increases, the clearance between the bushing and the work decreases. In general, the more accurate the size of the machined hole, the smaller the chip generated. The smaller the chip generated, the closer the bottom of the bushing should be to the workpiece.

Thus, in Fig. 10.16a the bottom of the liner, or bushing, almost rests on the work. This drilling operation could be a finishing or sizing operation. In Fig. 10.16b

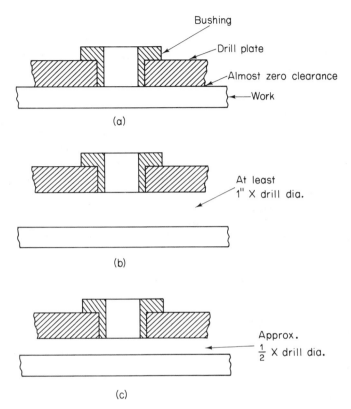

Figure 10.16

the chip produced is continuous. In Fig. 10.16c the chip produced is small, as is the case when using secondary operation drills or reamers. Renewable bushings, generally used in reaming operations, have very little clearance between the work and the bottom of the bushing, Fig. 10.14f.

In addition to the above, when bushings are used, the jig plate supporting these bushings must be at least $1\frac{1}{2}$ times as thick as the diameter of the drill. This is necessary to resist the cutting forces.

Because the purpose of the drill bushing and fixture is to locate the drill or reamer accurately, straying is not permitted. The closer the bottom of the bushing is to the workpiece, the less chance there is that the drill, or reamer, will wander. If the drill wanders, the clearance between the bushing and the work should be reduced. It should be remembered that the closer the clearance between the bushing and the work, the more chance that the chips will clog. In effect, the designer must make a compromise between the wandering of the cutting tool and the clogging effect while generating chips.

PROBLEMS

10.1. (a) Explain the construction of the hinge pin in Fig. 10.1a through c.
(b) Which construction is preferred?

10.2. (a) Describe each of the hinges in Fig. 10.1d through i.
(b) Why is it necessary to confine the hinge pins?

10.3. (a) List the eight types of screws and bolts discussed in this section.
(b) State a characteristic of each that could be used to identify them.

10.4. Make a freehand sketch of the various kinds of heads that are used with threaded fasteners.

10.5. (a) Make a sketch of the six types of setscrew points discussed in this chapter.
(b) State the purpose for which each may be used. (Check your answer with your instructor.)

10.6. (a) Make a sketch of the three types of faces that nuts may have.
(b) What is the fundamental difference between regular, slotted, jam, or castle nuts?

10.7. Is the plain flat washer considered a locking mechanism? Explain your answer.

10.8. Sketch the seven types of washers available. Explain their uses.

10.9. How is a bent washer used to lock a nut? Illustrate.

10.10. Describe two uses for a jam nut. Illustrate.

10.11. How many ways may a cotter pin be used to lock a nut in position? Draw a sketch for each method and explain your drawing.

10.12. Explain the locking devices shown in Fig. 10.7f and how they work.

10.13. Rest buttons may be installed in a fixture in at least two ways. What are they?

10.14. (a) What is the purpose of a shoulder screw?
(b) How does it function in a fixture?

10.15. (a) Make a sketch of a quarter- and a half-turn thumbscrew.
(b) How are they used in fixture design?

10.16. How does the adjusting head screw differ from the hand knob?

10.17. What is the difference between a handle, a handwheel, and a crank? Illustrate.

10.18. Show how eye-bolts are used in fixture design.

10.19. Make a sketch of a C-washer, a swing C-washer, and a spherical washer. How are they used? Illustrate.

10.20. Describe the use of spring-loaded plungers. Why is there a need for different types of tips?

10.21. Describe the purpose for which the various thumbscrews in Fig. 10.13 are used.

10.22. List the three types of indexing plungers discussed in this chapter.

10.23. What are jig legs? How are they used?

10.24. (a) What is a plain bushing?
(b) How is it used?

10.25. Describe the use of
(a) fixed and
(b) slip renewable bushings.

10.26. Describe the three ways in which liners may be secured in a fixture.

10.27. List the standard jig plate thickness.

10.28. Given a bushing number $\frac{3}{8}$–S–34–22, draw and dimension.

10.29. Given a bushing number 0.8750–L–42–60–U, calculate and dimension this bushing.

10.30. Explain the clearances in Fig. 10.16a through c.

10.31. What effect does the size of the chip generated during the drilling operation have on the clearance between the bottom of the bushing and the workpiece?

10.32. Figure 10.17 is a workpiece that is to have surface *A* machined. Design a fixture to hold the workpiece so that the surface can be machined. Use buttons as supports. You must take care of all six degrees of freedom.

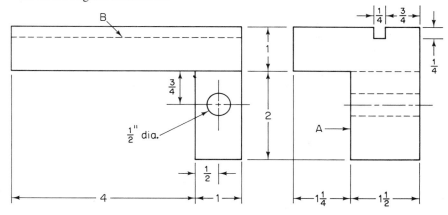

Figure 10.17

10.33. Design a fixture to cut the slot *B* on a milling machine. The purpose of this assignment is to test your ability to use the items in this chapter. Use Fig. 10.17.

10.34. You are to drill a hole into the workpiece, Fig. 10.18. Note that the surface of the work is not perpendicular to the required hole. Draw the bushing and show how you would mount the bushing into a drill plate.

10.35. Repeat Prob. 10.34 for Fig. 10.19.

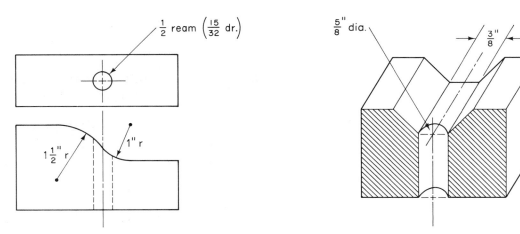

Figure 10.18   **Figure 10.19**

Chapter 11

Clamps and Clamping Principles

11.1 CLAMPING PRINCIPLES

In Chapter 9 the six degrees of freedom necessary for good fixture design were studied. It was pointed out at that time that these degrees of freedom had to be confined in one direction and that some type of clamping mechanism should hold the workpiece along the same line of action. That is, it should hold the work firmly against a locating or banking surface. There are several principles that should be understood to aid in the design of clamps. They are:

1. The clamp should press against a strong portion of the workpiece without distorting the workpiece.
2. It should hold the work firmly and transmit the clamping force through the work, directly over a fixed support point.
3. It should be capable of overcoming the maximum possible force exerted on it by the workpiece. If possible, the minimum clamping force applied by the clamp should be able to accomplish this.
4. A fixed locator or support should oppose the forces of cutting. The clamp should not be designed or required to resist the forces of cutting.
5. Simple clamps are always preferred to complicated clamps.
6. Clamps should be easy to operate, positioned, or loosened. They should never interfere with the positioning or removal of the workpiece.
7. Clamps should be designed so that they will retract enough to clear the workpiece.

8. Wherever possible, clamps should be designed so that they can be operated from the front, or the operator's side, of the fixture.

9. When side clamps are used, they should be designed so that they press down and in and tend to seat the work.

10. When cams or wedges are used, they should be designed so that loads or vibrations will tend to tighten rather than loosen them.

11. The clamp should not damage the workpiece. This is especially true if the surface to be clamped has already been machined. If the surface is to remain rough, such as a catalog surface, the clamp may be designed to bite into the workpiece surface.

12. When estimating the cutting forces that a clamp is to overcome, it should be remembered that these forces can increase because of uneven depths of cut, the dulling of cutters, variations in cutting rates or stock removal, or changes in the hardness of materials. It is best to overdesign the clamp if there is some doubt as to the accumulated forces that will be present.

11.2 TYPES OF CLAMPS

There are several general categories that can be used to group clamps: screw, level, cam, toggle, wedge, or latch.

Screw clamp.   The basic type of fixture clamp is the screw clamp. The simplest of all the screw type clamps is the threaded rod equipped with some method for tightening. Figure 11.1 shows two such screws used as clamps. Additional applications of the screw as a method of applying a force will be seen in the discussions that follow.

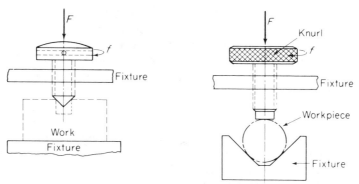

Figure 11.1

Lever clamp.   The lever as a clamp uses the physical principle of mechanical advantage. The three possible combinations of fulcrum, input, and output forces are shown in Fig. 11.2.

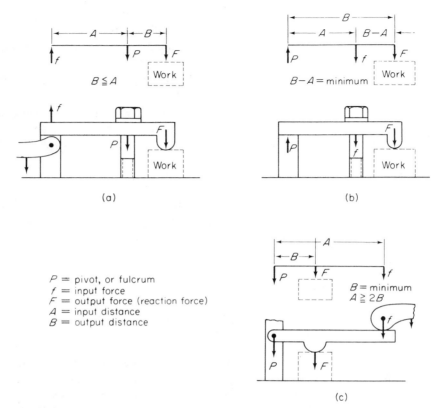

P = pivot, or fulcrum
f = input force
F = output force (reaction force)
A = input distance
B = output distance

Figure 11.2(a,b,c)

In Fig. 11.2a the distance B from the fulcrum to the point at which the clamp acts on the workpiece must be shorter than the distance A. Under these conditions the cam exerts an upward force on the clamp while the pivot bolt P puts a downward force on the clamp. Since f is attempting to rotate its end of the clamp clockwise about P, the other end of the clamp places a clockwise (down) force against the work. If distance B is less than distance A, the greater proportion of the applied force will act against the workpiece.

In Fig. 11.2b, if $B - A$ is less than A, the greater proportion of the input force will be transmitted to the workpiece at F. Therefore the input force f should be applied to the clamp as close to the point at which the output force F is applied as the workpiece permits. In this setup the bolt f is pulling down on the clamp thus applying a downward force on the work. Both act clockwise about the pivot P.

In Fig. 11.2c the force F should be as close to the fulcrum P as the workpiece size will permit, and the distance A should be at least twice the distance B. Note that the input force f is attempting to rotate the clamp about the pivot point P in a clockwise direction. These two actions cause the clamp to place a downward force F on the work.

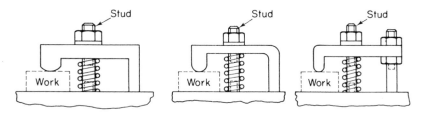

Figure 11.3

Figure 11.3 shows a few of the ways in which lever clamps may be used. The spring holds the clamp up while the work is being removed or put into the fixture.

In many instances the strap is provided with some method for *pushing* it out of the way so that the workpiece can be inserted or removed from the fixture. Figure 11.4 shows some of these types of straps. They are generally slotted and provided with guides so that they may be moved in a straight line.

Another type of lever clamp provides for a method of *swinging* it out of the way. This type of clamp is generally supported when it is being loosened. This may be done with a spring or a block. Several of these lever clamps, called *straps,* are shown in Fig. 11.5.

Figure 11.6 shows the use of a pad or swivel two- or three-point clamp. These

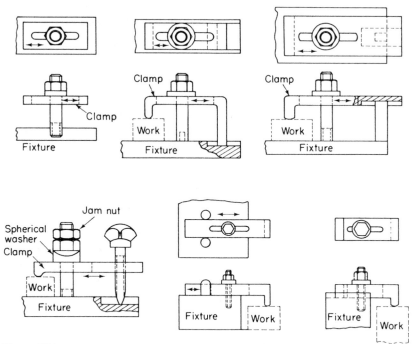

Figure 11.4

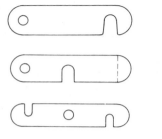

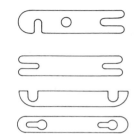

Figure 11.5

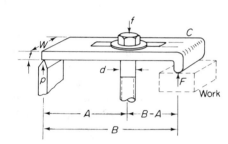

Figure 11.6   **Figure 11.7**

pressure pads or clamps are generally used with lids on fixtures as shown in Fig. 11.6. The same arrangement may be used with a hook clamp from Fig. 11.5.

11.3 STRAP CLAMP CALCULATIONS

Strap clamps are usually made to at least the same width as the washers under the head of the bolt used to tighten the clamp. The slots are made approximately $\frac{1}{16}$ in. wider than the diameter of the bolt. The width, W, of the clamp, Fig. 11.7, may be calculated as

$$W = 2.3d + 0.062$$

The thickness, t, of the clamp for a bolt diameter d is given by the equation

$$t = \sqrt{0.85dA\left(1 - \frac{A}{B}\right)}$$

d = bolt diameter
A = distance, pivot to bolt
B = span, pivot to workpiece
W = width of clamp
t = thickness of clamp

Example 11.1

In Fig. 11.7 the distance A is 3 in. and B is 5 in. Assume the bolt size to be $\frac{3}{4}$–10 NC. Calculate (a) the allowable width of the clamp and (b) the allowable thickness of the clamp.

Solution   (a) The allowable width of the clamp is

$$W = 2.3d + 0.062 = 2.3(0.750) + 0.062 \qquad A = 3 \text{ in.}$$
$$= 1.787 \qquad\qquad\qquad\qquad\qquad B = 5 \text{ in.}$$
$$\qquad\qquad\qquad\qquad\qquad\qquad\qquad\qquad d = 0.750 \text{ in.}$$

(b) The thickness of the clamp should be

$$t = \sqrt{0.85dA\left(1 - \frac{A}{B}\right)} = \sqrt{0.85(0.750)\,(3)\left(1 - \frac{3}{5}\right)}$$

$$= 0.875 = \tfrac{7}{8} \text{ in.}$$

The minimum clamping forces that can be exerted by the various bolt sizes are shown in Table 11.1.

TABLE 11.1   FORCES EXERTED BY BOLTS

Bolt Size, NC	Minimum Clamping Force[a] (lb) T.S. = 6000 psi
$\tfrac{1}{4}$–20	To 160
$\tfrac{5}{16}$–18	160–270
$\tfrac{3}{8}$–16	270–410
$\tfrac{7}{16}$–14	410–560
$\tfrac{1}{2}$–13	560–760
$\tfrac{9}{16}$–12	760–1000
$\tfrac{5}{8}$–11	1000–1210
$\tfrac{3}{4}$–10	1210–1810
$\tfrac{7}{8}$–9	1810–2520
1–8	2520–3300
$1\tfrac{1}{4}$–7	3300–5350

[a]Clamping force = 6000 × area at root diameter.

The load on the bolt is a function of the torque on the bolt and the diameter of the bolt. The equation is

$$T = \frac{df}{5} \qquad \begin{aligned} f &= \text{load on bolt, lb} \\ d &= \text{diameter of bolt, in.} \\ T &= \text{torque on bolt, in-lb} \end{aligned}$$

In Fig. 11.7 the moment on the strap at both contact points where the strap touches the work and the support points are equal. The moment equation can be derived using the vector diagram Fig. 11.8.

$$\Sigma\, F_y = F + P = f$$

$$\Sigma\, \overset{\frown}{M_p} = Af - BF = 0$$

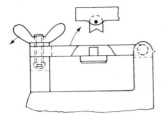

Figure 11.8

The force on the work is

$$F = \frac{Af}{B}$$

Substitute in equation

$$\Sigma F_y = P = f - F = f - \frac{Af}{B} = \frac{Bf - Af}{B} = \frac{f(B - A)}{B}$$

The moment M about point f is force times perpendicular distance to the line of action of that force. Thus

$$M = PA = \frac{f(B - A)}{B} \ (A)$$

Therefore,

$$M = \frac{fA(B - A)}{B}$$

f = load on bolt, lb
A = distance, bolt to support, in.
B = span, in.
M = moment on strap, lb-in.

The stress on the clamp is obtained by dividing the moment on the strap by the section modules. Thus

$$\text{sec. mod.} = \frac{(W - c) \ (t)^2}{6}$$

and

$$S = \frac{M}{\text{sec. mod.}}$$

W = width of strap, in.
c = width of slot, in.
t = thickness of strap, in.
S = stress on clamp, psi

The diameter of the bolt may be calculated from

$$d = 1.35 \sqrt{\frac{f}{S}}$$

S = working stress, psi
d = diameter of bolt, in.
f = axial load, lb

Example 11.2

Assume an 8-in.-long open-end wrench is used to tighten a $\frac{3}{4}$–10 bolt in Fig. 11.7 and that a force of 10 lb is exerted at the end of the wrench. Calculate (a) the necessary width of the clamp, (b) the thickness of the clamp, (c) the load on the bolt, (d) the moment on the strap, (e) the working stress on the clamp, (f) the safety factor if the ultimate stress of the clamping material is 65,000 psi, and (g) the maximum radial load that can be applied to this bolt.

Solution

(a) The width of the clamp should be

$$W = 2.3d + 0.062 = 2.3(0.750) + 0.062$$

$$= 1.787 \sim 1\tfrac{28}{32} \text{ in.}$$

(b) The thickness of the clamp should be

$$t = \sqrt{0.85dA\left(1 - \frac{A}{B}\right)} = \sqrt{0.85(0.750)\,(3.000)\left(1 - \frac{3.000}{5.000}\right)}$$

$$= 0.875 \text{ in.}$$

$l = 8$ in.
$d = \frac{3}{4}$ in.
$A = 3$ in.
$B = 5$ in.
$c = \frac{7}{8} + \frac{1}{16} = \frac{15}{16}$
$F = 10$ lb
$T = 8 \times 10$
$\quad = 80$ in-lb

(c) The load on the bolt from $T = \dfrac{df}{5}$,

$$f = \frac{5T}{d} = \frac{5(8 \times 10)}{0.750}$$

$$= 533 \text{ lb}$$

(d) The moment on the strap is

$$M = \frac{fA(B - A)}{B} = \frac{533 \times 3.000(5 - 3)}{5}$$

$$= 640 \text{ lb}$$

(e) The stress on the clamp is a function of the section modulus of the strap. Thus

$$\text{sec. mod.} = \frac{(W - c)(t)^2}{6} = \frac{(1\tfrac{25}{32} - 0.9375)(0.875)^2}{6}$$

$$= 0.108$$

The stress on the clamp is

$$S = \frac{M}{\text{sec. mod.}} = \frac{640}{0.108}$$

$$= 5926 \text{ psi}$$

(f) The safety factor is

$$\text{safety factor} = \frac{65,000}{5926} = 11$$

(g) The maximum radial force that can be placed on this clamp is

$$d = 1.35 \sqrt{\frac{f}{S}}$$

Solving for f yields

$$f = \frac{Sd^2}{1.35^2} = \frac{5926 \times 0.750^2}{1.35^2} = 1829 \text{ lb}$$

11.4 CAM ACTION CLAMPS

Cam clamps.   These types of clamps make use of the changing radial distance found in eccentrics and in cams.

Eccentrics.   These clamps, shown in Fig. 11.9a, are designed to be rotated by a lever at the end or center of the shaft. The movement of the handle causes motion due to the radius increase. This process allows for high forces to be exerted as a result of the large lever radius actuating a small rise, or increase of radius in the eccentric.

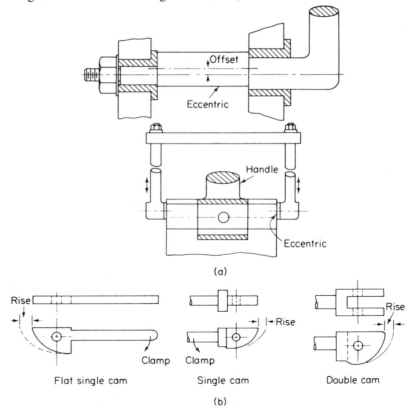

(a)

Flat single cam   Single cam   Double cam

(b)

Figure 11.9(a,b)

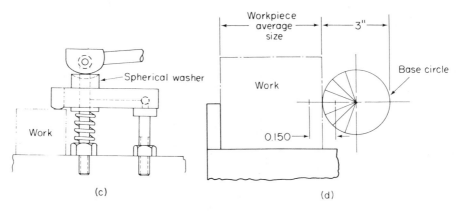

Figure 11.9(c,d)

There are several types of *cams* used with clamps shown in Fig. 11.9b. Figure 11.9c shows one of these cams used with a strap clamp and a spring to hold it against the cam as the workpiece is being inserted or removed from the fixture.

In some instances the tool designer is required to lay out a cam that will lock workpieces in jigs or fixtures. There are several considerations that must be kept in mind. First, the rise of the cam must be steep enough to lock the workpiece firmly. If it is made too steep, it is an inconvenience to the operator. If it is not made steep enough, it may not take care of the variations in workpiece sizes. Also, it must not be designed so that vibrations will work it loose. One procedure for the layout of a cam is to allow 0.001 in. per degree for every inch of radius of the base circle.

Example 11.3

A cam has a base circle of 3 in. and a total rise that takes place through 100°. The middle of the rise is to clamp the average block size. Design the cam.

Solution   The total rise is

$$\frac{3.000}{2} \times 0.001 \times 100 = 0.150 \text{ in.} \qquad D = 3000$$

Half of this rise is to be above the base circle and half below the base circle, as shown in Fig. 11.9d.

The 100° portion of the base circle in which the rise is to take place is divided into an equal number of parts. The rise is divided into the *same* number of parts. This is shown in Fig. 11.9e. In this case the base circle is divided into 10 parts (10° each) through 100° and the rise is divided into 10 parts of 0.015 in. each: five divisions above the base circle and five below. Starting at 0°, consecutive rise and degree divisions are intersected as shown in Fig. 11.9e. These intersections are connected to give the cam surface.

Another method used to lay out a clamping cam is to determine the rise needed through 100°. Since the average clamping action takes place at the midpoint of the 100° in Fig. 11.9f, the radius (r) of the average clamping circle is

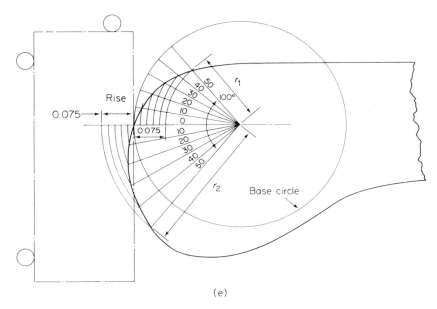

(e)

Figure 11.9(e)

$$r = \frac{r_1 + r_2}{2}$$

r = radius of average base circle
r_1 = radius at beginning of rise
r_2 = radius at end of rise

where r_1 is the radius at the beginning of the rise and r_2 is the radius at the end of the rise. Once the rise and the radius (r) of the average base circle are determined, the procedures of Ex. 11.3 are followed.

Example 11.4

In Ex. 11.3, calculate r_1 and r_2.

Solution   The total rise is

$$\Delta r = r_2 - r_1 = 0.150 \text{ in.} \qquad \Delta r = \text{total rise } 0.150 \text{ in.}$$

Thus the maximum and minimum radii are

$$r \pm \frac{\Delta r}{2}$$

Therefore, the medium radius is

$$r_1 = r - \frac{\Delta r}{2} = 1.500 - \frac{0.150}{2} = 1.425 \text{ in.} \qquad r = 1.500$$

and the maximum radius is

$$r_2 = r + \frac{\Delta r}{2} = 1.500 + \frac{0.150}{2} = 1.575 \text{ in.}$$

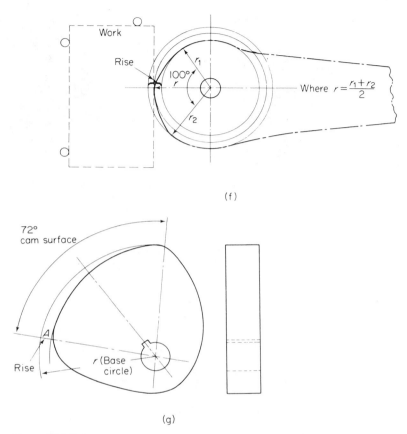

Figure 11.9(f,g)

If it is possible to purchase cams that will serve the designer's purpose, he should buy them. It is rarely less expensive to make cams than it is to buy them commercially. Figure 11.9g and Table 11.2 show such cams and their dimensions.

TABLE 11.2

A	r
$\frac{5}{32}$	$1\frac{5}{32}$
$\frac{13}{64}$	$1\frac{15}{32}$
$\frac{5}{16}$	$2\frac{7}{32}$
$\frac{13}{32}$	$2\frac{27}{32}$
$\frac{7}{16}$	$3\frac{1}{32}$

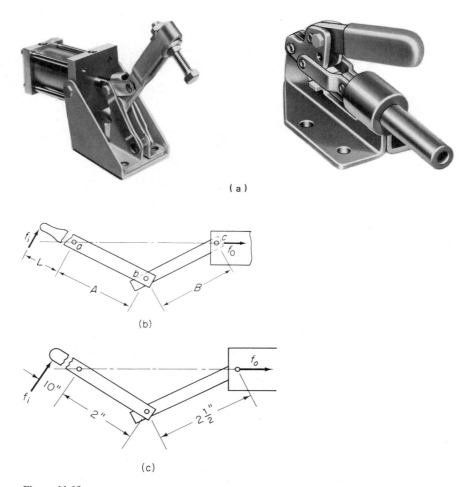

(a)

(b)

(c)

Figure 11.10

11.5 TOGGLE CLAMPS

Still another type of clamping mechanism is the toggle. The toggle clamps shown in Fig. 11.10a may be purchased as standard items. The range of holding forces that may be achieved with these clamps is from 25 to 1200 lb. These toggles may be adapted to many types of fixtures. They can be operated very quickly and can be moved out of the way, making loading and unloading of the fixture an easy matter. Generally, toggles have one fixed and one sliding point.

The pin diameter in Fig. 11.10b is

$$d = 2\sqrt{\frac{\mu f_o}{S}}$$

f_o = force, output
S = allowable shear stress
μ = coefficient of friction
d = diameter of pin

The coefficient of friction for steel pins may be taken as $\mu = 0.22$ for most toggles. The fixed pin is a and the movable pin is c.

Example 11.5

Assume that a material is used to make the pins for the toggle, Fig. 11.10c, to have an allowable shear stress of 5000 psi. If the toggle is to deliver a 600-lb force to the work, calculate the pin diameter.

Solution

$$d = 2\sqrt{\frac{\mu f_o}{S}} = 2\sqrt{\frac{0.22 \times 600}{5000}} \qquad \begin{aligned} \mu &= 0.22 \\ f_o &= 600 \text{ lb} \\ S &= 5000 \text{ lb} \end{aligned}$$

$$= 0.325 \sim \tfrac{5}{16} \text{ in.}$$

If it is required that the toggle deliver f_o pounds of force when it is closed, then

$$f_i = 4\mu d f_o \left(\frac{A + B}{LB}\right)$$

A = toggle distance pin A to B
B = toggle distance pin B to C
L = handle distance to pin A
f_i = input force

Example 11.6

A toggle clamp, Fig. 11.10c, is to deliver a 600-lb force to a workpiece in a fixture. Assume the pin diameter to be $\frac{1}{4}$ in. and the coefficient of friction 0.22. Calculate the force required at the end of the handle.

Solution

$$f = 4\mu d f_o \left(\frac{A + B}{LB}\right) \qquad \begin{aligned} L &= 10 \text{ in.} \\ A &= 1.5 \text{ in.} \\ B &= 2.5 \text{ in.} \end{aligned}$$

$$= 4(0.22)(0.250)(600)\left(\frac{1.5 + 2.5}{10 \times 2.5}\right) \qquad \begin{aligned} \mu &= 0.22 \\ d &= \tfrac{1}{4} \text{ in.} \end{aligned}$$

$$= 21.1 \text{ lb} \qquad\qquad\qquad\qquad\qquad f_o = 600 \text{ lb}$$

11.6 WEDGES AND LATCHES AS CLAMPS

Wedge clamps.   Wedges may be used to clamp workpieces into position in a fixture. They may be driven into fixtures, but are put to better use when pushed into place by a screw or some other mechanism. Thus, in Fig. 11.11a, the wedge is used effectively to force the workpiece up underneath and against the drill plate. In Fig. 11.11b the wedge is used to force the wedge end of the clamp up, which causes the clamping action.

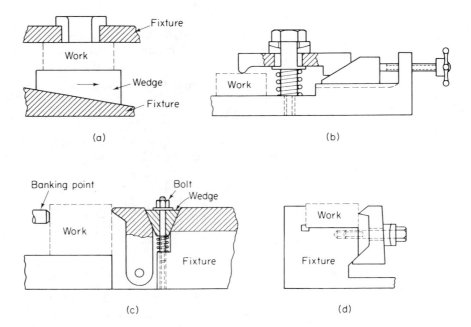

Figure 11.11(a,b,c,d)

In Fig. 11.11c the wedge is forced down with the bolt. The wedge forces the latch against the work. The work is held between the banking point and the latch. In Fig. 11.11d the wedge is used to pull the clamp down until it is tightened. This in turn forces the work down as well as back.

Latch clamps.   These clamping devices are quick acting and easy to operate. In general, this type of locking device is used to rapidly secure a hinged cover plate, which in turn seats the workpiece. The latch may be either a hook type, Fig. 11.12a,

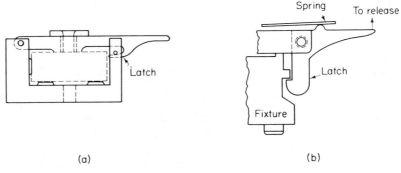

Figure 11.12(a,b)

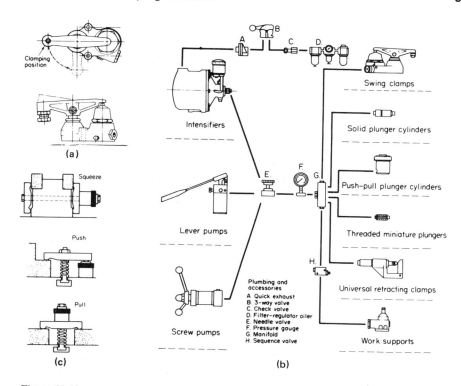

Figure 11.13

which relies on the cam action of the lever to secure the plate, or a self-locking hook, Fig. 11.12b, which snaps closed when the lid is closed.

11.7 OTHER CLAMPING METHODS

The *rack-and-pinion clamp* permits using the motion of the rack to transmit the clamping force. Thus the pinion when rotated, may be used to actuate the rack to clamp or release the workpiece in a fixture. A locking mechanism is usually required since the motion is reversible. Of course, if the rack is used to actuate a cam, a wedge, or a toggle, the locking mechanism is not needed because these mechanisms are themselves self-locking.

Pneumatic and *hydraulic clamping* is usually justified when production requirements are high. The energy expended by the operator is very small and the force exerted is even and uniform. The hydraulically operated swing clamp, Fig. 11.13a, allows the clamp to swing out of position when the pressure is released. Figure 11.13b shows a schematic hookup for several types of clamping operations. Figure 11.13c shows squeeze, pull, and push types of clamping actions.

PROBLEMS

11.1. What role does the clamp play in a fixture relative to the six degrees of freedom?

11.2. List the 12 rules which should be followed when designing clamps for fixtures. Explain each in your own words.

11.3. What is the simplest type of screw clamp? Illustrate its use.

11.4. Explain the force diagram in Fig. 11.2.

11.5. What is the purpose of the spring in Fig. 11.3?

11.6. Sketch at least five methods used to permit moving a clamp out of the way while the work is being inserted or removed from the fixture.

11.7. Draw several swing clamps (straps) and explain or illustrate how they may be used.

11.8. What is a pad? What is a swivel two-point clamp? Illustrate.

11.9. The distance from the pivot to the bolt in Fig. 11.7 is 5 in. The length of the clamp is 8 in. and a $\frac{3}{4}$–10 NC bolt is used to tighten the clamp. Calculate (a) the allowable width of the clamp and (b) the allowable thickness of the clamp.

11.10. Assume the bolt in Fig. 11.7 to be halfway between the pivot point P and the workpiece. If the distance A is 3 in., B is 6 in., and the bolt is $\frac{5}{8}$–11 NC, calculate (a) the allowable width of the clamp and (b) the allowable thickness of the clamp.

11.11. Assume the conditions of Prob. 11.9. What size bolt should be used if the clamp is $\frac{3}{4}$ in. thick?

11.12. Assume that an 8-in.-long wrench is used to tighten the bolt in Fig. 11.14 with a 6-lb force at the end of the wrench. Calculate (a) the width of the clamp, (b) the thickness of the clamp, (c) the load on the bolt, (d) the moment on the strap, (e) the working stress on the clamp, (f) the safety factor if the ultimate stress of the material is 50,000 psi, and (g) the maximum radial load that can be applied to this bolt.

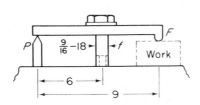

Figure 11.14

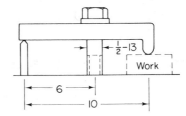

Figure 11.15

11.13. Assume a $5\frac{1}{2}$-in.-long open-end wrench is used to tighten a $\frac{1}{2}$–13 bolt in Fig. 11.15 and that a force of 12 lb is exerted at the end of the wrench. Calculate (a) the width of the clamp, (b) the thickness of the clamp, (c) the load on the bolt, (d) the moment on the strap, (e) the working stress on the clamp, (f) the safety factor if the ultimate stress of this clamp material is 65,000 psi, and (g) the maximum radial load that can be applied to this bolt.

11.14. What is the difference between an eccentric and a cam? Illustrate.

11.15. What are some of the considerations when designing an appropriate rise in a cam?

11.16. A cam is to have a base circle of 4 in. and a total rise through 120°. It is assumed that the middle of the rise is to clamp the average block size. Calculate the maximum and minimum radii and design the cam.

11.17. List several advantages of toggle clamps over other types of clamps.

11.18. Calculate the pin diameter of a toggle that is to deliver 600 lb of force to the work if the allowable shearing stress of the material to be used in the pin is 8000 psi, ($\mu = 0.22$.)

11.19. The ultimate stress of pin material in a toggle is 6000 psi. If the safety factor is 5 and the diameter of the pin is $\frac{1}{4}$ in., what force may be delivered to the work by this toggle? ($\mu = 0.22$.)

11.20. In Fig. 11.10b the following data are taken:

$$
\begin{aligned}
L &= 12 \text{ in.} & d &= \tfrac{3}{8} \text{ in.} \\
A &= 2\tfrac{3}{8} \text{ in.} & f_o &= 800 \text{ lb} \\
B &= 3 \text{ in.} & \mu &= 0.22
\end{aligned}
$$

Calculate the input force at the end of the handle.

11.21. Assume that a material is used to make the pins in a toggle, Fig. 11.16, to have an allowable shear stress of 6000 psi. If the toggle is to deliver a 400-lb force to the work, calculate the pin diameters.

11.22. A toggle clamp, Fig. 11.16 is to deliver a 400-lb force to a workpiece in a fixture. Assume the pin diameters to be $\frac{1}{4}$ in. and the coefficient of friction 0.22. Calculate the force required at the end of the handle.

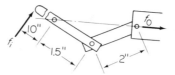

Figure 11.16

11.23. Draw three wedge clamps and show how they act to hold a workpiece.

11.24. Show how latches are used to secure work in fixtures.

11.25. Why is it *not* necessary to use a locking mechanism when a rack and pinion is used with a toggle? Explain.

11.26. List and discuss two advantages achieved over other methods if pneumatic or hydraulic devices are used to actuate clamps.

Chapter 12

Jigs and Fixtures

12.1 DEFINITIONS AND AXIOMS

The first part of this book was concerned with manufacturing costs and the various components that enter into the final determination of the manufacturing method and the holding device that needs to be designed. This chapter deals with the drawing together of these elements into the design of a holding device so that operations may be repeated within the allowable tolerances, finishes, and at a competitive cost. It should be evident that the number of possibilities and variations for designing jigs and fixtures is endless. Yet there are some very basic considerations which should be observed when planning a fixture.

It was pointed out that holding devices may be categorized as jigs or fixtures. A *jig* is a workpiece locating and holding device that positions and guides or controls a cutting tool. A *fixture* is a workpiece locating and holding device used with machine tools, inspection, welding, and assembly. It does not control the position of the tool or instrument that is being used. In many instances gage blocks may be provided for setting the depth or position of the cutting tool.

There are elements that need to be present in order to classify a device as a jig or a fixture. Jigs and fixtures must have a *structure* which maintains the relationship between its various elements. This structure may be a machine table, a plate, a standard machine attachment which has been altered, or a fabricated structure such as a box. Also present must be elements of the jig or fixture that *support* the work; and elements, called locators, that *position* the work on a repetitive basis. Once located and positioned, the work is *clamped* so that it will not move off the supports or locators. In the case of jigs a *tool-guiding* element must also be present.

The designer approaches the design project by first becoming fully aware of all critical dimensions and surfaces. If possible, he should also know how the part fits into the main assembly. Once he is familiar with the special functional requirements and dimensions, he collects the data that may have an effect on his design: data such as type of material, physical properties, weight, and the characteristics of the raw material. Other information that could affect the type of fixture to be built is the available budget, production quantities, and lot sizes.

Next, the designer determines whether fixtures already in the plant can be altered and used for his job. He must also consider the possibility and advisibility of using standard holding devices, such as vises, parallels, and clamps. Once it is decided that a series of operations are to be done with newly designed fixtures, the designer combines as many operations into one fixture as possible. The fixture design plan is then developed and the design carried through. After the fixture is built and becomes operational, changes may be suggested for future fixtures.

There are many rules that should be observed when designing a jig or a fixture.

1. The main frame of a jig or a fixture must be strong enough so it will deflect a minimum amount as a result of the forces of cutting, workpiece clamping, or clamping to the machine table. It must also have the mass necessary to prevent chatter. Some fixtures, such as milling fixtures, may need to be reinforced to take the end thrust of cutting.

2. Frames may be built up from simple sections which may then be either fastened with screws or welded. In some instances a combination of both welding and screws are used. Those parts of the frame that will remain for the life of the jig or fixture may be welded. Those that may need changing may be held with screws. Where the body of the jig or fixture has an intricate shape, it may be cast from a good grade of cast iron.

3. Provisions should be made so that workpieces can be located in the fixture only in the correct position. Some kind of obstruction should be built into the fixture which does not permit inserting a workpiece in the wrong position. The converse is also true. Obstruction, which may prevent workpieces from being properly placed into the fixture, should be removed.

4. Fixtures should be designed so that workpieces may be easily inserted or removed.

5. Clamping should be rapid and require the least amount of effort on the operator's part.

6. Clamps should be arranged so that they are readily available to the operator and may be easily positioned or removed. This should be possible without removing the clamp, but with the least amount of movement.

7. Wherever possible, clamps should be supported with springs so that the clamp is held against the bolt head.

8. If the clamp is to swing off the work, it should be permitted to swing only as far as is necessary to allow for the removal of the workpiece. An obstruction, such as a bumper pin or block, should be built into the fixture to control this movement.

9. All clamps, supports, and locators should be clearly visible to the operator and easily accessible for cleaning, positioning, or tightening.

10. All clamps and support points that need to be adjusted with a wrench should be of the same size. If it is at all possible, all clamps and adjustable support points should be capable of being operated from the front of the fixture.

11. Loose parts, such as shims, small wrenches, and bushings, should be fastened to the fixture with a chain; or there should be some method provided which will ensure that they will not be lost or misplaced.

12. Workpieces must be stable when they are placed into a fixture. The most favorable condition is to have three fixed support points. When positioning rough workpieces, such as sand castings, the maximum number of *fixed* points is three. In some instances, it is advisable to provide a means for adjusting these fixed support points to take care of wear or change in design. This does not refer to adjustments that need to be made for each workpiece, but rather, the adjustments that might have to be made between job lots.

13. If the workpiece is smooth and flat, more than three points may be used. In general, all support points beyond the original three fixed support points should be capable of being adjusted. In some instances this adjustment should be sensitive to "feel." In other instances, they should be self-registering.

14. Support points should be set as far apart as is consistent with the nature of the work. The purpose is to provide maximum stability.

15. To ensure stability the three fixed support points should circumscribe the center of gravity of the workpiece.

16. The contact surface area of the support points should be as small as possible without causing damage to the workpiece as a result of clamping or work forces.

17. Clamping should take place above the fixed support points or inside the three-support-point triangle. Special care should be taken that the clamping forces do not lift the workpiece off the support points or away from the banking points, or that they do not distort the workpiece. All side clamps should press down.

18. Support points, and other parts, should be designed so that they may be easily replaced if they break or wear.

19. Support points should provide for easy chip removal or storage during the time that the workpiece is in the fixture. This can be done by making the support points longer.

20. In the case of jigs, provision must be made for easy chip escape. This must be possible so that the storage of chips does not interfere with the operation, and that their removal during the operation does not interfere with the cutting process.

21. Tolerances of the workpiece should be as liberal as interchangeability will permit. The closer the tolerances of the workpiece, the greater the cost of the fixture and of the manufacturing of the workpiece. Simple fixtures are always preferred to complicated fixtures.

Figure 12.1

22. All provisions of safety to the operator should be observed when designing a fixture, even if that means less output.

Figure 12.1 shows a cast fixture that demonstrates many of the rules stated above.

12.2 LATHE FIXTURE

Although round workpieces are generally machined in a lathe, the discussion of the various methods for clamping round workpieces may be applied to attachments used with other machine tools. When using a lathe the workpiece must be held and also driven.

Figure 12.2a shows a round piece of work mounted in a three-jaw universal chuck. The workpiece is self-centering. Figure 12.2b shows a workpiece mounted in a four-jaw independent chuck. These are standard attachments for a lathe. Collets are also available as standard lathe attachments. It should be noted that round plugs, and star disks made from flat steel may be used as banking surfaces for the workpieces so that depth of shoulders may be maintained. Plugs may be fitted into the spindle of the lathe as a stop for collets, as shown in Fig. 12.2c.

These chucks may also be used to grip internal diameter. The jaws are provided with contour gripping surfaces to hold internal diameters. A chuck may also be used in conjunction with the steady rest. This is necessary when work is to be done on the end of the workpiece.

Probably the most reliable holding devices on a lathe are the centers. The headstock center is powered but usually is not relied upon to do the driving. The tailstock center may be stationary, hardened, and ground, it may be carbide tipped, or it may be a ball-bearing center. The 60° center in a ball-bearing center is interchangeable with other forms or included angles. Figure 12.2d shows several ways of using them.

Figure 12.2e shows the use of a chuck and a live center, and Fig. 12.2f shows the use of a center, dog, driving plate, and steady rest. Rawhide is used to tie the workpiece to the center.

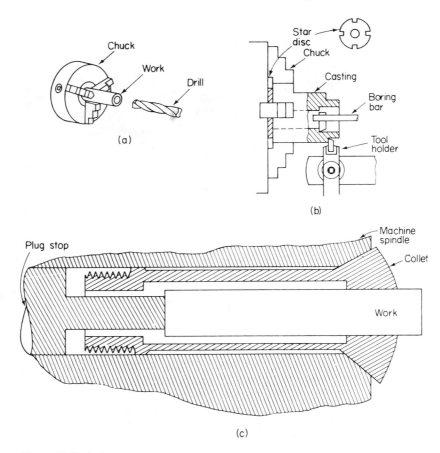

Figure 12.2(a,b,c)

Plain mandrels, Fig. 12.2g, and the expanding type, Fig. 12.2h, are standard methods for mounting round objects that have holes. Variations of these are tapers that expand a shell so that they grip the work internally or externally, as shown in Fig. 12.2i. Another type of fixture is the arbor and nut, shown in Fig. 12.2j.

Face plates are larger versions of drive plates. They have additional slots in their faces so that attachments or fixtures may be mounted. Figure 12.2k shows four jaws mounted on the surface of a face plate which converts it into a large four-jaw chuck. Figure 12.2l shows a disk mounted on a face plate and being bored. Figure 12.2m shows an angle plate mounted on a face plate. The work is ready for boring.

12.3 DRILL JIG

Simple drill jigs for round stock may be made from standard V-blocks. Figure 12.3a shows a V-block adapted for drilling a hole in round a rod.

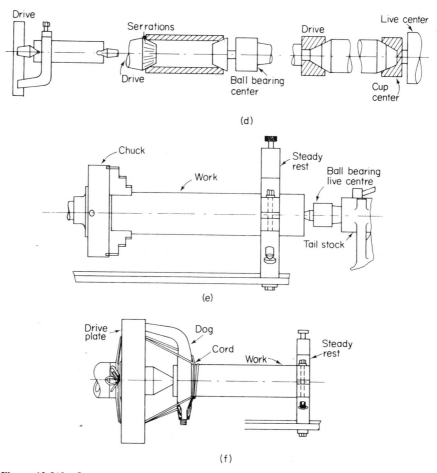

(d)

(e)

(f)

Figure 12.2(d,e,f)

Figure 12.3b shows a standard vise adapted for drilling two holes into a flat plate. The supports are screwed to the solid and movable jaws. They are recessed to receive the workpiece. The drill plate is pinned after it has been located so that it may be removed and replaced without disturbing the location of the bushings. It should be noted that a horizontal V-block could have been fastened to the solid jaw. The vise would then have been converted for drilling round stock. Horizontal V-blocks introduce a lateral location error about the centerline of the workpiece. This design may be used only if the tolerances permit.

Sometimes slow-thermosetting plastic materials are poured around an irregularly shaped object and allowed to set. After setting, the jaws of the vise will have the matching shape of the workpiece.

Another approach is to make two matching halves of a container. The irreg-

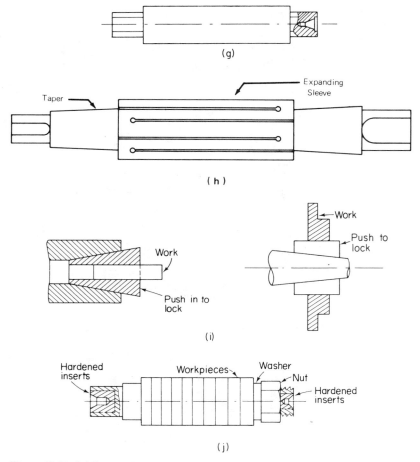

Figure 12.2(g,h,i,j)

ularly shaped object is centered in the container and a low-melting-point alloy is poured into the container. After setting, the faces of the container are machined to produce clearance so that the workpiece may be clamped, see Fig. 12.3c. The two halves are then mounted in a vise. The vise is then equipped with components for drilling, milling, and so on. Fig. 12.3d shows a type of "pump" lock jig which may be fitted with bushings and used for drilling. Figure 12.3e shows a solid block that has been bored to accept a round workpiece for drilling two holes at right angles to each other.

At times box fixtures may be built up and welded into a box as shown in Fig. 12.3f. The rails are welded to the base, the pads to the side and back. The box is welded, normalized, and machined.

A more expensive but more accurate method is to fasten the sides and elements with screws and dowels. This avoids warping and provides accuracies not possible

(k)

(l)

Weight

Face plate

Angle plate

Work

(m)

Figure 12.2(k,l,m)

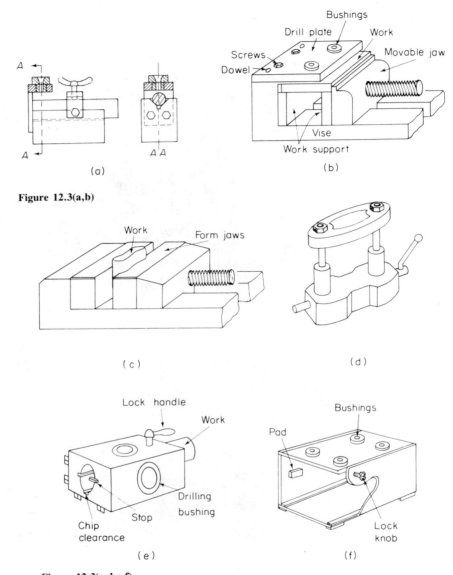

Figure 12.3(a,b)

Figure 12.3(c,d,e,f)

with welds. Furthermore, if it should become necessary to replace the rails or pads, it can be done rather easily. Still another refinement is to hinge the top or side plate. This makes it easier to insert and remove workpieces.

12.4 TYPES OF MILLING OPERATIONS

Milling fixutres may be classified according to some characteristic method of locating and moving the workpiece relative to the cutter. These methods may be classified as single-piece, string, abreast, multiple, reciprocal, box, progressive, transfer base, index, rotary, and contour milling. These are *methods* of milling. The method to be used when production milling is used must be determined before the fixture design plan can be formulated. These methods may be used singly or in a combination.

Single-piece milling is the simplest of all milling operations. The operation consists of locating the workpiece in a fixture or on the milling table, rapidly advancing the table to locate the workpiece very near the cutter, setting the feed to start, completing the cut, rapidly returning the table to its starting position, and removing the machined workpiece. The cycles are shown in Fig. 12.4a.

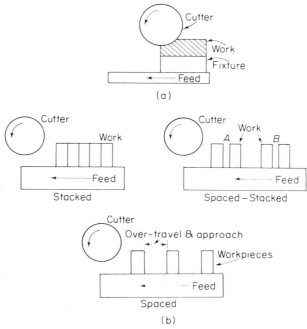

Figure 12.4(a,b)

String milling is that operation in which two or more workpieces are placed in a row in the direction of the feed, Fig. 12.4b. The workpieces may be stacked as on a mandrel. The entire cycle is run before unloading and reloading takes place. The

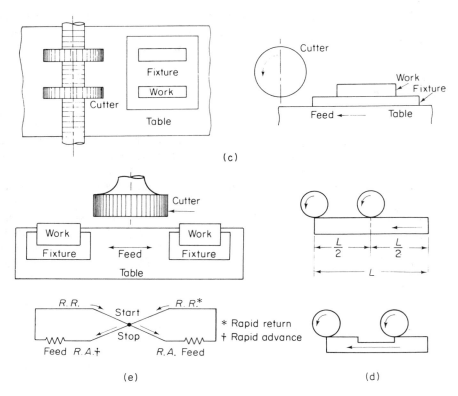

Figure 12.4(c,d,e)

string milling operation may be spaced so that the first is finished before the second starts. Several rows, side by side, may be machined in this manner. Also shown in Fig. 12.4b is a combination of grouping and spacing of workpieces. This permits loading one set while the other set is being machined. A is machined, B is machined, and A is loaded. At the completion of B, the table is dropped, returns to the starting position, and cut A is started. The cycle is repeated.

   Abreast milling, Fig. 12.4c, requires that the pieces be stacked side-by-side, or aligned in rows perpendicular to the direction of movement of the table. This saves cycle time.

   Cycle time can also be saved by using the principle of *multiple milling.* If two cutters are used longitudinally and set a distance of one-half the length of the cut, the table travel will be reduced by one-half, as shown in Fig. 12.4d.

   Reciprocal milling, Fig. 12.4e, carries on the milling operation in both longitudinal directions. A right- and a left-hand fixture may be required. An antibacklash arrangement may be necessary because climb milling may take place at the left end of the table. During face milling this is not present. This differs from string milling in that the cutter in string milling is always milling from left to right, whereas in reciprocal milling, the cutter is milling from right to left and then from left to right.

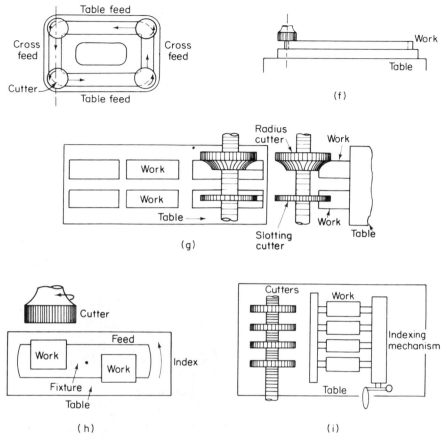

Figure 12.4(f,g,h,i)

Box milling, Fig. 12.4f, is an operation that coordinates the longitudinal and cross slides into one cycle of the milling operation. Shown is the continuous milling cycle around the outer rim of a casting.

Progressive milling is a method whereby several similar and different operations are done simultaneously or successively. Several of the same workpieces may be arranged abreast and in line, each having a different operation performed on it. Figure 12.4g shows the workpiece setup to cut a radius in one line, and a saw cut in a second line on the same fixture.

Transfer base milling, Fig. 12.4h, requires that two fixtures be mounted on an index table or base so that while the workpiece in one fixture is being cut, the other is being loaded. While the cutter is repositioning, the base is being indexed ready to cut the next piece. This process is similar to reciprocal milling except that the indexing time needs to be added to the total milling time.

Index milling, Fig. 12.4i, is the case where several pieces are having the same

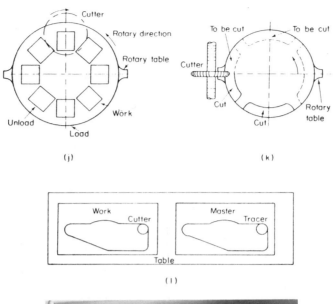

(j)

(k)

(l)

(m)

Figure 12.4(j,k,l,m)

cut made on each piece at the same time. The pieces then index manually or automatically and the operation is repeated on all pieces.

Rotary milling, Fig. 12.4j, is the milling operation where workpieces are mounted directly on a rotary table, in a fixture that is mounted on a rotary table, or the case where the rotary mechanism is built into the fixture. The rotation of the table may be continuous so that it may be loaded and unloaded while the cutting operation is taking place.

Sweep milling, Fig. 12.4k, is a type of rotary milling where only a portion of the workpiece is being rotary milled. The metal is being swept out for a portion of the cycle.

Contour milling, Fig. 12.4l is that operation where the cutter is controlled by a master or template and a tracer.

Many of these operations may be combined for purposes of production. A combination of progressive, abreast, and string milling is shown in Fig. 12.4m.

12.5 THE MILLING FIXTURE

Figure 12.5a shows a series of step blocks, parallel blocks, and various types of clamps which when used in conjunction with parallel bars and bolts produce a wide range of fixture possibilities. These are generally used for short-run setups, or for positioning and holding large workpieces where it would be impractical or very expensive to build a fixture.

The milling vise shown in Fig. 12.5b is fitted with special jaws to hold the workpiece and a setting gage for setting the depth of the cutter. This arrangement has incorporated in its construction all the elements necessary for good fixture design. Attachments such as indexing heads and rotary tables may also be adapted to provide a method of indexing workpieces.

Figure 12.1 shows a fixture used to mill cylinder heads. This has a cast body, clamps, a gage block, locator pins, and rest blocks. This fixture permits three cutters to operate simultaneously. The clamps are supported and can be moved out of the way so that the workpiece can easily be removed. The two locator pins fit into two predrilled and reamed holes, and position the workpiece so that the location of these machined surfaces to other dimensional requirements may be maintained. A gaging or setting block is provided for locating the depth of one cutter. The others are set at the appropriate spacing during the setup. The bottom of the fixture is keyed to the slot in the machine table, the pins locate the workpiece, and a shim, shown in the drawing, Fig. 12.5c, sets the lateral position of the cutters.

Figure 12.5d shows a shell mill and fixture used to mill a cast iron part. The fixture, clamps, and banking points have been designed so that they do not interfere with the upper surfaces which are to be machined.

Figure 12.5e is another fixture that exposes the sides and upper surfaces to be machined by a cluster of inserted cutters.

12.6. GRINDING OF FIXTURES

Grinding fixtures may be those designed for cylindrical or surface grinding machines. Fixtures to be used with cylindrical grinders are essentially the same as those used on lathes. Those used with surface grinders serve essentially the same function as those used with milling machines. The one major difference is that grinding is generally a finishing operation to higher accuracies than would be required from the lathe or miller. The main purposes for grinding a part are that high degrees of accuracy and high finish may be achieved by grinding. Therefore, the grinding fixtures must not

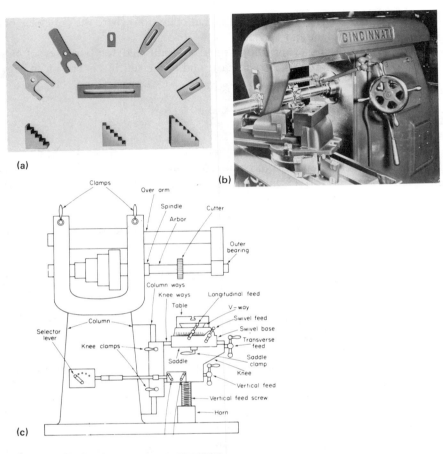

(a)

(b)

(c)

Clamps
Over arm
Spindle
Cutter
Arbor
Outer bearing
Column ways
Knee ways
Longitudinal feed
Table
V-way
Swivel feed
Swivel base
Transverse feed
Column
Selector lever
Knee clamps
Saddle
Saddle clamp
Knee
Vertical feed
Vertical feed screw
Horn

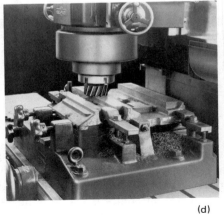

(d)

(e)

Figure 12.5

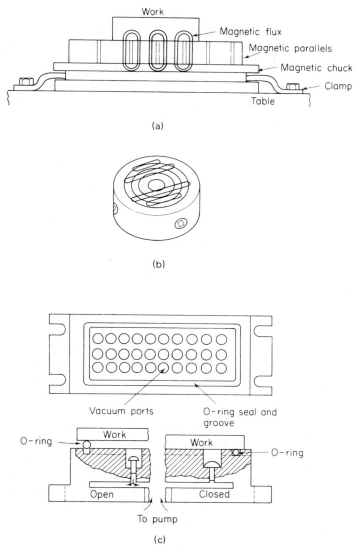

Figure 12.6

only locate the workpiece to greater accuracy as far as location is concerned compared to a milling fixture, but the fixture must also have greater rigidity.

Magnetic chucks are generally used to hold the work directly or to hold a fixture that holds the work. Vises may be held by the magnetic chuck. The vise then secures the workpiece. Sometimes a laminated structure of ferrous and nonferrous materials is used to separate the magnetic flux so that it will hold the work in position as shown in Fig. 12.6a.

If the workpieces are nonmagnetic, a vise, fixture, or epoxy tool material may be used to hold the work in position. It should be noted that the magnetic chuck on a surface grinder may be removed and a fixture may be clamped in its place. Cylindrical grinders may also be equipped with rotary magnetic chucks, Fig. 12.6b.

In some cases vacuum chucks may be used to hold work. The suction created between the workpiece and the rubber O-ring draws the workpiece down to the solid chuck surface. A vacuum pump that delivers about 12 psiv is used to drop the pressure in the chamfer between the chuck and the workpiece. Atmospheric pressure forces the workpiece down (see Fig. 12.6c).

PROBLEMS

12.1. Define (**a**) a jig and (**b**) a fixture. (**c**) What is the purpose of a gage block when used with a fixture? Illustrate with an example.

12.2. Define and illustrate the purpose of a structure when it is referred to as a jig or fixture.

12.3. Explain and illustrate the purposes of supports and locators in fixtures.

12.4. Discuss the five elements of a jig. Select an object in your room and make a sketch of a jig including all five elements.

12.5. A designer is given a print of a workpiece. (**a**) What must he do before he can begin designing a fixture? (**b**) What information and data must be collected? (**c**) Once it is decided which operations are to be included, what does the designer do? (**d**) Does the designer have any responsibilities for his design once the fixture is operational? What are they?

12.6. Discuss the design and fabrication of a jig or fixture frame.

12.7. In some instances the design of obstructions in fixtures serves a useful purpose; in others, they are detrimental. Explain.

12.8. There are six rules in fixture design which apply to clamps and their operations. State each and make a sketch showing what is meant or explain what is meant.

12.9. How are loose pieces, needed with a fixture, secured so that they are not mislaid or lost?

12.10. There are seven rules that refer to support points. State and explain them.

12.11. "Tolerances of the workpiece should be as liberal as interchangeability will permit." How does this sentence affect fixture design?

12.12. Design a stop to be used with a three- or a four-jaw chuck so that shoulder depths in successive pieces of work may be maintained within the allotted tolerances.

12.13. How may a ball-bearing tailstock center be altered to support a large piece of tubing in a lathe? Illustrate.

12.14. Sketch the use of an expanding mandrel to grip (**a**) a gear blank that has an accurately reamed hole or (**b**) an accurately ground shaft.

12.15. (**a**) How does an arbor differ from a mandrel? (**b**) Design a nutted arbor which is to be used to cut teeth in a gear blank.

12.16. A pillar block is to have a bearing hole bored, Fig. 12.7, on a lathe using an angle plate. Sketch the fixture.

12.17. Assume in Fig. 12.3a that the V-block had been turned 90°. What error would have been introduced in the design?

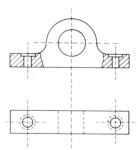

Figure 12.7

12.18. Why is the drill plate doweled into position when used as shown in Fig. 12.3b?

12.19. Discuss the use of epoxy or low-melting-point alloys in the design of clamping jaws which are to clamp irregularly shaped objects.

12.20. What are the advantages of fastening the components of a fixture with screws rather than with welds?

12.21. (a) Describe string milling. (b) Discuss the purpose of stacked and spaced workpieces in string milling.

12.22. Describe abreast and multiple milling. How does each save cycle time?

12.23. Describe reciprocal milling. How does it differ from string milling?

12.24. Describe box milling.

12.25. Describe progressive milling.

12.26. (a) Describe transfer base and index milling. (b) How do they differ? (c) How do these milling operations differ from rotary milling?

12.27. Describe rotary and sweep milling. How do they differ?

12.28. What is contour milling? Describe the process.

12.29. Check some of the design magazines in your library for a fixture and describe the various components of that fixture.

12.30. (a) What are the essential similarities between milling and grinding fixtures? (b) What are the differences?

12.31. (a) Explain the various ways in which magnetic chucks may be used on surface grinders and (b) on cylindrical grinders. (c) What standard parts may be used as fixtures with a magnetic chuck?

12.32. Describe the operation of a vacuum chuck for holding work.

Chapter 13

Design of Single-Point Tools

13.1 PURPOSE OF CUTTING TOOLS

The first of the single-point cutting tools to be used on a machine was developed by Wilkinson when he developed his boring machine so that he could machine Watt's steam engine in 1775. In 1800, Mardslay developed the screw-cutting lathe. Soon after that the planer, milling, and drilling machines came into existence.

The purpose of the cutting tool is to remove metal under controlled conditions. Therefore, the tool must be harder than the material that it is to cut. The second requirement is that the workpiece offer resistance to cutting. It is this resistance to the applied cutting forces that makes possible the cutting action of the tool. Under these conditions the metal on the workpiece compresses, flows, and separates from the base material. The efficiency with which a tool will remove metal from the base material is a function of the relative hardness between the work and tool materials, the sharpness of the tool, the tool geometry, the forces applied, and so on.

Single-point tools are simply made so that they have one edge cutting, as in a lathe or shaper. In some instances many single-point cutting tools may be mounted in a boring bar. In other cases these tools may be mounted in a disk, as in a fly cutter. Of course, the next step is the milling cutter, with its multiple teeth and disk manufactured from one piece. A drill bit is actually a two-point cutter. It quickly becomes evident that the design of all cutting tools will draw heavily on the theory developed for single-point cutting.

340

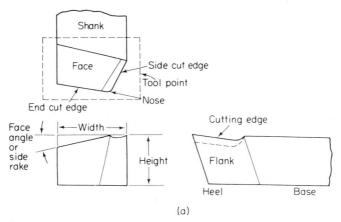

Figure 13.1(a)

13.2 TOOL NOMENCLATURE

The *elements* of a tool are shown in Fig. 13.1a and are defined as follows:

The *size* of the tool is given by stating the width (W), height (H), and overall length (L) of the tool in inches. Thus $\frac{3}{4} W \times \frac{1}{2} H \times 6 L$ indicates a $\frac{3}{4}$-in. width, $\frac{1}{2}$ in. height, and 6-in. length.

The *shank* is that part of the tool which is held in a holder. It is the main body of a tool.

The *base* is the bottom part of the shank. It takes the tangential forces of cutting.

The *heel* is that part of the front of the tool blank adjacent to the intersection of the base and the flank.

The *face* is the surface of the tool against which the chip impinges as it is separated from the work. Sometimes a short land is ground at the cutting edge of the face. The face angle is always greater than the land angle. Sometimes the land angle is negative ($-4°$) and the face angle is $15°$. In this case the angle would be stated $-4(15)$.

The tool *point* is all that part of the tool which is shaped to produce the cutting edges and the face.

The *cutting edge* is the edge of the face that separates the chip from the work. It consists of the side cutting edge, the nose, and the end cutting edge.

The *nose* is the corner, arc, or chamfer which joins the side-cutting and end-cutting edge.

The *flank* of the tool is the surface below the cutting edge.

The *neck* is a small cross section of the shank behind the point. A boring tool has a reduced cross section behind the cutting point.

The *tool angles,* Fig. 13.1b, have been standardized by the American National Standards Institute (ANSI). The *signature* for single-point tools is listed in the order

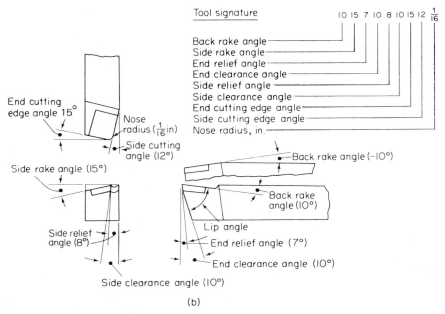

Figure 13.1(b)

rake angles, relief angles, cutting edge angles, and *nose radius.* If the *clearance angles* are to be included, they follow each of the relief angles, as shown in Fig. 13.1b. Generally, they are not included in the signature. The standard signature when applied to Fig. 13.1b would be 10, 15, 7, 8, 15, 12, $\frac{1}{16}$.

The tool angles are defined as follows (Fig. 13.1b):

The *back rake angle* is the angle made by the face and the base of the tool in a plane parallel to the longitudinal centerline of the shank. If it slopes down as shown in Fig. 13.1b and tends to reduce the lip angle, it is said to be a positive rake. If it slopes up, tending to increase the lip angle, it is said to be a negative rake.

The *side rake angle* is the angle between the face of the tool and the base in a plane perpendicular to the centerline of the shank.

The *end relief angle* is the angle between a plane perpendicular to the base and the end flank. It prevents rubbing between the work and the end of the flank.

The *end clearance angle* is the angle between the plane perpendicular to the base of the tool and the end of the flank. It is a secondary angle just below and larger than the end relief.

The *side relief angle* is the angle made by the flank of the tool and a plane perpendicular to the base just under the side cutting edge.

The *side clearance angle* is the angle between the flank of the tool and a plane perpendicular to the base just under the side cutting edge. It is a secondary angle just below and larger than the side relief angle.

The *end cutting edge angle* is the angle between the face of the tool and a plane perpendicular to the side of the shank.

The *side cutting edge angle* is the angle between the side of the shank of the tool and the line at the intersection of the flank and the face of the tool.

The *nose radius* is the radius that connects the end with the side cutting edge. In some instances the end and side cutting edges are connected by a chamfer.

13.3 IDENTIFICATION NUMBERS

The ANSI standards notation may have up to 12 positions. The selection of the tool insert is dependent on the machine used, the characteristics of the material to be machined, the tolerance and finish required, the cutting feed, the speed and depth of cut, the design of the holder, and the type of operation to be performed. When describing a tool insert, the notations shown in Table 13.1 should be used in the signature that will describe that insert.

TABLE 13.1

First position: shape

A	Parallelogram 85°		M	Diamond 86° (rhombic)	
B	Parallelogram 82°		O	Octagon	
C	Diamond	80° (rhombic)	P	Pentagon	
D	Diamond	55° (rhombic)	R	Round	
E	Diamond	75° (rhombic)	S	Square	
H	Hexagon		T	Triangle (rhombic)	
K	Parallelogram 55°		V	Diamond 35°	
L	Rectangular		W	Trigon 80°	

Second position: clearance

N	0°	G	30°
A	3°	H	0–11°*
B	5°	J	0–14°
C	7°	K	0–17°
P	11°	L	0–20°
D	15°	M	11–14°
E	20°	R	11–17°
F	25°	S	11–20°

*Primary and secondary

Third position: tolerance (in.)

	Cutting point	Thickness		Cutting point	Thickness
A	0.0002	0.001	H	0.0005	0.001
B	0.0002	0.005	J	0.0002	0.001
C	0.005	0.001	K	0.0005	0.001
D	0.005	0.005	L	0.001	0.001
E	0.001	0.001	M	0.002	0.005
F	0.0002	0.001	R	0.005	0.005
G	0.001	0.005	S	0.002	0.001

(continued)

TABLE 13.1 (*Continued*)

Fourth position: type

A	Hole	J	Same as H, chip groove both sides
B	Hole, one countersink	K	Smaller than a $\frac{1}{4}$ IC with hole and chip groove both sides
C	Hole, two countersinks		
D	Smaller than $\frac{1}{4}$-in. inscribed circle (IC) with hole	L	Same as K, except no hole
E	Smaller than $\frac{1}{4}$-in. IC	M	With hole and chip grooves on one top rake surface
F	Chip groove, both sides, no hole, 0 top rake, no land	P	With hole and 10 positive chip breaker, both sides
G	Same as F, with hole	R	No hole, chip groove one surface
H	Hole, one countersink and chip groove on one top	S	With hole, 20 chip breaker one side

Fifth position: size of inscribed circle

Regular polygon or diamond
  IC is equal to or greater than $\frac{1}{4}$ in., the number of $\frac{1}{8}$ in. in IC
  IC is less than $\frac{1}{4}$ in., the number of $\frac{1}{32}$ in. in IC
Rectangle and parallelogram (use two digits)
  1st digit: number of $\frac{1}{8}$ in. in width
  2nd digit: number of $\frac{1}{4}$ in. in length

Sixth position: thickness

Regular polygon or diamond
  IC of $\frac{1}{4}$ in. and over, then the number of $\frac{1}{16}$ in. in thickness
  IC less than $\frac{1}{4}$ in., then the number of $\frac{1}{32}$ in. in thickness
Rectangle and parallelogram
  Use width dimension instead of IC

Seventh position: cutting point configuration

0	sharp corners	A	Square insert 45° chamfer
1	$\frac{1}{64}$ radius	D	Square insert 30° chamfer
2	$\frac{1}{32}$ radius	E	Square insert 15° chamfer
3	$\frac{3}{64}$ radius	F	Square insert 3° chamfer
4	$\frac{1}{16}$ radius	K	Square insert 30° double chamfer
6	$\frac{3}{32}$ radius	L	Square insert 15° double chamfer
8	$\frac{1}{8}$ radius	M	Square insert 3° double chamfer
		N	Truncated triangle
		P	Flatted corner triangle

Eighth position: special cutting point configuration

Only used following letter in seventh position. Number of $\frac{1}{16}$ in. in primary faceted face.

Ninth position: hand

R Right   L left

Tenth position: edge and surface preparation

A	Honed, 0.0005–0.003 in.	C	Honed, 0.005–0.007 in.
B	Honed, 0.003–0.005 in.	J	Polished 4 in., rake surface only

TABLE 13.1 (*Continued*)

Eleventh position: special cutting edge conditions

S indicates the insert is tipped with a composite material (boron nitride, diamond, etc.).

Twelfth position: special cutting edge effective length

Used when following the letter S. Indicates the number of $\frac{1}{16}$ in. in length of the composite material along the tip of the cutting edge.

Source: SME, *Tool and Manufacturing Engineers Handbook,* Vol. 1, *Machining* (Dearborn, Mich., Society of Manufacturing Engineers, 1979), pp. 8–36, 8–37); ANSI Standard B94.4.

Inserts have also been standardized and identification numbers assigned to them so that their characteristics may be easily determined. Figure 13.2 shows several inserts that may be secured in holders, see Fig. 13.6h. The identification numbers consist of numbers and letters. The position of the number, or letter, in the identification number is important. Thus

1	2	3	4	5	6	7
R	H	L	M	5	4	2
shape	clearance	tolerance	type	inscribed circle in $(\frac{1}{32}, \frac{1}{8}, \frac{1}{4})$	thickness $(\frac{1}{32}, \frac{1}{8}, \frac{1}{4})$	radius $(\frac{2}{32} = \frac{1}{16})$

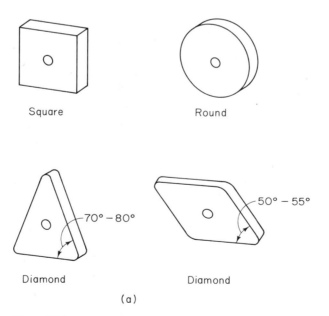

Square   Round

Diamond   Diamond

(a)

Figure 13.2

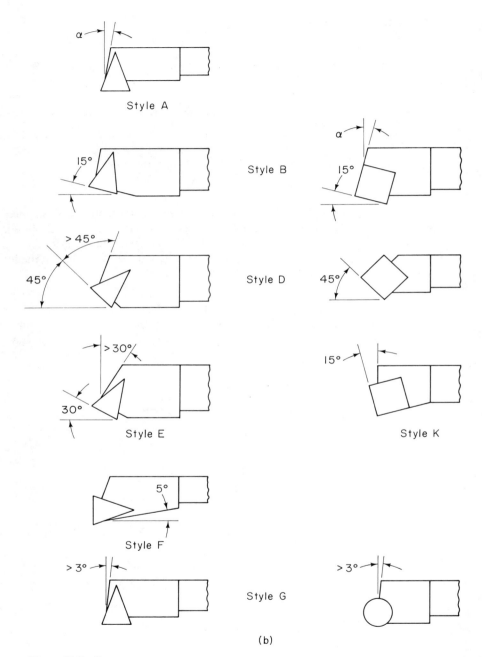

Figure 13.2 Cont.

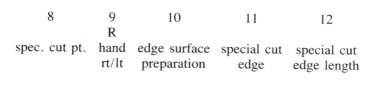

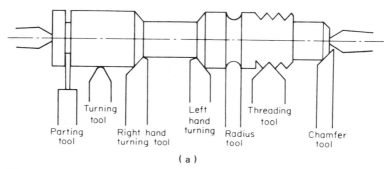

(a)

Figure 13.3(a)

13.4 SELECTION OF ANGLES

Figure 13.3a shows the various types of tool bit grinds. It is to be noted that a right-hand turning tool operates from the right to the left. It is also important that the edges of the tool involved in the cutting operation be the only edges permitted to touch the work. All other faces should be ground to recede so that they will not interfere with the cutting action. The degree to which they are ground away is a compromise between free cutting and strength of the point of the tool. The same may be said for the face of the tool which creates the rake.

As a general rule increasing the rake angles will reduce the cutting forces. It also results in better surface finishes and increases the tool life. Control of the direction in which chips are caused to flow, called the "true rake," results from a control of the rake angles. Figure 13.3b shows the three types of rake angles: positive, negative, and zero rake. It should be noted that excessive positive rake angle will weaken the cutting end of the tool. Sometimes a slight land, about equal to the feed, is added to a high positive rake angle.

If conditions are right, the sliding of chips over the face of the tool will cause cratering in the tool face back of the cutting edge. This action has the effect of creating a "natural" rake angle.

Positive rakes are used when cutting low tensile strength and nonferrous materials, long small-diameter shafts, or materials that work-harden while being machined. Negative rakes are used for machining high-tensile-strength materials, heavy feeds, and interrupted cuts. It is the rare case that a negative rake will be used with high-speed or cast tool bits. They are generally reserved for carbide tools cutting materials such as gray, malleable, or cast iron, cast steels, hot-work die steels, tool steels, or plain carbon steels.

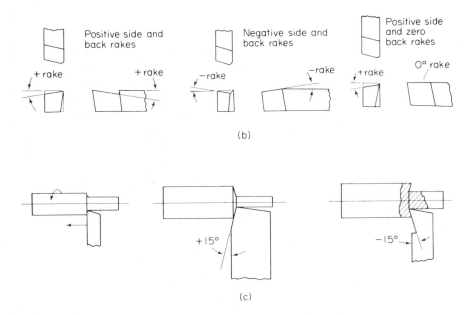

(b)

(c)

Figure 13.3(b,c)

When relief angles are increased, they reduce the cutting forces at the cutting edge for low-tensile-strength materials. For high-tensile-strength materials relief angles of more than 15° will weaken the cutting edge. If the relief angles are too small a vertical flat land may develop rapidly just under the cutting edge. The effect of such a land is to destroy all relief and clearance with the quick destruction of the tool. It should be remembered that the relief angles are provided to ensure that the cutting edge will be free to cut as the relative movement between the work and the tool takes place. If greater relief of the cutting edge is needed, a secondary relief, called clearance, should be provided. This is generally required when the heel of the tool is likely to rub on the work. Such a condition could occur when boring a hole.

The side cutting edge angle is generally ground to an angle of 15° where there are no special requirements on the workpiece. If a square shoulder is needed, this angle must be 0°. Some of the advantages of increasing this angle are an increase in tool life, control of the chip movement, and less power requirements. As the angle is increased the chip thickness is decreased and the width of the chip is increased. Since the forces of cutting are reduced, greater feeds may be used. However, thin cross sections will deflect under this type of cutting action because the resultant vector forces are no longer parallel to the centerline of the work. Figure 13.3c shows the effect of a 0° side cutting edge angle and a 15° cutting edge angle. Note that side cutting angles may also be negative. This type of tool is sometimes used to face the shoulder of the work after turning.

End cutting-edge angles are generally ground between 4 and 15° to avoid "dragging." Sometimes a land the length of the feed is provided to produce smooth finish.

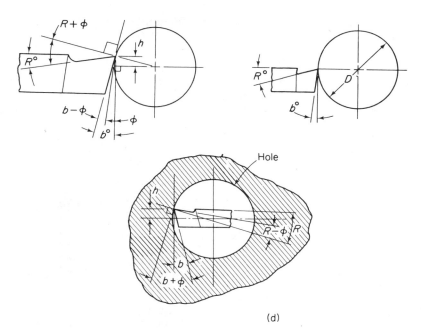

Figure 13.3(d)

The nose radius is provided to increase the finish and strength of the cutting tip of the tool. Small radii will produce smooth finishes and are used on thin cross sections of work. Large radii strengthen the tool and are used on cast iron and castings where the cuts are interrupted.

Tables 13.2 through 13.5 give the various angles and nose radii for high-speed steel tools, cast alloy steel, and carbide single-point toolbits.

It should be understood that tools generally operate with the cutting point at the centerline of the workpiece. At times they are raised above center. This has the effect of increasing the top rake and decreasing the front relief when machining external diameters, and the reverse effect when boring holes, Fig. 13.3d.

The correction angle ϕ may be calculated from the equation

$$\sin \phi = \frac{2h}{D} \qquad \begin{array}{l} \phi = \text{correction angle} \\ D = \text{diameter of shaft} \\ h = \text{offset} \end{array}$$

where h is $(+)$ if the offset is above the centerline, and $(-)$ if below.

The general equations from Fig. 13.3d (the shaft) are

$$R' = R + \phi \qquad \begin{array}{l} R = \text{actual rake} \\ R' = \text{effective rake} \end{array}$$

and

$$\qquad\qquad\qquad\qquad \begin{array}{l} b = \text{actual front relief} \\ b' = \text{effective front relief} \end{array}$$

$$b' = b - \phi \qquad \phi = \text{correction angle}$$

TABLE 13.2 RECOMMENDED ANGLES FOR HSS SINGLE-POINT TOOL BITS

Material	Relief (deg)		Rake (deg)	
	Side	End	Back	Side
Aluminum	12–14	8–10	30–35	14–16
Bakelite	10–12	8–10	0	0
Brass				
Free cut	10–12	8–10	0	1–3
Yellow, red	8–10	8–10	0	−2–(−4)
Bronze				
Cast, commercial	8–10	8–10	0	−2 to (−4)
Free cut	8–10	8–10	0	2 to 4
Phosphor hard	8–10	6–8	0	0
Cast-iron, gray	8–10	6–8	3–5	10–12
Copper				
Commercial	12–14	12–14	14–16	18–20
Hard	8–10	6–8	0	0
Soft	10–12	8–10	0–2	0
Fiber	14–16	12–14	0–2	0
Formica	14–16	10–12	14–16	10–12
Micarta	14–16	10–12	14–16	10–12
Monel and nickel	14–16	12–14	8–10	12–14
Nickel iron	14–16	10–12	6–8	12–14
Nickel silver	10–12	10–12	8–10	0 to (−2)
Rubber hard	18–20	14–16	0–(−2)	0 to (−2)
Steel				
HSS, alloy	7–9	6–8	15–7	8–10
High-C, st. st.	7–9	6–8	5–7	8–10
SAE:				
1020, 1035, 1040	8–10	8–10	10–12	10–12
1045, 1095	7–9	8–10	10–12	10–12
1112, 1120	7–9	7–9	12–14	12–14
1314, 1315	7–9	7–9	12–14	14–16
1335	7–9	7–9	12–14	14–16
2315, 2320	7–9	7–9	8–10	10–12
2330, 2335, 2340	7–9	7–9	8–10	10–12
2345, 2350	7–9	7–9	6–8	8–10
3115, 3120, 3130	7–9	7–9	8–10	10–12
3135, 3140	7–9	7–9	8–10	8–10
3250, 4140, 4340	7–9	7–9	6–8	8–10
6140, 6145	7–9	7–9	6–8	8–10

Source: Lathrobe Steel Co.

TABLE 13.3 RECOMMENDED NOSE RADII FOR SINGLE-POINT TOOLS

Type	Cross Section Dimension (in.)	Radius Dimension (in.)
Square	$\frac{3}{8}$ or less	$\frac{1}{64}$
	$\frac{3}{8}$ to $1\frac{1}{4}$	$\frac{1}{32}$
	Over $1\frac{1}{4}$	$\frac{1}{16}$
Rectangle	$\frac{1}{2} \times 1$ to $1 \times 1\frac{1}{2}$	$\frac{1}{32}$
	1×2 and $1\frac{1}{2} \times 2$	$\frac{1}{16}$

Source: Tool Engr., p. 21–7, Table 21.1.

TABLE 13.4 RECOMMENDED ANGLES FOR CAST ALLOY TOOLS[a]

Material	Rake (deg)		Relief (deg)		Cutting Edge (deg)	
	Back	Side	Side	End	Side	End
Steel	8–20[b]	8–20[b]	7	7	10	15
Cast steel	8	8	5	5	10	10
Cast iron	0	4	5	5	10	10
Bronze	4	4	5	5	10	10
Stainless steel	8–20[b]	8–20[b]	7	7	10	15

[a]Haynes Stellite 98M2 turning tool.
[b]Boring tools use the same rake but greater relief to clean work.

TABLE 13.5 RECOMMENDED ANGLES FOR CARBIDE SINGLE-POINT TOOLS

Material	Relief (deg)		Rake (deg)	
	End	Side	Back	Side
Aluminum and magnesium alloy	6–10	6–10	0–10	10–20
Brass and bronze	6–8	6–8	0–(−5)	8–(−5)
Cast iron	5–8	5–8	0–(−7)	6–(−7)
Copper	6–8	6–8	0–4	15–20
Steel				
Up to 1020	5–10	5–10	0–(−7)	6–(−7)
1025 and over	5–8	5–8	0–(−7)	6–(−7)
Alloy	5–8	5–8	0–(−7)	6–(−7)
Free machining, 1100 and 1800 series	5–10	5–10	0–(−7)	6–(−7)
Stainless				
Austenitic	5–10	5–10	0–(−7)	6–(−7)
Hardenable	5–8	5–8	0–(−7)	6–(−7)
High-nickel alloy: monel, inconel, etc.	5–10	5–10	0–(−3)	6–10
Titanium alloy	5–8	5–8	0–(−5)	6–(−5)

Example 13.1

A tool bit has a rake angle of 8° and a front relief angle of 6°. The diameter of the shaft is $1\frac{1}{2}$ in. and the offset is above the centerline by 0.030 in. Calculate the effect on the front relief angle and the rake angle if the rake angle is (a) positive and (b) negative.

Solution The correction angle is

$$\sin \phi = \frac{2h}{D} = \frac{2 \times (0.030)}{1.500} = 0.040$$

$$\phi = 2.3°$$

$R = 8°$
$b = 6°$
$D = 1.500$ in.
$h = 0.030$ in.
$\phi =$ correction angle

(a) For a positive rake angle (see Fig. 13.4a)

$$R' = R + \phi = 8 + 2.3$$
$$= 10.3°$$

$$b' = b - \phi = 6 - 2.3$$
$$= 3.7°$$

$\phi = 2.3°$
$R = 8°$
$b = 6°$

(b) For a negative rake angle (see Fig. 13.4b)

$$R' = R + \phi = -8 + 2.3$$
$$= -5.7°$$

$$b' = b - \phi = 6 - 2.3$$
$$= 3.7°$$

$\phi = 2.3°$
$R = -8°$
$b = 6°$

Example 13.2

Assume the same data as in Ex. 13.1 except that the tool bit is dropped below the centerline of the workpiece by 0.030 in.

Solution The correction angle is the same as in Ex. 13.1, except that it is below the centerline and taken as negative.

$$\phi = -2.3$$

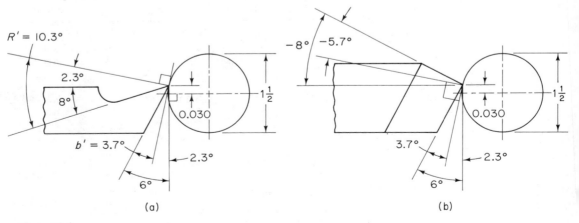

(a)   (b)

Figure 13.4

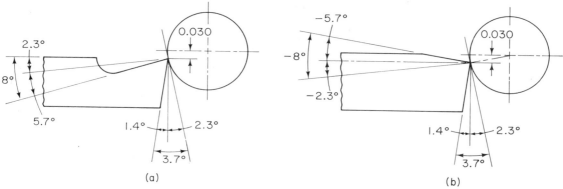

Figure 13.5

(a) For the positive rake angle, Fig. 13.5a,

$$R' = R + \phi = 8 + (-2.3)$$
$$= 5.7°$$

$$\phi = -2.3°$$
$$R = 8°$$

$$b' = b - \phi = 6 - (-2.3)$$
$$= 8.3°$$

$$b = 6°$$

(b) For a negative rake angle, Fig. 13.5b,

$$R' = R + \phi = -8 + (-2.3)$$
$$= -10.3°$$

$$\phi = 2.3°$$
$$R = -8°$$

$$b' = b - \phi = 6 - (-2.3)$$
$$= 8.3°$$

$$b = 6°$$

13.5 CHIP BREAKERS

Besides controlling the flow of chips with rake and cutting angles, it is also possible to use chip breakers. This is done by grinding grooves into the face of tools, or by fastening small pieces of steel to the surface of the face of the tool. As the chips are removed from the workpiece, their movement is interfered with by the chip breaker. They break into small pieces and are easily removed. The distance that the obstruction is set back from the cutting edge determines the type of chip control achieved. Table 13.6 shows some recommended parallel-type chip breakers.

There are several ways to grind chip breakers into the surface of tool bits to achieve desired effects. Figure 13.6a through f shows various types of chip breakers. Mechanical chip breakers serve the same purpose and may be used with inserts as shown in Fig. 13.6g. Figure 13.6h shows the chip breaker and locking system for a carbide insert system. Referring to Fig. 13.6i, it is possible to estimate the chip path that results from the width and height of the chip breaker. A chip breaker directs the chip into its curling pattern. The actual breaking of the chip is a secondary result,

TABLE 13.6 CHIP BREAKERS—PARALLEL AND ANGULAR

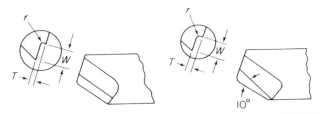

	Feed	0.006–0.012	0.013–0.017	0.018–0.025	0.028–0.040	Over 0.040
	R	0.010–0.025	0.035–0.065	0.035–0.065	0.035–0.065	0.035–0.065
Depth of Cut (in.)	T	0.010	0.015	0.020	0.030	0.030
$\frac{1}{64}$–$\frac{3}{64}$	W	$\frac{1}{16}$	$\frac{5}{64}$	$\frac{7}{64}$	$\frac{1}{8}$	
$\frac{1}{16}$–$\frac{1}{4}$	W	$\frac{3}{32}$	$\frac{1}{8}$	$\frac{5}{32}$	$\frac{3}{16}$	$\frac{3}{16}$
$\frac{5}{16}$–$\frac{1}{2}$	W	$\frac{1}{8}$	$\frac{5}{32}$	$\frac{3}{16}$	$\frac{3}{16}$	$\frac{3}{16}$
$\frac{9}{16}$–$\frac{3}{4}$	W	$\frac{5}{32}$	$\frac{3}{16}$	$\frac{3}{16}$	$\frac{3}{16}$	$\frac{3}{16}$
Over $\frac{3}{4}$	W	$\frac{3}{16}$	$\frac{3}{16}$	$\frac{3}{16}$	$\frac{3}{16}$	$\frac{1}{4}$

generally when it encounters some kind of obstruction. The equation for calculating the radius of the chip paths is

$$R = \frac{1}{2T}(W^2 + T^2)$$

R = radius of chip., in.
W = width of chip breaker, in.
T = height of chip breaker, in.

Example 13.3

Calculate the width of a chip breaker if its height is 0.015 in. and the radius of curvature of the chip removed is to be $\frac{5}{8}$ in.

Solution Solve the equation above for W:

$$W = \sqrt{2TR - T^2}$$
$$= \sqrt{2(0.015)(0.625) - 0.015^2}$$
$$= 0.136 \sim \frac{1}{8} \text{ in.}$$

$T = 0.015$ in.
$R = \frac{5}{8}$ in.

The feed is the most important single element in determining the results after the chip breaker dimensions have been determined. If the feed is increased, the chip breaking forces will be greater than when the feed is decreased. The converse is also true. A tighter radius R will produce a stronger chip breaking action. The effectiveness of a chip breaker depends on the feed and the radius of curvature of the chip flow. Finally, by varying the feed, any chip breaker can be made to produce almost any type of chip.

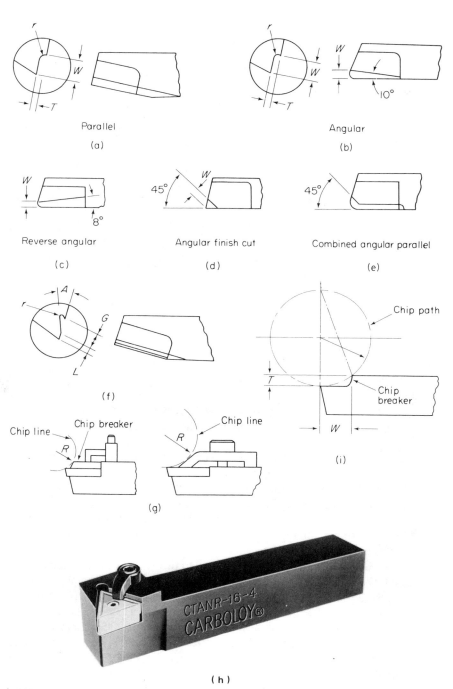

Parallel

(a)

Angular

(b)

Reverse angular

(c)

Angular finish cut

(d)

Combined angular parallel

(e)

(f)

Chip path

Chip breaker

(i)

Chip line

Chip breaker

Chip line

(g)

CTANR-16-4
CARBOLOY®

(h)

Figure 13.6

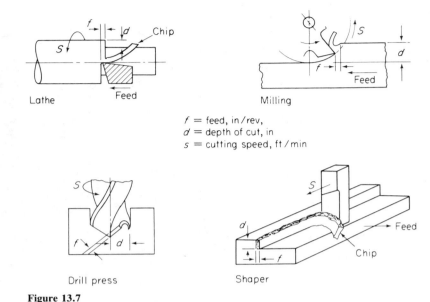

Lathe   Feed   Milling

f = feed, in/rev,
d = depth of cut, in
s = cutting speed, ft/min

Drill press   Shaper

Figure 13.7

13.6 CHIP FORMATION

The form that the chip takes is a result of such factors as the tool geometry, the feed, depth of cut, cutting speed, tool material, and the workpiece material. These factors by themselves or in combination also affect the horsepower required, the force of cutting, tool life, tool wear, finish, the heat generated, and the dimensional stability of the workpiece.

As indicated earlier in this chapter, the cutting tool must be harder than the material being cut. There must be a resistance to cutting, and there must be an interference to the motion of the tool by the workpiece. In Fig. 13.7 are shown the formation of chips resulting from the four basic machining operations.

There are three basic types of chips: continuous, discontinuous, and built-up. These are shown in Fig. 13.8. The *continuous chip,* Fig. 13.8a is obtained when cutting ductile material. The chip is severely deformed by the action of the cutting forces. Slip planes form. The resistance to further slip increases and slip stops. This is called *work hardness.* Continuous formation of new slip planes, and their transfer to successive shear planes and further work hardening results. If the process occurs and continues before fracture takes place, the material is said to be *ductile.* Since the chip has work-hardened and because of the orientation of the shear planes, the resulting chip will be shorter than the length of the material that went into the production of the chip before it was cut from the base material. It will also be more brittle than the original material because it has work hardened. However, the volume of the material removed from the work of the chip will remain the same. Therefore, since the length of the chip is shorter, and if the width is considered to be unchanged, the thickness

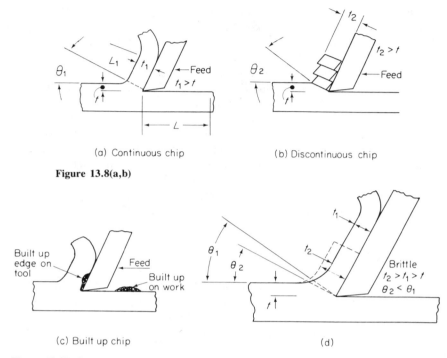

(a) Continuous chip   (b) Discontinuous chip

Figure 13.8(a,b)

(c) Built up chip   (d)

Figure 13.8(c,d)

must increase to maintain the volume of the material removed after cutting. In Fig. 13.8a, if the ratio r is about 1:2, it will yield good finish results for ductile cutting and continuous chips. The relationship is

$$r = \frac{t}{t_1} = \frac{L_1}{L}$$

r = ratio
t = chip thickness before cutting
t_1 = chip thickness after cutting
L = length of chip before cutting
L_1 = length of chip after cutting

If the chip is *discontinuous*, Fig. 13.8b, the mechanism is such that the fracture occurs before very many new slip planes form. That is, the work hardening cannot be overcome by plasticity in the material and therefore the chip separates from the base material. Because the shearing force is high, a crack develops early in the cutting process and small pieces fracture from the base material. This is characteristic of gray cast iron.

A *built-up chip*, Fig. 13.8c, occurs when the friction between the chip and the tool face is high. If the friction force is high, seizure will occur at the face of the tool. This will result in the high generating of localized heat causing the material to weld to the face of the tool. When the weld is stronger than the ultimate shear strength of the material being cut, it separates from the base material and "piles up" at the cutting edge of the tool. If the pile-up becomes excessive, it will break off, part remaining with the surface of the workpiece and part being carried off by the chip. A very undesirable surface condition results. It should be noted that a slight buildup at the cutting edge is sometimes desirable since it protects the cutting surface.

The shear angle θ_1, Fig. 13.8a, is larger than the shear angle θ_2, Fig. 13.8b. Since the material is more ductile in continuous chip formation, the material will pile up more and the shear angle will increase. Brittle materials do not pile up as much for the same stock removed. Thus if t is constant in both cuts, the shear angle for the brittle material will be less than the shear angle for ductile material.

It should be pointed out very quickly that larger shear angles require less cutting forces than smaller shear angles because the smaller the shear angle the greater the thickness of the material which the fracture plane needs to traverse. This is shown in Fig. 13.8d. The shear angle is a function of the rake of the tool, the material being cut, the tool material, and the friction at the face of the tool.

13.7 TOOL FAILURE

Tool failure may occur because of *structural failure* of the tool holder, the workholder, the machine, or the production requirements. It may occur because of a failure of the *geometry* of the cutting tool or it may result from *tool wear*. Failure has also occurred when the tool ceases to produce the workpieces to the requirements of the print.

Structural defect. If tool failure results from a *structure defect*, it may result from a lack of rigidity of the tool holder, the workholding fixture, the machine, or the workpiece. Under the worst possible conditions, any of these might break and become inoperative. Under the most favorable conditions, if any of the above are present, vibration (chatter) may occur and a poor finish and failure to hold size may result. Increasing the mass of a fixture or supporting an overhanging tool may reduce the amplitude and dampen the resonant frequency of the vibrations.

Failure of geometry. Earlier in the chapter we had discussed *tool geometry* and its effect on the cutting operation. The rakes and the cutting angles may be either positive or negative, depending on the job at hand. The relief and clearance angles must always be positive. The shape of the tool must fit the requirements of the job. Violations of any of these rules will result in lowered production, poor finish, failure to hold size, increased power requirements, and so on. Any of these failures could be considered to be tool failures. Either structural defects, or tool geometry error may cause chipping of the cutting edge. Thus any changes in tool geometry that will produce less friction, less chip distortion, and may be ground with larger shearing angles will produce longer life in the tool.

Tool wear. *Wear* is probably the most misunderstood reason for tool failure. Certainly, a failure to apply the proper geometry in tools could cause excessive wear

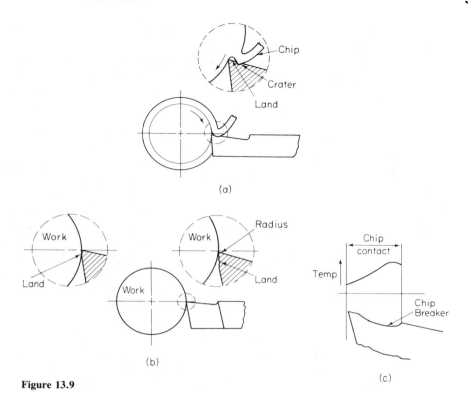

Figure 13.9

in tools. If a lathe tool is considered, there are two surfaces on the tool which are subject to wear. These two types of wear are face and flank wear.

The *face* of a tool is that part of the tool over which the chip slides. As indicated the sliding wears a cavity into the face a short distance in back of the cutting edge. This is shown in Fig. 13.9a. As the cutting proceeds the cavity gets wider and deeper. The land gets smaller as cutting progresses, until it disappears. The next phase is a breakthrough of the cutting edge, which creates a notch effect. The result can be that the tool breaks down suddenly because of the stress concentration at the notches; or the notch can become the focal point for a buildup of material. In either case the tool fails suddenly.

Flank wear is the eroding away of the flank of the tool immediately under the cutting edge. It is the part of the tool that takes the full effect of the feed. If the side relief is too small for the amount of feed, the wear will be excessive. If the feed is decreased, cutting time may be excessive. If the tool is reground to increase the side relief, the support under the cutting edge may be weakened. A compromise of feed and side relief must be achieved for optimum production and longest tool life. Figure 13.9b shows the wearing away of the flank of the tool bit to create a land. This wearing starts at the cutting edge and gets wider until the operating tool forces and the heat generated cause tool failure. When cutting highly abrasive material, such as cast iron, the flank wears into a land and a radius appears at the cutting edge as shown in the insert, Fig. 13.9b.

If a parting tool is used over a period of time, the front relief will also wear into a small land under the cutting edge. It should be noted that since the nose radius is the connecting cutting edge between the side and front relief and since it may be the only edge that cuts during the finishing operation, it can develop a worn land, which could result in tool breakdown. In cases where a pointed nose zero radius is used, the nose itself may fail before the flank or face wears. It is for the latter reason that nose radius failure is sometimes considered a third type of wear.

Many factors contribute to wear in a tool bit. The most important single factor that influences tool wear is *temperature*. Figure 13.9c shows the increase in the temperature of the tool point as the distance increases from the tip.

These temperatures are affected by the abrasive action of the work material against the tool material, the change of composition and diffusion as the critical temperature is reached, the chemical decomposition that takes place, the welding of work material to the tool material because of minute irregularities at the surface of the tool, and the plastic deformation of the work and the tool materials. Temperature increases result from the sum of all these conditions. Diffusion of heat and chemical decomposition progress very slowly or not at all until a critical temperature is reached.

Abrasive action results from hard particles or foreign matter in the structure of the material being cut. These particles remove minute pieces of material from the tool bit material. At high speeds the impact factor of work material against the tool material may dislodge a greater number of particles per unit time, thus contributing to the shortening of the tool breakdown time.

As the heat reaches a critical temperature, alloying of the component elements in the tool material seems to take place at the tip of the tool. This alloying effect results in a change of the hardness of the tool material. As a result it may be rubbed away, or caused to build up an edge which is eventually torn loose by the relative movement between the work and the tool.

Another action which contributes to the wearing mechanism is that of the decomposition of the tool material which results at the tool–workpiece interface from a localized chemical reaction. In the case of carbides it may weaken the bonding material. In the case of other tool material, it may form structurally weak components which are subject to faster wear.

All tool surfaces have minute irregularities. It is probable that these projections shear off small pieces of work material. Since this is more likely to happen at low cutting speeds and therefore at low temperatures, it is thought that the work hardening which resulted when the chip was cut from the workpiece is retained. These pieces then cling to the projection buildup, and eventually small pieces of material are dislodged. The contribution of this mechanism is greatest when the temperature of the system is below the critical temperature. Plastic deformation of the tool material and the work material at the point of contact contributes to the generation of heat because the deformation of the tool point has the same effect as a slightly dull tool point.

TABLE 13.7 Machinability

Material AISI	Rating (%)	Material AISI	Rating (%)	Material AISI	Rating (%)	Material	Rating	Material	Rating
1008	50	1108	80	4023	70	Cast iron		Brass	
1009	50	1109	80	4032	65	Hard	50	Rod	200
1010	55	1111	95	4037	60	Medium	65	Yellow	200
1012	55	1112	100	4040	55	Wrought iron	50	Leaded	150–600
1015	50	1113	135	4130	60	Stainless		Bronze	
1018	70	1115	80	4135	60	Austenite	25	Leaded	200–500
1020	65	1118	85	4140	55	Free machining	50	Zinc	200
1025	65	1119	100	4150	50	Ni-resistant	30	Mn bronze	40
1030	70	1132	75	4340	60	Mn steel	30	Copper	
1040	60	1137	70	4640	60	Tool steel	30	Cast	70
1045	55	1140	70	5150	50	HSS	30	Rolled	60
1050	45	2330	45	6120	50	High C–high Cr	25	Nickel	20
1070	55	2340	40	6152	45			Monel	
1080	45	3130	45	8650	‚50	Nonferrous		Cast	35
1085	45	3140	50	9260	45	Aluminum		Rolled	45
1090	45	3150	50			2-S	300–1500	K-	50
1095	45					11-S	500–200	Inconel	45
						17-S	300–1500	Everdur	60

Source: American Machinist, p. 12–71.

13.8 TOOL LIFE

Machinability refers to the ease with which various materials may be machined when related to AISI B1112 cold-rolled, or cold-drawn, steel, and when machined with a high-speed-steel cutter. These ratings depend on a great many factors. However, in general, material that has a 50% machinability rating will need to be machined at approximately one-half the cutting speed as a material which is rated 100% to maintain the same tool life. Table 13.7 shows a list of materials and their machinability rating relative to AISI 1112, which has a ratio of 100.

The ratings may be used to convert the cutting speed for one tool material to the cutting speed of another tool material. It is also possible to relate the tool life for one kind of tool material to another. The equation is

$$\frac{R_2}{R_1} = \frac{S_2}{S_1} = \frac{T_2}{T_1}$$

Material 1	Material 2
R_1 = rating (Table 13.7)	R_2 = rating
S_1 = cutting speed	S_2 = cutting speed
T_1 = tool life	T_2 = tool life

Example 13.4

Assume that an AISI 1040 material is machined with a cutting speed of 90 ft/min for 3 hr of tool life. A substitution of 4140 material is made using the same tool material. (a) To maintain the same tool life, what must be the cutting speed? (b) What should be the tool life if the cutting speed were held constant?

Solution   From Table 13.7, the rating for a 1040 steel is

$$R_1 = 60 \quad \begin{aligned} T_1 &= 3 \text{ hr} \\ S_1 &= 90 \text{ ft/min} \\ &\text{material } 1040 \end{aligned}$$

For a 4140 steel, the rating is

$$R_2 = 55$$

(a) The new cutting speed that will give 3 hr of tool life is

$$S_2 = \frac{R_2}{R_1}(S_1) = \frac{55}{60} \times 90$$

$$= 82.5 \text{ ft/min}$$

(b) The new tool life would be

$$T_2 = \frac{R_2}{R_1}(T_1) = \frac{55}{60} \times 3$$

$$= 2.75 \text{ hr}$$

It should be remembered that tool life is affected by many variables. Those related to the *material* being cut are: the microstructure, tensile strength, hardness, degree to which the material cold works, the rigidity of the work, and the shape and size of the work. A second set are those related to the *machining variables:* the cutting speed, feed, depth of cut, tool material, tool form, condition of the machine, and the conditions under which the tool engages and disengages from the work. A third set of variables are those related to the *machining conditions:* the temperature of the work and tool, the ability of the system to dissipate heat, the chip geometry, the forces required to remove a chip, the degree to which the chip hardens during removal, the feed rate, and in some cases, chip disposal.

It should also be understood that tool life and tool failure are synonymous and that tool failure must therefore be defined before tool life can be judged. There are many ways to define tool failure. Obviously, if the tool will no longer cut, it has failed. Probably the most noticeable evidence of failure is when the finish begins to change. The next most noticeable change that takes place is a dimensional change. Sometimes an increase in the cutting forces or feed forces will indicate that a land has been worn on the flank of the tool and the cutting relief has been destroyed. Sometimes a burnishing band will appear on the work if the tool is failing.

It is a judgment decision as to how long a tool should be permitted to operate after preliminary indications that the tool is beginning to fail. The tool engineer must weigh the factors of the economics of stopping production and regrinding the tool versus allowing it to continue to operate. In the case of disposables, the usable tool time is longer.

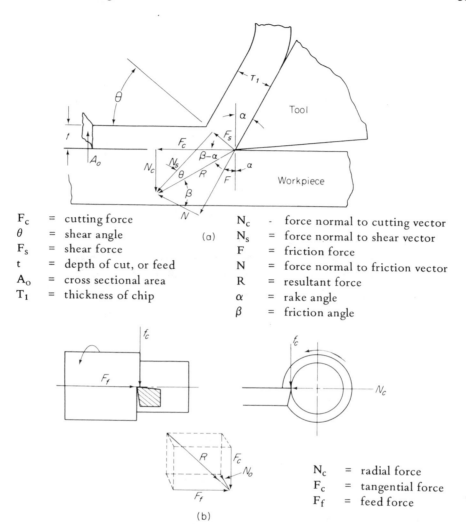

F_c	=	cutting force	N_c	-	force normal to cutting vector
θ	=	shear angle	N_s	=	force normal to shear vector
F_s	=	shear force	F	=	friction force
t	=	depth of cut, or feed	N	=	force normal to friction vector
A_o	=	cross sectional area	R	=	resultant force
T_1	=	thickness of chip	α	=	rake angle
			β	=	friction angle

(a)

N_c	=	radial force
F_c	=	tangential force
F_f	=	feed force

(b)

Figure 13.10(a,b)

13.9 CUTTING FORCES

One of the easiest ways to study the forces that enter into the cutting operation is to analyze the cutting action of a tool on a shaper during the action of cutting a keyway. This is called *orthogonal cutting*. Orthogonal cutting is two-dimensional cutting, Fig. 13.10a. A lathe tool for three-dimensional cutting is shown in Fig. 13.10b.

The forces operating in orthogonal cutting when producing a continuous chip are shown in Fig. 13.10a.

Earlier it was pointed out that the cutting ratio could be determined from the ratio

$$r = \frac{t}{t_1} = \frac{L_1}{L}$$

r = cutting ratio
t = chip thickness before cutting
t_1 = chip thickness after cutting
L = length of chip before cutting
L_1 = length of chip after cutting

Once this ratio is determined the shear angle $\theta°$ can be determined from

$$\tan \theta = \frac{r \cos \alpha}{1 - r \sin \alpha}$$

θ = shear angle (deg)
r = cutting ratio
α = rake angle

Example 13.5

Assume orthogonal cutting. The depth of cut is 0.075 in., with a chip thickness after cutting of 0.090 in. If the tool bit rake is 8°, calculate (a) the cutting ratio and (b) the shear angle.

Solution   (a) The cutting ratio is

$$r = \frac{t}{t_1} = \frac{0.075}{0.090} = 0.833$$

t = 0.075 in.
t_1 = 0.090 in.
α = 8°

(b) The shear angle is

$$\tan \theta = \frac{0.833 \cos 8°}{1 - 0.833(\sin 8°)}$$

$$= \frac{0.833(0.990)}{1 - 0.833(0.139)}$$

$$= 0.933$$

$$\theta = 43°$$

Table 13.8 shows some of the useful equations for calculating mechanical properties related to orthogonal cutting and Fig. 13.10a.

Power is the product of a force times a displacement. The horsepower needed at the cutting edge may be calculated in the following manner.

1. The forces operating on a tool are generally three-dimensional, as shown in Fig. 13.10b. The force F_f is the force that develops as a result of the feed. It operates parallel to the lines of action of the cutter and in the direction of the feed. The feed has a linear velocity which is very slow. The feed produces a force that is about one-half that of the force produced as a result of the rpm of the workpiece. But because the feed velocity is small, it reduces the overall horsepower to the point where it becomes insignificant. The equation that gives the horsepower at the cutter as a result of the feed is

TABLE 13.8 ORTHOGONAL CUTTING AND CONTINUOUS CHIPS

Quantity	Equation[a]
Coefficient of friction	$\mu = \dfrac{N_c + F_c \tan \alpha}{F_c - N_c \tan \alpha}$
Friction force	$F = N_c \cos \alpha + F_c \sin \alpha$
Mean shear strength	$S_s = \dfrac{F_c \sin \theta \cos \theta - N_c \sin^2 \theta}{A_o}$
Work done in shear	$W_s = S_s \left[\cot \theta + \tan(\theta - \alpha)\right]$
Work done in overcoming friction	$W_f = \dfrac{F}{A_o}\left[\dfrac{\sin \theta}{\cos(\theta - \alpha)}\right]$
Total work done in cutting	$W_n = \dfrac{N_c}{A_o}$

[a] μ = coefficient of friction

F = friction force; force component acting between tool face and sliding chip, lb

F_c = cutting force; force component acting in a direction of tool travel

N_c = thrust force; force component acting in a direction perpendicular to surface generated, lb

S_s = mean shear stress on shear plane; mean shear strength of material cut, psi

W_f = work done in overcoming friction between chip and tool per unit volume of metal removed, in.-lb./in.^{3}

W_n = total work done in cutting per unit volume of metal removed, in. = lb./in.^{3}

W_s = work done in shearing of metal per unit volume of metal removed, in.-lb/in.^{3}

α = effective rake angle of tool as measured in a plane perpendicular to its cutting edge, deg

θ = shear angle; angle between shear plane and surface being generated, deg

Source: ASTME, Tool Engineers Handbook, 2nd ed. (New York: McGraw-Hill, 1959).

$$Hp_c = \frac{F_f v_f}{33,000}$$

HP_c = horsepower at the cutter
F_f = feed force
v_f = longitudinal velocity

2. The radial force (N_c) operating on a lathe tool is about one fourth of the cutting force (F_c). However, since it has no velocity* toward the centerline of the work (it maintains a constant depth), the horsepower is zero. The equation for the horsepower of the radial forces is

$$HP_c = \frac{N_c v_c}{33,000}$$

HP_c = horsepower radial force
N_c = radial force
v_c = radial velocity

*A parting tool would have a radial velocity.

3. The cutting force, F_c, is the force that develops as a result of the rpm of the work, while the latter is in contact with the tool. This force is substantial and constitutes almost all the horsepower consumed at the cutting edge.

$$HP_c = \frac{F_c(S)}{33,000}$$

HP_c = horsepower at the cutter
F_c = cutting force, lb
S = cutting speed, ft/min

4. The shearing force F_s may be calculated from

$$F_s = S_s A_o \csc \theta$$

5. The friction angle β may be calculated from the coefficient of friction μ. Thus

$$\tan \beta = \mu$$

6. The cutting force F_c may be calculated from

$$F_c = \frac{F_s \cos (\beta - \alpha)}{\cos (\theta + \beta - \alpha)}$$

F_s = shearing force
α = rake angle
θ = shear angle
β = friction angle
S_s = shear strength of material
A_o = cross-sectional area of chip
μ = coefficient of friction

Table 13.9 shows several values for shear strength and coefficient of friction.

Example 13.6

Use the data from Ex. 13.5 and the following data:

Type of material   1045 hot-rolled steel
Shear strength   90,000 psi
Width of cut $\frac{1}{4}$ in.
Cutting speed   90 ft/min
Coefficient of friction 0.96

Calculate (a) the shearing force, (b) the friction angle, (c) the cutting force F_c, and (d) the horsepower at the cutting tool.

Solution

(a) The shearing force is

$$F_s = S_s A_o \csc \theta = 90,000(0.075 \times 0.250) \csc 43°$$

$$= 2474 \text{ lb}$$

S_s = 90,000 psi
t = 0.075 in.
θ = 43°
μ = 8°
S = 90 ft/min

TABLE 13.9 Mechanical Properties

Material AISI No.	BHN	Shear St. (× 1000 psi)	Coef. of Friction	Material AISI No.	BHN	Shear St. (× 1000 psi)	Coef. of Friction
1010	103	68	1.32	4340	210	93	1.06
1020	109	74	0.94	52100	186	73	1.11
1045 (HRS)	190	92	0.96				
1045 (CRS)	213	95	0.94	Stainless steel			
1095	185	95	0.89	303	162	106	0.80
1113	170	72	0.53	304	139	118	1.18
1340	192	92	1.10	410	217	88	0.87
2340	197	91	0.99	416	215	83	0.82
3130	169	83	0.94	430	156	81	0.92
3140	185	83	1.12				
3150	197	88	1.13				

Source: ASTME, *Tool Engineers Handbook,* 2nd ed. (New York: McGraw-Hill, 1959), Table 18–3.

(b) The friction angle is

$$\tan \beta = \mu = 0.96$$

$$\beta = 43.8$$

(c) The cutting force F_c is

$$F_c = \frac{2474 \cos (43.8 - 8)}{\cos (43 + 43.8 - 8)} = \frac{2007}{0.194}$$

$$= 10{,}345 \text{ lb}$$

(d) The horsepower at the tool is

$$HP_c = \frac{F_c S}{33{,}000} = \frac{10{,}345 \times 90}{33{,}000}$$

$$= 28 \text{ hp}$$

The horsepower at the motor (HP_m) is in many instances as important as the horsepower at the cutter. This is approached through an analysis of the cubic inches of material removed per minute from the workpiece. Since the volume of metal that can be removed each minute automatically takes into account all the variables, it becomes an efficient method for calculating horsepower. This concept is one of unit horsepower and is defined as the power required to remove 1 in.^{3} of metal per minute. Tables are available which show these values and also show correction factors for different feed rates.

The procedure is to calculate the horsepower at the cutter using the volume method. Table 13.10 lists the unit horsepower values for many materials.

$$HP_c = UHP \times C \times V$$

V = volume, in./min
C = correction factor
UHP = unit horsepower
HP_c = horsepower at cutter

TABLE 13.10 UNIT HORSEPOWER VALUES

(a) Ferrous Metals and Alloys

	BHN					
	150–175	176–200	201–250	251–300	301–350	351–400
Materials	Unit Horsepower (UHP)					
AISI						
1010–1025	0.58	0.67				
1030–1055	0.58	0.67	0.80	0.96		
1060–1095			0.75	0.88	1.00	
1112–1120	0.50					
1314–1340	0.42	0.46	0.50			
1330–1350		0.67	0.75	0.92	1.10	
2015–2115	0.67					
2315–2335	0.54	0.58	0.62	0.75	0.92	
2340–2350		0.50	0.58	0.70	0.83	1.00
2512–2515	0.50	0.58	0.67	0.80	0.92	
3115–3130	0.50	0.58	0.70	0.83	1.00	
3160–3450		0.50	0.62	0.75	0.87	1.00
4130–4345		0.46	0.58	0.70	0.83	1.00
4615–4820	0.46	0.50	0.58	0.70	0.83	0.87
5120–5150	0.46	0.50	0.62	0.75	0.87	1.00
52100		0.58	0.67	0.83	1.00	
6115–6140	0.46	0.54	0.67	0.83	1.00	
6145–6195		0.70	0.83	1.00	1.20	1.30
Plain cast iron	0.30	0.33	0.42	0.50		
Alloy cast iron	0.30	0.42	0.52			
Malleable cast iron	0.42					
Cast steel	0.62	0.67	0.80			

(b) High-Temperature Alloys

Material	BHN	UHP	Material	BHN	UHP	Material	BHN	UHP
A286	165	0.82	Hastelloy B	230	1.10	T_i–150A	340	0.65
A286	285	0.93	Inco 700	300	1.12	U–500	375	1.10
Chromoloy	200	0.78	Inco 702	230	1.10	4340	200	0.78
Chromoloy	310	1.18	M-252	230	1.10	4340	340	0.93
			M-252	310	1.20			

(c) Nonferrous Metals and Alloys

Material	UHP	Material	UHP	Material	UHP
Brass		Bronze		Aluminum	
Hard	0.83	Hard	0.83	Cast	0.25
Medium	0.50	Med.	0.50	Hard	0.33
Soft	0.33	Soft	0.33	Monel	1.00
Free-machining	0.25	Copper	0.90	Zinc (die cast)	0.25

Source: Carboloy Application Data Manual, Metallurgical Products Department, General Electric Co.; by permission.

Figure 13.11

The correction factor can be obtained from the graph in Fig. 13.11. If the tool has a lead angle, the correction factor should be reduced by

10% for a lead of 30°

30% for a lead of 45°

50% for a lead of 60°

30% if the depth of cut is less than the nose radius

TABLE 13.11 MACHINE TOOL EFFICIENCY

(a) Nonmilling Machines

Type	Efficiency (%)
Direct-drive spindle	90
One-belt drive	85
Two-belt drive	70
Geared head	70

(b) Milling Machines

Rated HP	Efficiency (%)
3	40
5	48
7.5	52
10	52
15	55
20	60
25	65
30	70
40	75
50	80

Source: ASTME, *Tool Engineers Handbook,* 2nd ed. (New York: McGraw-Hill, 1959).

The horsepower at the machine will be higher than the horsepower at the cutter because much energy is lost due to friction in the machine. The efficiency of various machine tools is listed in Table 13.11. These are approximate values and will vary as conditions change in the machine.

Example 13.7

Using the volume method, calculate (a) the horsepower at the cutter and (b) the horsepower at the motor given the following data:

Depth of cut	0.062 in.
Cutting speed	120 ft/min
Feed	0.030 in./rev.
Material	3130, BHN 169
Cutting tool	HSS with 45° side cutting-edge angle
Lathe	gear head

Solution

(a) The volume of metal removed per minute

$$V = 12fdS = 12(0.030)(0.062)(120) \qquad f = 0.030 \text{ in./rev.}$$
$$\qquad\qquad d = 0.062 \text{ in.}$$
$$= 2.68 \text{ in./min} \qquad\qquad S = 120 \text{ ft/min}$$

The unit horsepower from Table 13.10, 3130 steel, BHN 169 is $UHP = 0.05$. The lead of 45° will reduce the thickness of the chip. The feed should be reduced by 30%. Therefore,

$$0.030 - (0.030 \times 0.30) = 0.021 \text{ in./rev.}$$

The correction factor for this feed, from Fig. 13.11, is

$$C = 0.91$$

The horsepower at the cutter is

$$HP_c = UHP \times C \times V = 0.5 \times 0.91 \times 2.68 \qquad UHP = 0.5$$
$$\qquad\qquad\qquad\qquad\qquad\qquad\qquad C = 0.91$$
$$= 1.22 \qquad\qquad\qquad\qquad\qquad\qquad\qquad V = 2.68 \text{ in.}^3/\text{min}$$

(b) Efficiency from Table 13.11 for a gear motor is

$$\text{efficiency} = 70\%$$

The horsepower at the motor is

$$HP_m = \frac{HP_c}{\text{efficiency}} = \frac{1.22}{0.70}$$

$$= 1.74$$

13.10 CARBIDE TOOLS

At one time carbide tools were all made by brazing (silver) a slug of carbide material into a premachined tool shank, Fig. 13.12a. Solid tools have the disadvantage that

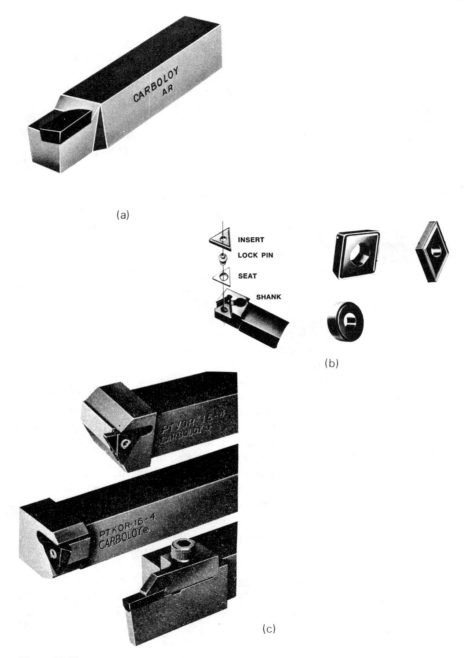

(a)

(b)

(c)

Figure 13.12

they must be discarded if they fracture or cannot be reground. Regrinding is a costly and time-consuming process, and stress may be set up in the carbide because of differences in the coefficient of expansion between the tool shank and carbide when the tool heats up.

The latest development is the disposable insert. These carbide inserts may be purchased in a large variety of shapes and sizes. Rake is built into the insert holder, or into the insert itself. Figure 13.12b shows an exploded view of one such holder and a variety of available inserts. Holders and insert shapes may be purchased to turn, thread, or bore almost any desired shape. They may have positive, zero, or negative rake. They may be rotated and in some instances (negative rakes) turned upside down and used. Once all possible rotations have been exhausted, they are thrown away.

These inserts may be indexed to within 1° and 0.0005 in. so that the indexing operation may be done accurately and swiftly. A plain shim is usually provided under the insert to take up shock and to provide a firm seat for the carbide. The clamp for the insert is used to hold the insert in position and to act as a chip breaker. Its purpose is not to take the thrust of the cutting forces. Figure 13.12b shows the lock pin clamp, Fig. 13.6h the bridge clamp type, and Fig. 13.12c the screw as a locking device. Also shown in Fig. 13.12c are unique holders and inserts which may be used for threading or grooving.

Everything that has been said about chip breakers for high-speed-steel cutters applies to carbide tips (see Sec. 13.5). Their use with carbides is very important because of the use of greater cutting speeds. The higher speeds produce a ductility that does not give the chips a chance to curl and break up. Mechanical, ground, or pressed chip breakers aid the curling process by placing an obstruction in the path of the flow of chips. Diamond wheels must be used to grind chip breakers into the face of the tool about one feed distance back of the cutting edge. Mechanical chip breakers are adjustable, so that as the depth of cut, feed, or cutting speed is changed, the distance of the obstruction back of the cutting edge may be changed. Pressed-in grooves are manufactured into the insert and are not to be reground.

Boring bars, which may be fitted with disposable inserts, are shown in Fig. 13.13. The boring bar shown in Fig. 13.13a is standard and integral with the shank. Figure 13.13b is an adjustable boring bar. It is manufactured with interchangeable heads that may be adjusted within ranges of $\frac{3}{8}$ to $1\frac{3}{8}$ in. These heads are provided with lead angles of -5 to $+15°$. Rake angles may range from -10 to $+8\frac{1}{2}°$ side rake in combination with $-6°$ to $+6°$ back rake angles. Figure 13.13c shows several of these combinations.

Figure 13.13d shows one of the advantages of the use of a negative rake. It can be seen that with negative rake the cutting forces are directed toward the main body of the shank, whereas with positive rake these forces are directed forward at the cutting edge. Therefore, when interrupted cuts are taken, negative rakes should be used. It is to be noted that in the side movement of the tool, the side rake is the cutting rake and the back rake is the control angle for determining the chip direction. In plunge cutting the reverse is true.

Also, a positive lead angle means that for a given depth of cut, the nose of the tool is the last edge to start cutting. The first contact is made back from the point, as

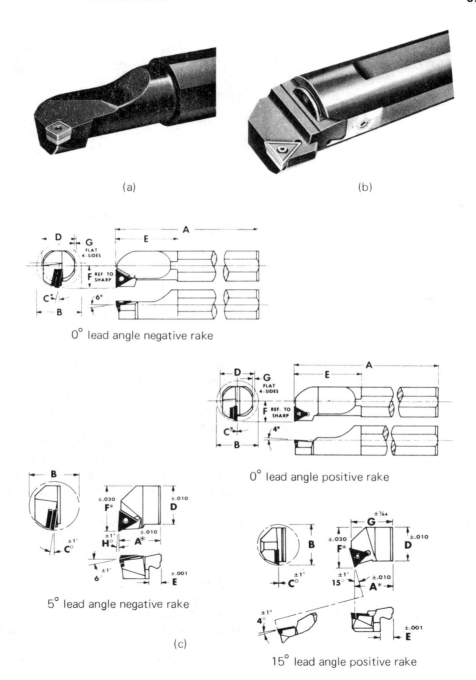

(a)

(b)

0° lead angle negative rake

0° lead angle positive rake

5° lead angle negative rake

15° lead angle positive rake

(c)

Figure 13.13(a,b,c)

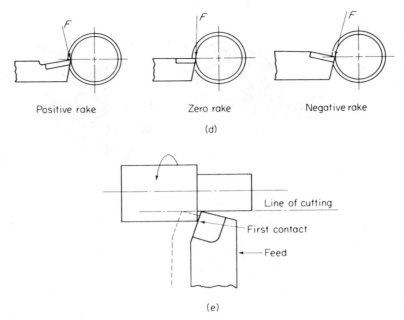

Figure 13.13(d,e)

shown in Fig. 13.13e. It should be noted that the cutting forces required to remove 1 in.³ of metal with a carbide tool are greater by about 15% then for high-speed steel, and that a great deal more heat is generated. The additional power requirements generally do not present a problem. The additional heat generated can be controlled by providing a sufficient coolant directed at the cutting edge.

Problems that may appear when using carbides are poor finish, chatter, chipped cutting edges, rapid wear of cutting edges, or poor breaking up of chips.

Poor finish, such as flaking, torn metal, or rough surfaces, may result if the nose radius is too large or if the cutting speed is too slow. The latter permits a built-up edge to form.

Chatter may result from a dull tool, too small an end or side cutting edge angle, too long a surface contact with the tool, as in the case of form tools or large-nose radii, excessive unsupported tool overchange, poorly operating chip breaker, or not a high-enough feed rate.

Chapped cutting edges may result from slow cutting speeds which permit built-up edges to form miniature grinding cracks which act as lines of stress. Concentration may be removed by boring, too slow a feed on rough or scale surfaces, cuts not deep enough on scale surfaces, improper chip breakers, or improper use of coolants which produce alternate heating and cooling.

Rapid wear of the cutting edge may result from light feeds, poor relief angles, or excessive cutting speeds.

Poor breaking of chips may result from a chip breaker that is winding the chip too tightly or too loosely, because the groove may be too narrow or too wide, respectively. A heavy flow of coolant sometimes makes it easier to break chips. A large radius may prevent proper chip breaking. The depth of the groove, as a chip breaker, may be incorrect. A deep groove winds a chip tight.

It should be made clear that these troubles may arise from the use of the wrong type of tool material, design, or grade of carbide. It is also possible that the causes of the trouble may be in the machine tool itself, or in the manner in which the machine tool and the cutting tool are used.

13.11 BORING TOOLS

Boring operations may take place with single-tool or multiple-tool cutting. Figure 13.14a shows the various rake and clearance angles for a boring tool. The side rake is used to direct the chip flow toward the center of the bored hole. The back rake is usually used to control the type of cut to be achieved when related to a particular material. It is obvious that the combination of side and back rakes creates a compound angle that affects the direction of chip flow. However, once the back rake is determined, the side rake is used to direct the chip flow.

It is important that the flow direction of the chips be carefully controlled. Chips that are permitted to curl into the cutting edge of the tool bit may crack it, or mar the surface finish. Chips will also mar the finished surface if they are directed into the wall of the workpiece. Stringers could easily wrap around the boring bat and interfere with the size or finish of the hole and may even break the boring bar.

The geometry of the tool cutting edge is important to the cutting operation. Once a satisfactory grind is established, it should be duplicated as accurately as possible. In Fig. 13.14b, c, and d are shown three types of cutting tips. The radius tip is the standard grind. If this produces chatter, or poor finish, it should be changed to a chamfer or positive lead angle. The chamfer and lead angle tool bits also are more stable than the radius.

The tool bit in boring may be set above the centerline of the work so that the bore clearance angle b is increased. The result of this increase makes it possible to decrease *angle b* and thus strengthen the cutting tip. The positive back rake, Fig. 13.14e, when raised above the centerline of the work a distance h, will cause an effective 0° rake. In Fig. 13.14f a tool bit with a 0° back rake angle when set above center creates an effective negative rake angle. Thus if the bore radius and the back rake angle of the tool are known, the height to which the tool must be set above center may be calculated from the equation

$$\sin \phi_r = \frac{2h}{D} \qquad \begin{aligned} \phi_r &= \text{back rake angle} \\ D &= \text{diameter of bore} \\ h &= \text{height above center} \end{aligned}$$

Example 13.8

Given a 3-in. bore diameter and a back rake angle of 7°, calculate the height above center of the tool to maintain 0° effective rake.

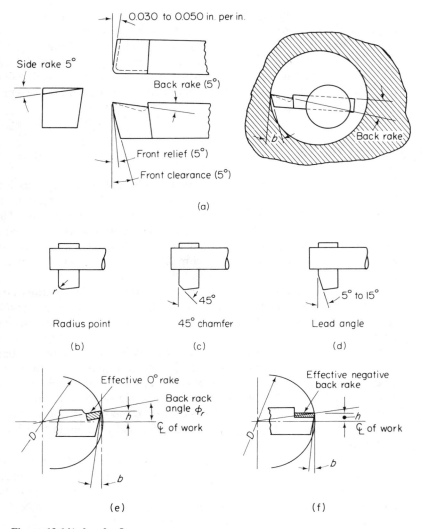

Figure 13.14(a,b,c,d,e,f)

Solution

$$h = \frac{D}{2} \sin \phi_r = \frac{3.0}{2} \times \sin 7° = 1.5 \times 0.122 \qquad \begin{aligned} D &= 3.0 \text{ in.} \\ \phi_r &= 7 \end{aligned}$$
$$= 0.182 \sim \tfrac{3}{16} \text{ in.}$$

If the tool is set at an *angle b°* as shown in Fig. 13.14g, the bore clearance is determined from angle *B°*. If the tool bit is ground with a clearance *angle b°* and an end-cutting edge of *E°*, the effective bore clearance *B°* becomes

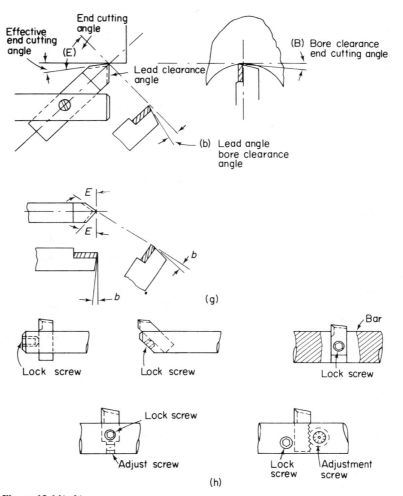

Figure 13.14(g,h)

$$\tan B° = \cos E° - \tan b°$$

$E°$ = end-cutting angle
$b°$ = actual bore clearance angle
$B°$ = effective bore clearance angle
(tool at an angle)

Example 13.9

Assume that a boring tool bit is ground with an end-cutting angle of 20° and one actual bore clearance angle on the tool of 7°. What is the effective bore clearance between the end-cutting face and the bore?

Solution

$$\tan B° = \cos E° \tan b° = \cos 20° \times \tan 7° = 0.940 \times 0.123$$

$$= 0.115$$

$$B° = 6.58$$

$E° = 20°$
$b° = 7°$

Figure 13.14h shows various types of boring bars for mounting single point tools for machining holes. Some lock into position and others have fine-adjustment screws and locking devices.

PROBLEMS

13.1. (a) How does the cutting action operate? (b) What effect does the relative hardness of the tool have on the cutting operation?

13.2. Make a drawing—three views—of a tool bit and label all the essential parts.

13.3. (a) Distinguish between the land angle and the face angle of a tool bit. (b) How would you interpret a symbol − 3(12) face angle?

13.4. (a) What is a tool point? (b) What does it include?

13.5. (a) Distinguish between the side-cutting edge and the end-cutting edge on a tool bit. (b) What is the flank?

13.6. List the order in which the various parts of a tool bit are given in the signature.

13.7. Explain each term of the tool signature

$$-8, 12, 10, 8, 0, 15, \tfrac{1}{32}$$

13.8. Explain each term of the tool signature

$$5, 0, 12, 8, 5, -6, \tfrac{3}{8}$$

13.9. Explain each of the position letters and numbers in the notation in Sec. 13.3 for tool bit inserts.

13.10. (a) Define back rake and distinguish between positive and negative rake. (b) Repeat for side rake.

13.11. (a) What is the difference between the terms "relief" and "clearance"? (b) Illustrate for front and side relief and clearance.

13.12. Define the edge- and side-cutting angles. Illustrate.

13.13. In general, what is the effect on the cutting operation when the rake angle is increased?

13.14. Draw the three possible types of rake angles.

13.15. What is "cratering" of a tool bit? Is it desirable? Explain.

13.16. Under what general conditions would positive or negative rake be used?

13.17. Assume too little relief on a tool bit. What effect does this have on the cutting action? Explain.

13.18. Show and explain why a front clearance angle would be necessary when boring a hole.

13.19. What are some of the advantages of increasing the side-cutting edge angle?

13.20. Make a sketch and show that an increase of the side-cutting edge angle produces a decrease in chip thickness and an increase in chip width.

13.21. When should a small nose radius or a large nose radius be used?

13.22. Design a $\frac{3}{8}$-HSS tool bit for cutting a square shoulder on aluminum. Refer to Tables 13.2 through 13.5.

13.23. Repeat Prob. 13.22 for a carbide tool bit.

13.24. Design a tool bit that is to cut a 15° angle on a shoulder at the end of a cylindrical cut, assuming an end-cutting edge angle of 5°. The tool is $\frac{5}{8}$ in. square. The tool material is (**a**) HSS, (**b**) cast steel, and (**c**) carbide. The material to be cut is 4140 steel, and the tool is set into the machine at an angle of 70° with the centerline of the workpiece.

13.25. A tool bit has a rake angle of 15°, a front relief angle of 10°, and is offset 0.030 in. above the centerline of a $1\frac{1}{2}$-in.-diameter workpiece. What is the effect on (**a**) the front relief angle and the positive rake angle and (**b**) the front relief angle and the negative rake angle?

13.26. Assume the same conditions as in Prob. 13.25 except that the tool bit is 0.030 in. below center.

13.27. Assume that the tool in Prob. 13.25 is used to bore a hole. Repeat the calculations of Prob. 13.25.

13.28. Assume that the tool in Prob. 13.26 is used to bore a hole. Repeat the calculations of Prob. 13.26.

13.29. A tool bit has a rake angle of 10° and a front relief angle of 8°. If the diameter of the shaft is 1 in. and the offset is *above* the centerline by 0.040 in., calculate the effect on the front relief angle and the rake angle if the rake angle is (**a**) positive and (**b**) negative.

13.30. Assume the same data as in Prob. 13.29 except that the tool bit is dropped below the centerline of the workpiece by 0.040 in.

13.31. (**a**) Make sketches of six different ground types of chip breakers. (**b**) Describe the purpose of chip breakers.

13.32. The depth of cut is to be $\frac{1}{4}$ in. Indicate the radius, the depth of the chip breaker, and the width of the chip breaker when (**a**) the feed is 0.015 in. and (**b**) 0.030 in.

13.33. The depth of cut is $\frac{1}{2}$ in. and the feed is 0.007 in. Indicate the (**a**) radius, (**b**) depth of the chip breaker, and (**c**) width of the chip.

13.34. Describe the use of mechanical chip breakers. What are some of the special considerations when designing these chip breakers?

13.35. Calculate the width W of the chip breakers if its height is 0.010 in. and the radius of curvature of the chip is to be 2 in. Explain this.

13.36. If the height of the chip breaker is to be 0.015 in. and the width $\frac{1}{8}$ in., what will the radius of curvature of the chip be?

13.37. Discuss the effect of "feed" on chip breaking.

13.38. Calculate the width of a chip breaker if its height is 0.020 in. and the radius of curvature of the chip removed is to be $\frac{3}{4}$ in.

13.39. (**a**) What factors affect chip formation? (**b**) What else do these factors affect during the cutting operation?

13.40. Make sketches of the four basic machining operations and label the parts of the chips that result from feed and depth of cut.

13.41. Discuss the formation of (**a**) continuous chips, (**b**) discontinuous chips, and (**c**) built-up chips. Support your explanation with sketches.

13.42. Discuss the relationships between shear angle, material thickness, and ductility during the cutting operation.

13.43. List the many reasons that may be responsible for tool failure.

13.44. Discuss ''chatter'' in terms of at least one of the reasons that will produce it. How may the cause of chatter be eliminated?

13.45. How does tool geometry affect tool life?

13.46. Tool wear affects tool life. Discuss the effect of (**a**) face wear and (**b**) flank wear of a lathe tool on tool life. (**c**) Why is nose radius wear considered a third type of failure?

13.47. Figure 13.9c shows the temperature curve as related to the distance from the cutting edge. Explain this curve.

13.48. Explain the effects of the following on a tool bit: (**a**) abrasive action, (**b**) heat increase, (**c**) decomposition of tool material, and (**d**) plastic deformation.

13.49. List the variables related to tool life which arise from (**a**) the material being cut, (**b**) the machining variables, and (**c**) the machining conditions.

13.50. Discuss the indicators of tool failure.

13.51. (**a**) What is orthogonal cutting? (**b**) Make a sketch of this type of cutting and allow all the vector forces operating. (**c**) Use a sketch and show how this type of cutting differs from the cutting done by a carbide tool.

13.52. Given a depth of cut of 0.125 in. and a rake angle of 10° which results in a chip thickness of 0.160 in., calculate (**a**) the cutting ratio and (**b**) the shear angle.

13.53. Why is it permissible in orthogonal cutting to omit the feed and radial forces from the horsepower calculations?

13.54. Use the data from Prob. 13.52 and the following:

Type of material	304 stainless steel
Shear strength	118,000 psi
Width of cut	$\frac{1}{2}$ in.
Cutting speed	40 ft/min
Coef. of friction	1.18

Calculate (**a**) the shearing force, (**b**) the friction angle, (**c**) the cutting force, and (**d**) the horsepower.

13.55. Use the volume method and calculate (**a**) the horsepower at the cutter and (**b**) at the motor given the data: cutting speed 40 ft/min, feed 0.020 in./rev, depth of cut 0.125 in., material 200 BHN cast steel, a HSS tool with a 20° side-cutting edge angle, and a gear head drive lathe. The radius of the tool is $\frac{1}{4}$ in.

13.56. AISI 4150 material is machined at a cutting speed of 80 ft/min with a tool that has a life of 4 hr. If 2340 material is substituted and the same tool life is expected, what cutting speed should be used?

13.57. Assume a 4340 material, a tool life of 6 hr, and a cutting speed of 90 ft/min. If the same cutting speed is maintained, (**a**) what tool life can be expected if a 1080 steel is substituted and (**b**) what cutting speed can be used if the tool life is constant?

13.58. Assume that an AISI 1018 material is machined at a cutting speed of 120 ft/min for 5 hr of tool life. A substitution of 1050 material is made using the same tool material. (**a**) What must the cutting speed be to maintain the same tool life? (**b**) What would be the tool life if the cutting speed were held constant?

13.59. Assume orthogonal cutting, a depth of cut of 0.050 in., and a chip thickness after cutting of

0.080 in. If the tool bit has a rake angle of 12°, calculate (**a**) the cutting ratio and (**b**) the shear angle.

13.60. Use the data from Prob. 13.59 and the following data:

Type of material	1045 hot-rolled steel
Shear strength	92,000 psi
Width of cut	$\frac{3}{8}$ in.
Cutting speed	100 ft/min
Coefficient of friction	0.96

Calculate (**a**) the shearing force, (**b**) the friction angle, (**c**) the cutting force F_c, and (**d**) the horsepower at the cutting tool.

13.61. Using the volume method, calculate (**a**) the horsepower at the cutter and (**b**) the horsepower at the motor given the following data:

Depth of cut	0.050 in.
Cutting speed	100 ft/min
Feed	0.040 in./rev
Material	1045, BHN 180
Cutting tool	HSS with 30° side cutting-edge angle
Lathe	single belt

13.62. What effect does the heat have on a brazed carbide tool?

13.63. Discuss the uses and advantages of carbide inserts.

13.64. Why are shims used with carbide inserts?

13.65. What is the dual purpose of a clamp that is to be used with an insert? Why should it *not* be used to absorb thrust during the cutting operation?

13.66. (**a**) How many types of chip breakers may be used with carbides? What are they? (**b**) Why are they important when using carbides? (**c**) How far back of the cutting edge should a chip breaker begin? (**d**) What determines this distance?

13.67. (**a**) What effect do the back and side rake have in controlling the chip flow? (**b**) Why is a lead angle advisable with carbides? (**c**) When is a negative rake advisable with a carbide tool?

13.68. (**a**) What problems may arise when using carbides? (**b**) List the probable causes for each.

13.69. How are the rakes and chip breakers used with boring tools to control the chip direction?

13.70. What can be done to the tip of a boring tool to help achieve good results?

13.71. What effect does setting the boring tool above center have on the rake angle?

13.72. A tool bit has a back rake of 12°. Calculate the height above center to which the tool must be set to maintain a 0° rake. The bore is 3 in. in diameter.

13.73. A tool bit is ground with an end-cutting angle of 25° and an actual bore clearance angle of 10°. What is the effective bore clearance between the end-cutting face and the bore, Fig. 13.14g?

13.74. Given a $2\frac{1}{2}$-in. bore diameter and a back rake angle of 6°, calculate the height above center of the tool to maintain 0° effective rake.

13.75. Assume that a boring tool bit is ground with an end-cutting angle of 30° and one actual bore clearance angle on the tool of 8°. What is the effective bore clearance between the end-cutting face and the bore?

13.76. Draw and describe at least three ways of mounting a tool bit in a boring bar.

Chapter 14

Multiple-Point Tools

14.1 MULTIPLE TOOL BITS

In addition to single-point cutting, various combinations of single-point tools may be clustered in holders to make roughing operations easier. Figure 14.1a shows a tool setup where the cutting forces are equalized on blades x and y. Each blade removes an equal amount of metal, blade x cutting the smaller diameter B at depth d_x slightly ahead of blade y. Blade y cuts the larger diameter at depth d_y but removes the same size chip from the hole. Sometimes the rough hole A is bored to receive a hardened and ground pilot at the end of the boring bar. This acts as a support for the end of the boring bar.

Figure 14.1b shows two single-point tools mounted in a bar for line boring two holes in one setup. Figure 14.1c and d shows block tool holders. They have the advantage over single-toolbits because the block may be removed, mounted in a fixture between centers, ground to size, and placed back in the boring bar. The repositioning of the block in the boring bar is very accurate and saves setup time. As a matter of fact, it is accurate enough so that the roughing or finishing diameters are generally engraved on the block once they are determined.

Another design is shown in Fig. 14.1e, which uses a cartridge. It may be used with a block or in a multiple-head cutting bar.

14.2 MILLING CUTTERS

Milling cutters are classified according to the manner in which they are used, such as gear cutters, radius cutters, slitting saw, and so on. They also may be classified

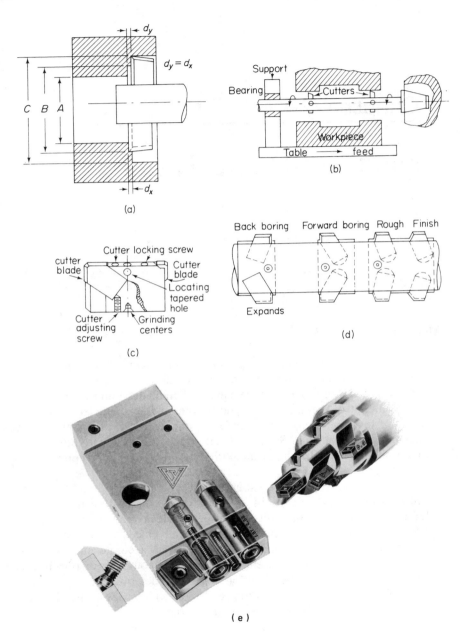

Figure 14.1

according to the method in which they are mounted, such as shank or arbor-type cutters. In some instances they are classified according to some construction feature, such as inserted tooth, carbide, or face mill.

Figure 14.2a shows a plain milling cutter which has straight teeth on the periphery and no cutting edge on the face of the cutter. A single-side milling cutter will cut on the periphery as well as one side. A double-side milling cutter will cut on the periphery as well as on both faces. This type of cutter is shown in Fig. 14.2b.

Plain milling cutters may have helical teeth. If, when viewed from the end of the cutter, the helix rotates clockwise away from the viewer, the helix is said to be *right-hand*. If it rotates counterclockwise away from the viewer, it is said to be *left-hand*.

Figure 14.2c shows a plain milling cutter with the helical teeth. The helix angle of the cutter shown is 18° and is right-handed. Other cutters have helix angles of 45° and 52°; the former for roughing, the latter for finish cutting. The large helix angle will produce fine finishes because of the shearing action of the teeth.

Sometimes a pair of cutters will be matched with opposite helix angles to equalize the thrust of cutting. Figure 14.2d shows a stagger-tooth side-milling cutter. In this case adjacent teeth have opposite helix angles.

Figure 14.2e shows a pair of interlocking cutters. With this arrangement it is possible to insert spacers between the two cutters and thus control the width of the slot being cut without developing a ridge in the center of the slot. As in the earlier example, the opposite helix angles equalize the thrust.

Slitting saws may be ground concave on both sides, or have chip clearance. They may also have stagger teeth. Figure 14.2f shows a plain slitting saw for cutting steel, brass, and so on. Figure 14.2g shows a single-angle cutter. The teeth are much finer than in the slitting saw.

Figure 14.2h shows a double-angle cutter. Figure 14.2i shows a corner-rounding cutter. Figure 14.2j shows a concave-radius cutter and Fig. 14.2k shows a convex-radius cutter. They are used to cut angles and radii.

A Woodruff key cutter used to cut keyways is shown in Fig. 14.2l. Figure 14.2m is a T-slot cutter which is small but similar to the side-milling cutters and usually has stagger teeth. Dovertail cutters, Fig. 14.2n, are used to cut dovetails, as the name indicates.

Shell mills, Fig. 14.3a, are mounted on the type of arbors shown. When they are large they may have inserted teeth and are called *face milling cutters.*

Solid end mills may have two cutting edges, Fig. 14.3c, and are called *two-flute* end mills, or multiple cutting edges when they are called three-fluted, four-fluted, and so on, end mills. They may also be center-cutting end mills, in which case the cutting edges will meet in the center, as shown in Fig. 14.3c, or they may be center relieved, as shown in Fig. 14.3b. Other end mills may have radius, or ball ends, and still others may have tapered ends. They may be purchased with straight, standard taper, or interlocking tapered shanks.

A *fly cutter* is generally a single cutter mounted in a disk or arbor. Sometimes two cutters are mounted in the disk. As it revolves and sweeps the work, it cuts the material. This is an effective method that may be used for cutting some types of

Figure 14.2(a,b,c,d,e,f,g,h,i,j,k)

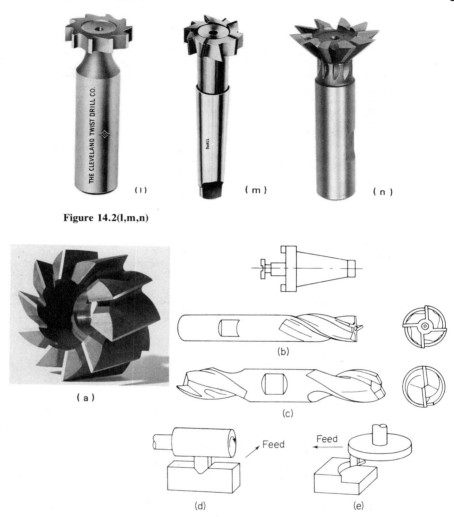

Figure 14.2(l,m,n)

Figure 14.3

contours or for surface milling at a very low cost. Figure 14.3d shows a fly cutter milling a small V-slot or multiple contours. Figure 14.3e shows a fly cutter milling a flat surface. Most cutters may be purchased with inserted blades, Fig. 14.4a, or inserted carbides, Fig. 14.4b. Cutters for making gears, taps, reamers, and so on, are special-purpose cutters.

14.3 NOMENCLATURE

Figure 14.5a shows a double-side milling cutter. It is mounted on an arbor and caused to rotate while in contact with the work. The action may be as shown in Fig. 14.5b

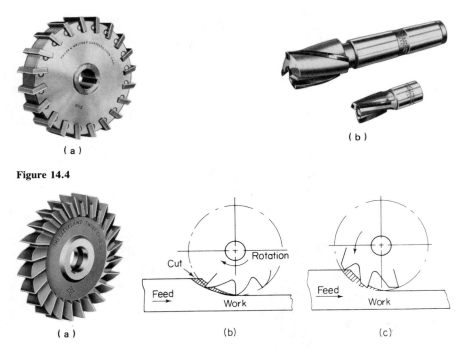

Figure 14.4

(a)   (b)

(a)   (b)   (c)

Figure 14.5(a,b,c)

and is referred to as *conventional* or *up milling*. If the cutter tooth action is down on the work as shown in Fig. 14.5c, it is called *climb* or *down milling*. A face mill is shown in Fig. 14.5d and it can be seen that one half the cutter is operating like conventional milling, whereas the other half of the cutting is similar to down milling. The chip generated during conventional milling is very thin at the beginning of the cut and maximum where the chip leaves the work, as shown in Fig. 14.5b. In down milling, Fig. 14.5c, the maximum chip thickness is obtained when the tooth contacts the work. Chip thickness decreases at the point where the tooth leaves the work. Finishes achieved with down milling are generally better than for conventional milling, because in the former the chip breaks almost tangent to the finish surface, Fig. 14.5e.

If a tooth of a milling cutter is to operate effectively, the tooth edge must be free to operate without any interference from other tooth parts. This is accomplished by grinding a *primary* relief angle of about 3 to 5°, as shown in Fig. 14.6a. To avoid the possibility of the *heel* of the tooth touching the work, a *primary* and sometimes a *secondary clearance* angle is ground as shown in Fig. 14.6d. The primary clearance angle is generally ground so that it is twice the relief angle. The secondary clearance angle is generally ground to 35°.

One of the most important parts of a milling cutter tooth is the *land*. Depending on the diameter of the cutter and the number of teeth, the land of a cutter tooth is ground to a width of $\frac{1}{32}$ to $\frac{1}{16}$ in. The land is that part of the tooth which supports the

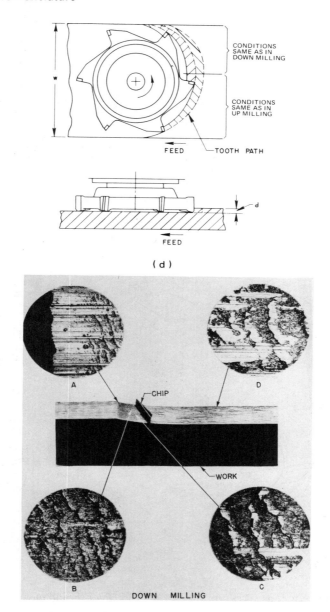

(d)

DOWN MILLING

Figure 14.5(d,e)

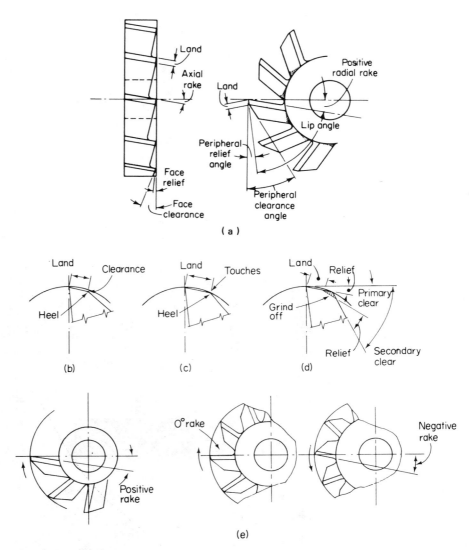

Figure 14.6(a,b,c,d,e)

cutting edge, as shown in Fig. 14.6b. The *heel* should clear the work so that the cutter tooth is free to cut. If the tooth becomes dull, the land is reground to sharpen the cutter. Each time the land is ground, it becomes wider, until it touches the work as shown in Fig. 14.6c. This interferes with the cutting action and the heel must be ground away. This creates a primary clearance angle, as shown in Fig. 14.6d.

Figure 14.6a shows two types of rake that provide a shearing cutting action. *Axial rake* is that angle which the face of the tooth makes with the axis of the cutter. *Radial rake* is the angle made by the face of the tooth and the radius of the cutter.

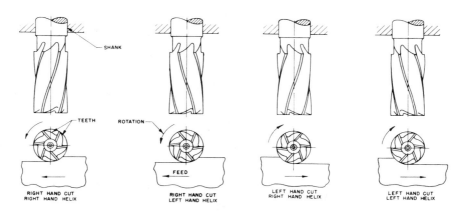

Figure 14.6(f)

Figure 14.6a shows both types of rake on a half-side milling cutter. Figure 14.6e shows positive, zero, and negative radial rake angles.

In addition to the tooth characteristics just discussed, it is possible to develop four combinations of the helix angle and the face cut. They are shown in Fig. 14.6f.

In the same manner that a *signature* was developed in Chap. 13 for tool bits, a signature has been developed for a milling cutter. This is shown in Fig. 14.7. Figure 14.8 shows the nomenclature for a four-flute end mill.

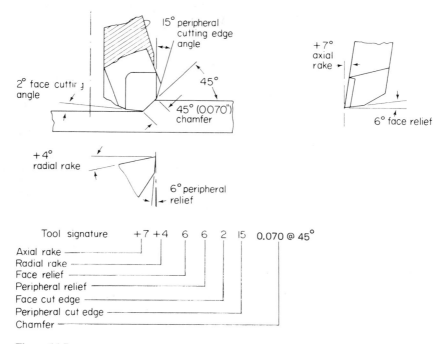

Figure 14.7

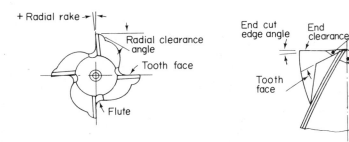

Figure 14.8

14.4 POWER REQUIRED FOR MILLING

The total horsepower used when milling is the sum of the horsepowers required to do the actual cutting and the power loss to the machine. The power at the cutter should reflect the power required to remove 1 in.^{3} of metal per minute, or the volume (V). This power per unit volume is affected by the feed, the cutting speed, and the structure of the material. Its hardness, tensile strength, and ductility also affect the power requirements. The configuration of the cutting tool, such as rakes, clearances, corner radii, or angles, also affect the power requirements. All of these factors are reflected in a K-factor which has been determined experimentally for different materials under different operational conditions. The K-factors in Table 14.1 are average values.

TABLE 14.1 K-FACTORS

Material	K	Material	K
Magnesium	4.0	Stainless	
Aluminum	4.0	Free machining	1.0
Copper	2.0	Other	0.6
Brass	2.5	Titanium	
Bronze	2.0	Under 100,000 psi	0.8
Malleable iron	1.0	100,000–135,000 psi	0.6
Cast iron		135,000 and over	0.4
Ferrite	1.5	High-tensile alloys	
Pearlitic	1.0	180,000–220,000 psi	0.5
Chilled	0.6	220,000–260,000 psi	0.4
Steel		260,000–300,000 psi	0.3
Up to 150 BHN	0.7	High-temperature alloys	
300 BHN	0.6	Ferritic low alloys	0.6
400 BHN	0.5	Austenitic alloys	0.5
500 BHN	0.4	Nickel-based alloys	0.4
		Cobalt-based alloys	0.4

Source: National Twist Drill & Tool Co., Lexington, S.C., Catalog No. 69, p. 218.

The horsepower at the cutter to remove 1 in.³ of metal is

$$HP_c = \frac{V}{K}$$

HP_c = horsepower at cutter
V = volume of metal, in.³/min
K = power factor

If y is the width of a chip, x is the depth of a cut, and F is the feed, the volume of the chip is

$$V = yxF$$

F = feed, in./min
x = depth of cut, in.
y = width of chip, in.

It is important to note that a better definition for x and y *for purposes of this equation* is to define y *parallel* to the centerline of the cutter, and x *perpendicular* to the centerline of the cutter. This is illustrated for a face mill, Figure 14.9a, and a peripheral mill, Fig. 14.9b.

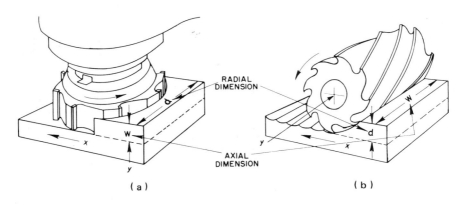

(a)   (b)

Figure 14.9(a,b)

The feed F in inches per minute may be converted to feed per tooth F_t. Recommended feeds F_t are given in Table 14.2.

$$F = F_t nN$$ F_t = feed per tooth (chip thickness, in.)

Therefore,

$$HP_c = \frac{V}{K} = \frac{yxF}{K} = \frac{yx(F_t nN)}{K}$$

n = number of teeth
N = rpm
K = constant (Table 14.1)

It should be noted that N may be calculated using the method of Chap. 6. Thus

$$N = \frac{12S}{\pi D}$$

S = cutting speed, ft/min
D = diameter of cutter, in.
N = rpm

TABLE 14.2 Suggested Feed per Tooth for Cutters

Material	Face Mills	Helical Mills	Slotting and Side Mills	End Mills	Form Relieved Cutters	Circular Saws
		High-Speed-Steel Milling Cutters				
Plastics	0.013	0.010	0.008	0.007	0.004	0.003
Magnesium and alloys	0.022	0.018	0.013	0.011	0.007	0.005
Aluminum and alloys	0.022	0.018	0.013	0.011	0.007	0.005
Free-cutting brasses and bronzes	0.022	0.018	0.013	0.011	0.007	0.005
Medium brasses and bronzes	0.014	0.011	0.008	0.007	0.004	0.003
Hard brasses and bronzes	0.009	0.007	0.006	0.005	0.003	0.002
Copper	0.012	0.010	0.007	0.006	0.004	0.003
Cast iron, soft (150–180 BH)	0.016	0.013	0.009	0.008	0.005	0.004
Cast iron, medium (180–220 BH)	0.013	0.010	0.007	0.007	0.004	0.003
Cast iron, hard (220–300 BH)	0.011	0.008	0.006	0.006	0.003	0.003
Malleable iron	0.012	0.010	0.007	0.006	0.004	0.003
Cast steel	0.012	0.010	0.007	0.006	0.004	0.003
Low-carbon steel, free machining	0.012	0.010	0.007	0.006	0.004	0.003
Low-carbon steel	0.010	0.008	0.006	0.005	0.003	0.003
Medium-carbon steel	0.010	0.008	0.006	0.005	0.003	0.003
Alloy steel, annealed (180–220 BH)	0.008	0.007	0.005	0.004	0.003	0.002
Alloy steel, tough (220–300 BH)	0.006	0.005	0.004	0.003	0.002	0.002
Alloy steel, hard (300–400 BH)	0.004	0.003	0.003	0.002	0.002	0.001
Stainless steels, free machining	0.010	0.008	0.006	0.005	0.003	0.002
Stainless steels	0.006	0.005	0.004	0.003	0.002	0.002
Monel metals	0.008	0.007	0.005	0.004	0.003	0.002

Sintered Carbide Tipped Cutters

Material						
Plastics	0.015	0.012	0.009	0.007	0.005	0.004
Magnesium and alloys	0.020	0.016	0.012	0.010	0.006	0.005
Aluminum and alloys	0.020	0.016	0.012	0.010	0.006	0.005
Free-cutting brasses and bronzes	0.020	0.016	0.012	0.010	0.006	0.005
Medium brasses and bronzes	0.012	0.010	0.007	0.006	0.004	0.003
Hard brasses and bronzes	0.010	0.008	0.006	0.005	0.003	0.003
Copper	0.012	0.009	0.007	0.006	0.004	0.003
Cast iron, soft (150–180 BH)	0.020	0.016	0.012	0.010	0.006	0.005
Cast iron, medium (180–220 BH)	0.016	0.013	0.010	0.008	0.005	0.004
Cast iron, hard (220–300 BH)	0.012	0.010	0.007	0.006	0.004	0.003
Malleable iron	0.014	0.011	0.008	0.007	0.004	0.004
Cast steel	0.014	0.011	0.008	0.007	0.005	0.004
Low-carbon steel, free machining	0.016	0.013	0.009	0.008	0.005	0.004
Low-carbon steel	0.014	0.011	0.008	0.007	0.004	0.004
Medium-carbon steel	0.014	0.011	0.008	0.007	0.004	0.004
Alloy steel, annealed (180–220 BH)	0.014	0.011	0.008	0.007	0.004	0.004
Alloy steel, tough (220–300 BH)	0.012	0.010	0.007	0.006	0.004	0.003
Alloy steel, hard (300–400 BH)	0.010	0.008	0.006	0.005	0.003	0.003
Stainless steels, free machining	0.014	0.011	0.008	0.007	0.004	0.004
Stainless steels	0.010	0.008	0.006	0.005	0.003	0.003
Monel metals	0.010	0.008	0.006	0.005	0.003	0.003

Source: Cincinnati Milling Machine Co., *A Treatise on Milling and Milling Machines*, 3rd ed., 1951, Table 13–14, pp. 214–215.

TABLE 14.3 EFFICIENCY OF MACHINES

HP of Machine	Efficiency (%)	HP of Machine	Efficiency (%)
3	40	20	60
5	48	25	65
7.5	52	30	70
10	52	40	75
15	52	50	80

Source: Kearney and Trecker Corp., Milwaukee, WI.

Table 14.1 shows some representative K-factors. These factors include an allowance of about 25% for dull cutters, so that this allowance need not be considered separately. The efficiency of various milling machines is shown in Table 14.3. The equation for calculating the power at the motor (HP_m) is given by the equation

$$HP_m = \frac{HP_c}{E}$$

E = efficiency
HP_m = horsepower at motor
HP_c = horsepower at cutter

Example 14.1

The feed of a 12-carbide-tooth face mill is 0.014 in. per tooth at 250 rpm. The material cut is 250 BHN steel. If the depth of cut is ¼ in. and the width of the cut is 2 in., calculate (a) the horsepower at the cutter and (b) the horsepower at the motor.

Solution
(a) The required horsepower at the cutter is

$$HP_c = \frac{yxF_t nN}{K} = \frac{0.250 \times 2 \times 0.014 \times 12 \times 250}{0.7}$$
$$= 30$$

F_t = 0.014 in.
n = 12 teeth
N = 250 rpm
y = 0.250 in.
x = 2 in.
K = 0.7 (Table 14.1)
E = 70% (Table 14.3)

(b) The horsepower at the motor is

$$HP_m = \frac{HP_c}{E} = \frac{30}{0.70}$$
$$= 42.9$$

Earlier it was indicated that the choice of the corner angle or radius on the tooth of a cutter affects the required horsepower. For corner angles on cutting teeth the *calculated chip thickness* always equals the feed per tooth (F_t); whereas the *actual chip thickness* (t) is a function of the cosine of the chamfer angle. That is,

$$t = F_t \cos \alpha \qquad \begin{aligned} t &= \text{actual chip thickness} \\ F_t &= \text{feed} \\ \alpha &= \text{chamfer angle} \end{aligned}$$

This is shown in Fig. 14.10.

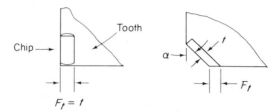

Figure 14.10

Example 14.2

What would be the horsepower in Ex. 14.1 if the cutter teeth had a 25° angle?

Solution

(a) The horsepower at the cutter is

$$HP_c = \frac{yxF_t(\cos \alpha)nN}{K} = \frac{0.250 \times 2 \times (0.014 \cos 25°)12 \times 250}{0.70}$$

$$= 27.2$$

(b) The horsepower at the motor is

$$HP_m = \frac{HP}{E} = \frac{27.2}{0.7}$$

$$= 38.8$$

14.5 THE DRILL PRESS

Many operations generally thought of as being reserved for other machines may be done on the drill press. However, as the name of this machine tool indicates, the drill bit, Fig. 14.11a, is the tool most commonly used for cutting. Twist drills may have two, three, or four flutes. The three- and four-fluted drill bits are generally reserved for enlarging irregularly shaped holes.

The nomenclature for the standard drill bit is given in Fig. 14.11b. These drill bits may be used, or adapted for use, on the rotary type of power tools. They are manufactured with straight, tapered, or special types of shanks for quick locking. The shank is usually soft to facilitate holding. The cutting portion of the drill is made from hardened high-carbon, holding-speed, or cobalt-alloy steels. The drill bit may also be made from carbide or fitted with carbide inserts. The flutes may be straight or have long leads or short leads. The helixes of most manufactured drills are right-hand.

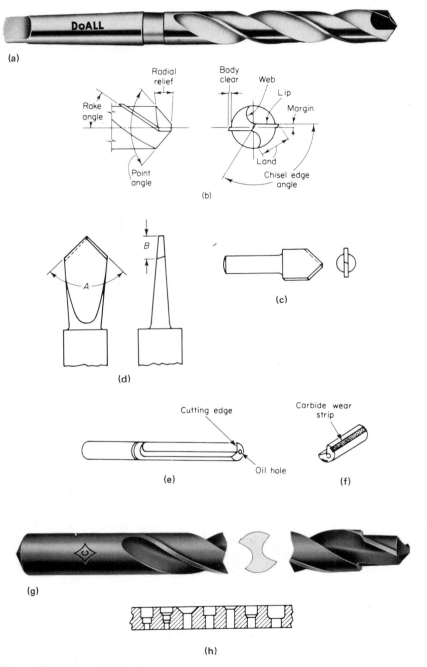

Figure 14.11(a,b,c,d,e,f,g,h)

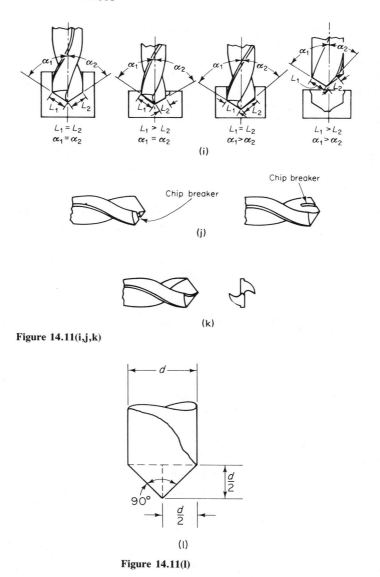

Figure 14.11(i,j,k)

Figure 14.11(l)

When viewed from the cutting edge, the helix winds around the drill in a clockwise direction. Thus the drill must rotate counterclockwise when viewed from the cutting end.

The high-helix (short lead) drill is used when it is necessary to clear chips rapidly from the hole being drilled. The helix angle is usually about 45°, which cuts down the friction between the flutes of the drill and the chips as they are being forced out of the hole. They are used for cutting deep holes in nonferrous materials.

Low-helix angle (long lead) drills are used for drilling nonmetals such as plastics and nonferrous materials. Anyone trying to drill through thin sheet metal has had the experience of the drill point's breaking through the bottom of the sheet while the lip of the drill is still cutting. If the workpiece is not clamped, the helix will catch the undrilled portion and climb the drill bit while it is rotating. To minimize this danger, low-lead helix angle or straight flute drill bits are used. They are manufactured with right- or left-hand helixes.

Still another design is the core drill. This drill usually has more than two flutes. Its purpose is to enlarge previously drilled holes, cored holes, or punched holes. There is no chisel edge on the core drill. The chamfered edges do the cutting. Since this action is similar to that of a reamer, these drills are sometimes used as roughing reamers. The lack of a chisel edge means that this drill cannot be used to drill into solid stock.

Another function of the core drill is to correct the location of predrilled holes. In this instance no chamfer is used on the lips, and the drill-point angle is ground flat (180°).

Sometimes the end of a core drill is a separate piece that may be changed when the size of the desired hole changes. These core drills are usually large and require that a pilot be used.

Flat drills, Fig. 14.11c, may be made from a drill rod. The end is forged flat, turned to the appropriate diameter and drill-point angle, heat treated, and ground with the proper lip clearance. This drill may be used when the appropriate drill size is not readily available.

Microdrills, Fig. 14.11d, are flat drills mounted in an alloy holder. These drills are made from carbide. They range from 0.001 to approximately 0.125 in. in diameter. They may be purchased in increments of 0.0001 in. in diameter. Since they break very easily, they must be cleared frequently.

The operating rpm of these drills should be about 2000 with feeds determined by trial and error. As the diameter of the drill decreases below 0.020 in., the rpm and feed should be decreased. Web thickness, rake angles, point angles, and pressure all affect the hole size that is produced. Regular point angles A equal to 135° are used to drill very small holes and to drill hard materials. Drill-point angles of 118° are used to drill steel. Drill points of 90° are used for drilling plastics and other soft materials. The clearance angle B may range from 8 to 12°. The harder or tougher the material to be drilled, the smaller is the angle B.

Gun drills, Fig. 14.11e, are used for drilling deep holes. They have one flute, usually parallel to the centerline of the drill. One of the lips extends beyond the end of the flute and has lip clearance which makes it possible for the drill to cut. Some method of creating a flow of water under pressure is provided. This is usually done through a long hole in the body of the drill. Thus the water flows through the drill to the cutting edge, keeps the drill cool, and washes the chips back through the flute. The V-flute is usually 110°, with the cutting edge having a radial clearance of 15°. These drills may have carbide inserts at the cutting edge as well as carbide wear strips on the outside of the drill, as shown in Fig. 14.11f.

It should be remembered that the drill cuts only at the end. Once started, the body of the drill acts as a guide for further drilling. Thus the drill is usually started through a bushing. If the drill is long, it is usually also supported with a steady rest. The wear strips prolong the life of the drill.

Where the hole to be drilled is large, an operation known as *trepanning* may be used. Trepanning does not remove all the material from the hole being drilled. It cuts a circular groove and leaves a core in the center of the hole. Since the core must enter the head (if the hole is long, the core must also enter the head holder), the head is made hollow and fastened to a drive tube. The oil is under pressure, and the chips are driven out through this drive tube.

Step drills are specially ground drills which are capable of drilling more than one diameter in one operation. A step drill is shown in Fig. 14.11g and some of the possible combinations are shown in Fig. 14.11h.

Conventional twist drills may be purchased in fractional, letter, number, or metric sizes. The fractional drills range from $\frac{1}{64}$ to $3\frac{1}{2}$ in. in diameter. The letter drill sizes are from A to Z, with the A drill equal to 0.234 in. in diameter and the Z drill equal to 0.413 in. in diameter. The number drills start with the number 1 drill, equal to 0.228 in. in diameter, and go to the number 80 drill, equal to 0.0135 in. in diameter. Since the letter drills pick up where the number drill sizes end and since the number and letter drill sizes fill many of the gaps between the fractional sizes, a wide variety of holes may be drilled between 0.0135 and 0.500 in. in diameter. For fractional drills between $\frac{1}{2}$ and $1\frac{1}{4}$ in., the drill diameters increase in increments of $\frac{1}{64}$ in.; between $1\frac{1}{4}$ and $1\frac{1}{2}$ in. the diameters increase in increments of $\frac{1}{32}$ in.; between $1\frac{1}{2}$ and $3\frac{1}{2}$ in. the diameters increase in increments of $\frac{1}{16}$ in.

It is very important that drills be properly ground. The drill point must be in the center of rotation of the drill. Figure 14.11i shows a drill point centered so that angle $\alpha_1 = \alpha_2$. Also, $L_1 = L_2$. If there is any variation of either the angle or the length of one side, a side thrust will be developed and the drilled hole will be larger than the drill diameter itself. The effect is shown in Fig. 14.11i.

It should also be mentioned that chip breakers, Fig. 14.11j, and web thinning, Fig. 14.11k, are possible. Chip breakers are used to aid in chip removal by breaking up the chip. Web thinning is used to restore the original web thickness after grinding. Web thinning may also be used to alter chip formation.

Drill points usually have an included angle of 118°. Drills used with numerically controlled machines are usually 90° points. This makes for easy calculations since the depth of the point is exactly equal to half the diameter of the drill (see Fig. 14.11*l*).

14.6 POWER REQUIRED FOR DRILLING

Power requirements for drilling need to reflect the *rpm* (*N*) of the spindle, the *force* (*P*) or thrust required to feed the drill into the work, and the *torque* (*M*) needed to rotate the drill. The torque and the rpm determine the power requirements; and the force determines the rigidity of the setup and the fixtures.

The power required to drill materials is given by the equation

$$HP_c = 1.6 \times 10^{-5}NM$$

M = torque, in.-lb
N = drill speed, rpm
HP_c = horsepower at cutter

The revolutions per minute (N) for the drill may be obtained by the method of Chap. 6. The equation is

$$N = \frac{12S}{\pi D}$$

S = cutting speed, ft/min
D = diameter of drill, in.

The torque (M) is obtained from the equation

$$M = KA(f^{0.8})(D^{1.8})$$

K = work material constant
A = torque constant
D = diameter of drill
f = feed, in./rev
M = torque, in.-lb

Combining equations and simplifying, the equation for horsepower becomes

$$HP_c = 1.6 \times 10^{-5} NM$$
$$= \frac{1.6 \times 10^{-5}(12S) \, KAf^{0.8}D^{0.8}}{\pi D}$$
$$= 6 \times 10^{-5} \, SKAf^{0.8}D^{0.8}$$

S = cutting speed, ft/min
K = work material constant
A = torque constant
f = feed, in./rev
D = diameter of drill, in.

The constant K is shown in Table 14.4, and the constant A in Table 14.5. It should be noted that the constants A, B, E are based on the ratio c/D, which is the ratio of the chisel edge length (c) to the diameter D of the drill. An easier measurement is the web thickness at the point (W) to the diameter (D) of the drill. For standard designs the c/D ratio that may be used is 0.18. Table 6.1 shows values of cutting speed (S) for various materials.

TABLE 14.4 WORK MATERIAL
CONSTANT K

Material	$K \times 10^3$
Steel	
200 BHN	24
300 BHN	31
400 BHN	34
Aluminum	7
Magnesium	4
Brass	14
Leaded brass	7

Source: National Twist Drill, Lexington, S.C., Catalog No. 69, Table IV, p. 211.

TABLE 14.5 TORQUE AND FORCE (THRUST) CONSTANTS BASED ON RATIO OF C/D AND W/D

c/D	Approx. w/D	Torque Const. A	Thrust Constant B	Thrust Constant E
0.03	0.025	1.000	1.100	0.001
0.05	0.045	1.005	1.140	0.003
0.08	0.070	1.015	1.200	0.006
0.10	0.085	1.020	1.235	0.010
0.13	0.110	1.040	1.270	0.017
0.15	0.130	1.080	1.310	0.022
0.18	0.155	1.085	1.355	0.030
0.20	0.175	1.105	1.380	0.040
0.25	0.220	1.155	1.445	0.065
0.30	0.260	1.235	1.500	0.090
0.35	0.300	1.310	1.575	0.120
0.40	0.350	1.395	1.620	0.160

Source: National Twist Drill, Catalog No. 69, Table I, p. 211.

The horsepower equation at the motor which allows for machine losses and drill dulling considers an efficiency (E) of 60% is

$$HP_m = \frac{HP_c}{E} \qquad \begin{array}{l} HP_m = \text{at motor} \\ HP_c = \text{at cutter} \\ E \;\;= \text{efficiency} \end{array}$$

Example 14.3

(a) Given a $\frac{3}{4}$-in.-diameter drill, a cutting speed of 200 ft/min, and a feed of 0.010 in./rev. (see Table 14.6, which gives several ranges of recommended feed rates for drills). The material to be drilled is aluminum. Calculate the horsepower at the drill point. (b) Calculate the horsepower at the motor. Make allowance for the dulling of the drill.

Solution

(a) The horsepower at the drill is

$$HP_c = 6 \times 10^{-5} S \, K \, A \, f^{0.8} D^{0.8} = 6 \times 10^{-5}(200)(7000)(1.085)(0.010)^{0.8}(0.75)^{0.8}$$
$$= 1.81$$

TABLE 14.6 RECOMMENDED FEED RATES FOR DRILLS

Diameter (in.)	Feed (in./rev)	Diameter (in.)	Feed (in./rev)
Under $\frac{1}{8}$	0.001–0.003	$\frac{1}{2}$–1	0.007–0.015
$\frac{1}{8}$–$\frac{1}{4}$	0.002–0.006	Over 1 in.	0.015–0.025
$\frac{1}{4}$–$\frac{1}{2}$	0.004–0.010		

Source: National Twist Drill, Catalog No. 69, p. 210.

(b) The horsepower allowing for the losses is

$$HP_m = \frac{HP_c}{E} = \frac{1.81}{0.60} = 3.00$$

It is important that all elements of the system involved in the drilling operation be rigid. Any lack of rigidity directly affects the results and may even result in breakage. The length of the drill and the web thickness may have a major effect on torsional stiffness. If the web thickness is increased, it should be thinned at the cutting tip. Axial stiffness becomes a problem for long drills, or where the feed is increased too much, and buckling will take place.

Another problem may arise when drilling materials that have high hardness work harden, or are very highly abrasive. Drills that are hard and those that work harden should be drilled under conditions where the drill bit has a continuous pressure placed on it. The drill bit must not be allowed to slide over material that work hardens. Low spindle speeds, shortened drill bits, and thick webs should be used. The feed should be maintained. High heat is generated when cutting the titanium alloys. Since there is very little work hardening present, the feed may be reduced.

Carbide drills are very resistant to abrasion at high feeds and spindle speeds. Breakage could be high because the carbide is brittle and is generally brazed to a high-speed steel shank. One of the most common abuses of carbide drills is the fact that they are permitted to get too dull. A carbide drill that has dulled a normal amount may be reground with little effort and the removal of a small amount of metal.

14.7 REAMERS

Reamers are cutting tools generally used in secondary operations to enlarge and size previously machined holes to accurate dimensions and high finish. They are designed to take small amounts of material out of a predrilled or prebored hole. Figure 14.12a shows a chucking reamer with straight flutes and a tapered shank. This may also be purchased with either straight or helical flutes. Other types are shell, tapered, and hand reamers. Straight and tapered reamers may also be of the solid or expansion variety. The expansion reamers may be expanded as much as 0.010 in. beyond the nominal size.

Hand reamers may also be of the solid or expansion type. A hand expansion reamer is shown in Fig. 14.12b. The taper at the end of the hand reamer is usually about 0.015 in. per inch of the length, with a 45° chamfer at the starting edge. If the flutes have a helix, it is usually left-handed to create an axial rake which will create a back pressure against the feed. The shank is about 0.005 in. smaller than the nominal size of the reamer. Figure 14.12c shows the nomenclature of a machine reamer, and Fig. 14.12d the nomenclature of a hand reamer.

Radial relief is provided for the land. But a small segment of the land is left without relief so that the reamer can maintain size when ground. It should be pointed out that the 45° chamfer is relieved and actually does the cutting during the reaming

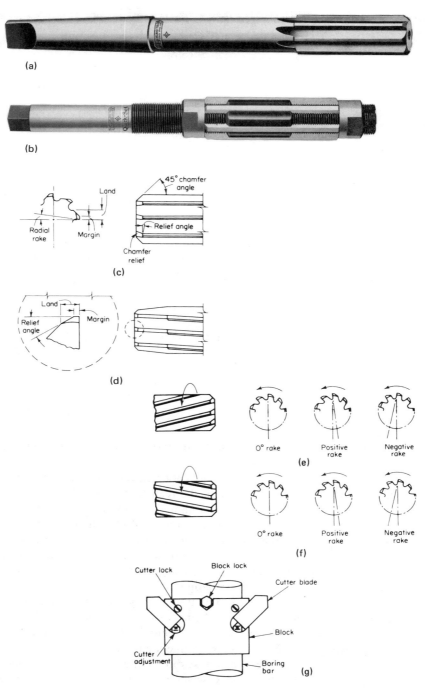

Figure 14.12(a,b,c,d,e,f,g)

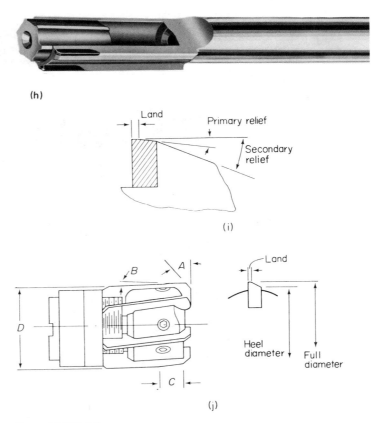

Figure 14.12(h,i,j)

operation. The body of the reamer is from 0.0005 to 0.001 in. smaller at the back than at the front to prevent drag.

Rose reamers have a land that is not relieved. This is to ensure that only the lead 45° does the cutting. The body of the reamer guides it as it reams the hole.

The feed used with reamers should be about three times that employed for drilling. The cutting speeds should be about 75% less than for drilling. The amount of stock removed depends on many factors, such as feeds, cutting speeds, type of material cut, setup, and so on. The amount of material removed should be enough to remove the feed helix from the former operation. This can be from 0.003 in. for hand reamers to $\frac{1}{64}$ in. for machine reaming.

Reamers have right-hand cutting. They may have left-hand, Fig. 14.12e; straight; or right-hand, Fig. 14.12f; helixes. They may also have positive, negative, or zero radial rake, shown in the end views of Fig. 14.12e and f.

Viewed from the end, these reamers cut when rotating counterclockwise. The right-hand helix twists in a clockwise direction when viewed from the end, as seen in Fig. 14.12f. The left-hand helix twists in a counterclockwise direction when viewed from the end, as shown in Fig. 14.12e. Thus the right-hand helix creates a positive axial rake and tends to pull the reamer into the work. This pulling action may be objectionable if there is any "play" in the machine spindle. This can be overcome by using a left-hand helix angle cutter which creates a negative axial rake, as shown in Fig. 14.12e.

Reamers may be either solid or expansion type. They may be purchased for machining one-diameter holes or tapered holes. They may be the rose type of reamer, with no margin relief, or the fluted type, in which the margin is relieved. If they are to be used for roughing, their flutes may be notched to break up the chip formation and thus permit greater stock removal.

Stock removal should not exceed $\frac{1}{32}$ in. for roughing and 0.005 to 0.010 in. for finishing. The removal of too much stock will result in oversize and rough holes. It may also cause damage to the reamer as well.

Combination, or step, reamers may also be specially purchased. A reamer with multiple diameters for line reaming is also a possibility.

It should be noted that reamers may be carbide tipped for longer life. Solid carbide reamers are usually reserved for reamers having diameters of less than $\frac{1}{4}$ in. All-carbide reamers need very careful handling. When used properly, they have the advantage of picking up less metal because they have high hardness and finish.

For reaming or boring large holes, inserted tooth blocks, Fig. 14.12g, are used. The body of the block is made from a fine alloy steel and heat treated. The block-locking device positions the block so that the reamer blades will be equidistant from the centerline of the boring bar when locked in position.

Carbide reamers are essentially the same as high-speed reamers. They have cutting edges made from carbide material, which makes them very highly wear resistant. All cutting is done by the chamfer.

Because carbide is brittle and because the body of the reamer must be rigid expansion reamers cannot be manufactured like high-speed expansion reamers. Figure 14.12h shows one type of expansion reamer. The drive plug is tapered 0.020 in. per inch, which permits adjustments of 0.001 in. The limit of adjustment is 0.003 to 0.005 in. The plug may then be removed and a 0.005-in. shim inserted. The plug is driven out through the hole in the shank. Because the teeth are not equally spaced, micrometer readings should be taken at the high diameter when setting the reamer size.

Land widths, Fig. 14.12i, range from 0.007 in. on small reamers to 0.020 in. on reamers $1\frac{1}{2}$ in. and over. Primary clearance angles are from 15° on small to 7° on large reamers. The chamfer on the lead is generally 6 to 8° with a 0.003 in. per inch of length back taper in the plane of the flutes, Fig. 14.12j.

Inserted blade reamers are adjustable and provide a practical method for manufacturing large reamers. Figure 14.12j shows the essential features of an inserted blade reamer. Angle A is generally maintained at 45° and is backed off for clearance in back of the cutting edge. The secondary lead angle B should be 3° for cast iron and nonferrous metals. Its length C should range from $\frac{3}{16}$ to $\frac{3}{8}$ in. For malleable iron and the steels angle B should be 5° and length C $\frac{1}{32}$ to $\frac{3}{32}$ in. It also should be backed off. The heel diameter should be about 0.005 in. less than the full diameter at the cutting end of the reamer. It should be backed off to leave a land width L of from 0.005 to

0.015 in. The back diameter D should be tapered smaller by an amount equal to 0.001 in. for each inch of length.

14.8 COUNTERSINKING, SPOTFACING, COUNTERBORING, BORING, AND TAPPING

Countersinking.   There are many kinds of countersinking tools. However, in almost all cases the operation is one of forming a chamfer of less than 90° with the centerline of a previously drilled hole. A countersinking tool and a countersunk hole are shown in Fig. 14.13a. Countersinks may be purchased with included angles of 45°, 53°, 60°, 78°, or 90°. They usually have three flutes. Six-flute countersinks are chatterless because of the uneven spacing of the flutes about the periphery.

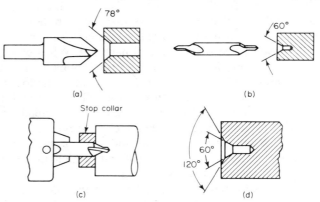

Figure 14.13

Figure 14.13b shows the combination drill and countersink used for centering work which is to be placed between centers of a lathe for turning or between the centers of a grinder when the work is to be ground. The included angle of the countersink is 60°.

Sometimes it becomes necessary to control the depth of the countersunk hole. Such an instance occurs in production grinding when a length dimension is to be maintained in grinding to a shoulder. In such cases some type of stop-ring must be affixed to the countersink to control the depth of the countersunk hole. A simple collar is shown affixed to the countersink in Fig. 14.13c. Countersinks fitted with micrometer stops may also be purchased.

In some instances, it is desirable to use a drill and countersink combination which has the 60° included angle and a secondary included angle of 120°. This serves the purpose of preserving the edge of the countersunk hole. This type of countersunk hole is shown in Fig. 14.13d.

Spotfacing.   Spotfacing is an operation in which a circular flat surface is machined perpendicular to a previously drilled hole. Spotfacing tools have pilots to

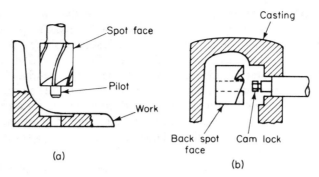

Figure 14.14

ensure concentricity of the spotfacing with the hole. Figure 14.14a shows a cast angle plate with a spotfacing at the surface of a bolt hole. This ensures a flat bearing surface for a washer or bolt at 90° with the hole.

Back spotfacing tools are also available. The holder usually has an eccentric or cam locking device. The holder is inserted into a hole to be back spotfaced. The spotfacer is placed on this holder in a reverse position. The rotation of the spotfacer holds it on the holder. This is shown in Fig. 14.14b.

Spotfacing tools may be of the shell type with an inserted blade, a solid blade that is integral with the shank, or a carbide tip. The shank may be a straight, tapered (usually Morse tapered), or pinlock type. The pilots may be integral with the spotfacing tool or interchangeable.

Counterboring.   The counterboring operation differs from the spotfacing operation in two essential ways: (1) the counterbore is deeper, and (2) its purpose is to enlarge a previously drilled hole for a given depth, terminating in a shoulder. The counterbored portion of the hole must be concentric to, and the shoulder is usually perpendicular to, the predrilled hole.

The counterboring tool is usually the same tool as that used for spotfacing. It must also be piloted unless used through a guide bushing in a fixture. Counterbores may be fitted with a stop collar to ensure proper depth, or they may be stepped to give multiple diameters.

They may be made from high-speed steel or have carbide tips or inserted blades. The clearance at the cutting edge is about 5° and the helix angle is 0 to 15°.

Boring.   Boring may also be done on a drill press. The purpose of boring on a drill press is usually to correct any runout created by the drilling operation before reaming. It is also possible that a hole may have to be bored on a drill press when an appropriate-sized drill or reamer is not available. In this instance a rigid boring bar with an inserted tool bit is used. A hole is drilled and then enlarged with the boring operation and, if necessary, then reamed. All the principles of clearance, rake, feed, and speed must be adhered to.

Tapping.   Tapping is another operation that may be done on a drill press. Tapping is a difficult operation under any conditions. The problems will relate to the amount of material to be removed with the first pass of the tap. This creates problems of torque, chip clearance, feeds and speeds. The feed is fixed by the lead of the tap.

To ensure starting the tap straight, especially when hand tapping is done, three types of taps are manufactured. A set of taps includes the starting (tapered), plug, and bottoming taps, The starting tap is tapered from the end for about six to eight threads, the plug tap is tapered for about three threads from the end, and the bottoming tap is merely ''backed off.'' The pitch diameter of all three taps is the same. The bottoming tap is used in blind holes.

Serial taps are different from the set of three taps just described. Serial taps come in a sequence of 1, 2, 3 and must be used in that order to achieve a full depth of thread. Each tap cuts more material than the previous tap. The number 1 tap can therefore be considered a roughing tap, and the number 3 tap, a finishing tap.

Taps are manufactured in fractional ($\frac{1}{2}$–13) and numbered (10 to 32) sizes. Tapered taps, such as pipe taps and acme taps, are also available as standard stock.

Gun taps are taps that have their cutting faces (in the flutes) cut back to give an axial rake, shown in Fig. 14.15a. This gives a shearing action to the cutting but, more important, causes the chips to move ahead of the tapping action. It should be obvious that such an action will clear the chips by forcing them deeper into the drilled hole. Gun taps should therefore be used only on through holes. This type of tap is also suitable for machine tapping.

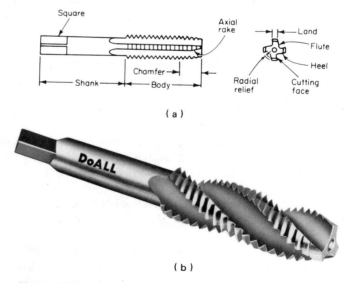

(a)

(b)

Figure 14.15

Spiral fluted taps, Fig. 14.15b, differ from gun taps in that the flutes have a helix about the same as a twist drill's. A tap of this type which cuts a right-hand lead should have a right-hand helix. This tap is excellent for cutting tough or soft materials such as stainless steel, copper, aluminum, and plastics.

Collapsible taps are manufactured so that they will collapse when the desired length of thread is cut. They may be collapsed by hand, in which case they must be reset, or they may be fitted with a trip ring which causes the chasers to retract when the plate is tripped. They may be made for turret lathes or other automatic equipment, where the tap does not revolve, or for machines such as a drill press, where the tap revolves. The chasers may be removed for sharpening, replaced, and adjusted rather easily. Collapsible die heads for cutting external threads are also available for use on drill presses, turret lathes, and automatic machines.

PROBLEMS

14.1. What is the advantage of taking a cut as shown in Fig. 14.1a over enlarging the bore diameter the same amount but taking the cut on one side of the boring bar?

14.2. What advantages are there to using block tool holders?

14.3. What determines the hand of the helix of a cutter?

14.4. Study the cutters in Fig. 14.2. Explain the general-purpose use of each cutter.

14.5. What is a fly cutter? Explain its use in cutting contours on a milling machine.

14.6. Select an inserted tooth carbide slab milling cutter and explain how the teeth are secured and adjusted before grinding.

14.7. Contrast conventional and up milling. What are the advantages of each? Why are surface finishes generally better when down milling?

14.8. Distinguish between relief, primary, and secondary clearance angles on a cutter tooth. What is the purpose of each?

14.9. Discuss the land of a cutter tooth, its width, the effect if it is too wide, and the function of the heel.

14.10. Define and illustrate (**a**) axial rake and (**b**) radial rake for a milling cutter.

14.11. Given a signature for a milling cutter of $-6, 5, 7, 6, 2, 18, 0.060$ at $45°$, draw the tooth, identify each angle, and insert the values.

14.12. Draw necessary views of a two-lip end mill and insert the names of all the tooth parts.

14.13. (**a**) Define the power requirements of a milling cutter. (**b**) Which factors are reflected in the value of K?

14.14. A 20-tooth cutter, 4 in. in diameter, is operating at 80 rpm with a feed of 0.625 in./rev. The material to be cut is medium cast iron. The width of the cut is $\frac{1}{2}$ in. and the depth is 0.050 in. Calculate the horsepower (**a**) at the cutter and (**b**) at the motor if the efficiency is 40%.

14.15. What would be the horsepowers in Prob. 14.14 if the cutter teeth have a corner angle of 20°?

14.16. The feed of an eight-tooth face mill is 0.013 in. per tooth at 200 rpm. The material cut is 300 BHN steel. If the depth of the cut is $\frac{1}{8}$ in. and width 4 in., calculate (**a**) the horsepower at the cutter and (**b**) the horsepower at the motor.

14.17. What would be the horsepowers in Prob. 14.16 if the cutter teeth had a 30° angle?

14.18. Make a sketch of a drill bit and label all parts.

14.19. Explain the action of a drill bit. What effect does a right-hand, straight, or left-hand helix have on the cutting action as a drill bit revolves?

14.20. What advantages does a short lead twist drill have over a long lead twist drill?

14.21. What is a core drill? When is it advisable to use a core drill?

14.22. When should a pilot be used with a core drill?

14.23. If a piece of drill rod is available, make a flat drill in your shop. Describe the procedure that you followed in making and heat treating the drill. Try it on your drill press and report the results to your instructor.

14.24. (a) What is a microdrill? (b) In what sizes are they available? (c) What are the operating speeds of these drills? (d) What point angles should be used on small holes in (1) hard materials, (2) steel, and (3) soft materials?

14.25. (a) Describe the action of a gun drill. (b) What is the characteristic of a gun drill that makes it possible to drill a straight hole?

14.26. What is trepanning?

14.27. Make several drawings of step drills that you have found in a textbook other than this one.

14.28. What is the range of diameters of (a) letter drills and (b) number drills?

14.29. Explain why imperfect holes are drilled (a) when the length of the cutting edge, L_1, is less than L_2; (b) when the lip angle α_1 is less than the lip angle α_2; and (c) when both L_1 and α_1 vary from L_2 and α_2. The notations refer to Fig. 14.11i.

14.30. Is it possible to use chip breakers on a drill bit? How?

14.31. When is web thinning used on a drill? Why is it important?

14.32. (a) What two factors enter into horsepower calculations of a drill bit? (b) What factors determine rigidity requirements?

14.33. Given a $1\frac{1}{2}$-in.-diameter drill, cutting speed of 90 ft/min, and a feed of 0.018 in./rev. The material to be dulled is steel, 300 BHN. Calculate the horsepower (a) at the drill point and (b) at the motor. Make a 75% allowance for dulling and their losses. Assume the web thickness is to be $\frac{5}{32}$ in.

14.34. (a) Given a 2-in.-diameter drill, a cutting speed of 150 ft/min, and a feed of 0.020 in./rev. (See Table 14.6, which gives several ranges of recommended feed rates for drills.) The material to be drilled is brass. Calculate the horsepower at the drill point. (b) Calculate the horsepower at the motor making an allowance for the dulling of the drill.

14.35. Discuss web thickness and length of drill as a consideration when designing or selecting a drill.

14.36. How can the problems of material hardness and abrasiveness be overcome in drill design?

14.37. Why is it important to resharpen carbide drills as soon as they begin to dull?

14.38. Given a 90° drill point. The diameter of the drill is $\frac{5}{8}$ in. What is the depth of the cone at the point where the drill starts to cut its full diameter? Explain your answer.

14.39. Why is the margin on a reamer ground without relief?

14.40. How does a rose reamer differ from a machine reamer?

14.41. How does a hand reamer differ from a machine reamer?

14.42. How much material should a reamer remove from a predrilled hole?

14.43. When are left-hand helixes used on reamers? Why is their use important?

14.44. What is a block reamer?

14.45. When finish reaming, how much material should be removed from a predrilled hole?

14.46. Describe the procedure for expanding a carbide expansion reamer, Fig. 14.12h.

14.47. Discuss the land, clearances, chamfer, and back taper of a reamer.

14.48. (a) Make a hand sketch of an inserted blade reamer and label all parts. (b) Indicate the range of sizes and angles for each part of the reamer in part (a).

14.49. (a) What is a combination drill and countersink? Make a sketch of this tool. (b) How can the depth to which this drill combination operates be controlled? (c) Why is it important to control this depth on a cylindrical grinder? Explain.

14.50. What is back spotfacing?

14.51. How does the counterbore operation differ from the spotfacing operation?

14.52. How is boring done on a drill press?

14.53. Describe the lead threads on a starting tap, a plug tap, and a bottoming tap.

14.54. How do serial taps differ from a set of starting-plug-bottoming taps?

Chapter 15

Piercing and Blanking Dies

15.1 PRESSES

There are various types of machines which are manufactured to apply forces, repeatedly, to workpieces so that the metal may be cut, shaped, or caused to flow into a desired pattern. These forces may be applied through the use of mechanical or hydraulic systems. The most commonly used press is the inclined press, sometimes referred to as the *open-back inclined press*, OBI, Fig. 15.1. The main features of this press are:

1. A *bed*, which is generally rectangular in shape. The center is open to allow finished parts or scrap to fall through as a result of gravity.

2. A *bolster plate* is fastened to the top of the bed. It is a steel plate to which the die set is fastened. It is from 2 to 5 in. thick. The openings through bolster plates have been standardized and are shown in Table 15.1.

3. The *ram*, at times called the slide, oscillates and provides the motion which moves the punch holder through its stroke. The position of this stroke may be altered. The length of the ram stroke is a function of the design of the machine. It should be noted that the shut height of a press is determined when the ram is in its *down* position and the adjustment in the maximum *up* position. This is measured from the top of the bolster to the slide.

4. *Knockouts* are mechanisms that are used to eject workpieces from the die set. The ram is built so that a knockout rod can be mounted and operated on the up stroke.

Figure 15.1

5. The *pitman arm* is the connecting link between the ram and the crank or main shaft.

6. The fly wheel has mass enough so that energy provided by a motor is stored, so that it will provide intermittent forces to the ram without appreciable loss of power.

Some presses are classified as *C-frame,* others as *straight-sided presses.* The C-frame presses have already been mentioned, and the important dimensions referred to in Table 15.1. The straight-sided presses are usually much larger than the C-frame presses. They have straight columns at both ends of the bed and are open in the front and rear.

Gear-driven presses may be single- or double-reduction. The single-reduction presses are the presses for shallow driving, or where average shearing forces are required. The double-reduction presses are usually very large. The stroke speed is usually slow but exerts high forces. In all cases the motor drives a flywheel, which in turn drives a crankshaft. The power is then transmitted from the crankshaft directly to the pitman; or from the crankshaft through a gear reduction and then to the pitman and ram.

Hydraulic presses utilize a cylinder and a double-action piston to drive the ram in both directions. The power is supplied by a hydraulic pump. The pressure exerted is a function of the cross-sectional area of the piston.

Presses may also be designated according to the number of slides that they have. Thus they may be called single-action, double-action, or triple-action presses. The single-action press has one ram. The double-action press has two rams, one inside the other. The triple-action press has three rams, one inside the other and a separate third ram. In all cases the rams have a specific job to perform and therefore must be synchronized with each other.

TABLE 15.1 STANDARD BOLSTER PLATE OPENINGS[a]

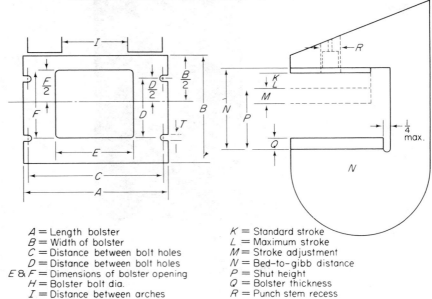

A = Length bolster
B = Width of bolster
C = Distance between bolt holes
D = Distance between bolt holes
E & F = Dimensions of bolster opening
H = Bolster bolt dia.
I = Distance between arches

K = Standard stroke
L = Maximum stroke
M = Stroke adjustment
N = Bed-to-gibb distance
P = Shut height
Q = Bolster thickness
R = Punch stem recess

Tonnage	A	B	C	D	E	F	H	I
22	20	12	18	$7\frac{1}{2}$	8	5	$\frac{3}{4}$	9
32	24	15	22	9	11	8	$\frac{3}{4}$	11
45	28	18	$25\frac{1}{2}$	$10\frac{1}{2}$	14	8	1	13
60	32	21	$29\frac{1}{2}$	12	16	11	1	15
75	36	24	33	18	18	14	$1\frac{1}{4}$	18
110	42	27	39	18	21	15	$1\frac{1}{4}$	21
150	50	30	47	18	21	17	$1\frac{1}{4}$	24
200	58	34	55	18	27	21	$1\frac{1}{4}$	27

Tonnage	K	L	M	N	P	Q	R	S
22	$2\frac{1}{2}$	4	2	$11\frac{1}{4}$	$8\frac{1}{2}$	$2\frac{1}{2}$	$1\frac{5}{8}$	$2\frac{1}{4}$
32	3	5	$2\frac{1}{2}$	$12\frac{3}{4}$	$9\frac{1}{2}$	$2\frac{1}{2}$	$1\frac{5}{8}$	$2\frac{1}{4}$
45	3	6	$2\frac{1}{2}$	$14\frac{1}{4}$	11	3	$2\frac{1}{8}$	3
60	$3\frac{1}{2}$	7	$2\frac{3}{4}$	$16\frac{3}{4}$	13	3	$2\frac{5}{8}$	3
75	4	8	3	$19\frac{1}{4}$	15	$3\frac{1}{2}$	$2\frac{5}{8}$	3
110	5	10	$3\frac{1}{2}$	$23\frac{1}{4}$	18	4	$3\frac{1}{8}$	3
150	6	12	4	$28\frac{1}{4}$	22	$4\frac{1}{2}$	$3\frac{1}{8}$	3
200	8	12	$4\frac{1}{2}$	$32\frac{1}{4}$	24	5	$3\frac{1}{8}$	3

[a]Mounting holes or slots in press bed, for bolting bolster plates to bed to conform to center-to-center distances on bolster plate. Jig-standard dimensions for OBI presses.
Source: ASTME, *Die Design Handbook* (New York: McGraw-Hill, 1965); by permission.

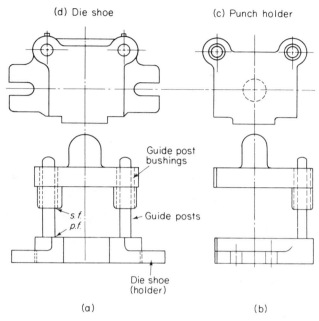

Figure 15.2(a,b,c,d)

15.2 DIE SETS

Die sets are used to hold and aid in the alignment of the cutting members. A die set and its components are shown in Fig. 15.2. The elevations have been drawn according to the conventions used by die designers. Figure 15.2a is the front elevation of a die set. It shows the punch holder (or shoe), die shoe (or holder), bushings, and guide posts. Figure 15.2b is the right elevation, and shows the front-to-back clearance available. The punch holder, Fig. 15.2c, is removed, rotated 180° as though it were the cover of a book so that it shows the view from the bottom of the punch holder and exposes the punches fastened to it. Note that the punch shank is dotted in the upper right view, Fig. 15.2c. Removing the punch holder exposes the top of the die shoe. The upper left view, Fig. 15.2d, shows the die shoe with all the die openings exposed.

Sometimes the front view, Fig. 15.2e, is split down the vertical centerline so that one half the die can be seen in the closed position, the other half in the open position. This technique enables the designer to evaluate interference. Crosshatching is used for clarity of the positions of the punches and dies. The punch holder and a die shoe are crosshatched with evenly spaced lines. The punch and dies are crosshatched with grouped lines.

Die sets may have no guide posts at all and rely on the machine ram for alignment or on pilots in the punches. They are called plain die sets and are not of high precision. They may also have shanks, or in the very large die sets fasten directly

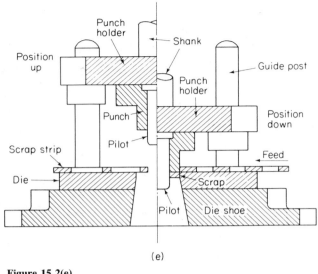

(e)

Figure 15.2(e)

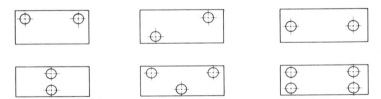

Figure 15.3

to the ram. Figure 15.3 shows various post positions, depending on the type of operation to be done. They are referred to as two-, three-, or four-post die sets. Table 15.2 shows dimensions for some standard back post die sets. There are many additional die sets available commercially.

15.3 THEORY OF CUTTING

The end result of a punch and die cutting operation is a partially finished, or semi-finished part, and scrap. When a punch cuts through unprocessed material and the slug that results is the workpiece, the operation is called *blanking*. If the slug that results is scrap, the operation is called *piercing*. In both cases the operations are called *shearing*. As far as the theory of cutting is concerned, these designations are purely arbitrary. In general, the results of a blanking operation are usually the *blank*, which is the workpiece, whereas the results of a piercing operation are usually scrap. These are shown in Fig. 15.4a.

The cutting operation that takes place when a punch cuts through a piece of

TABLE 15.2 BACK-POST DIE SET DIMENSIONS

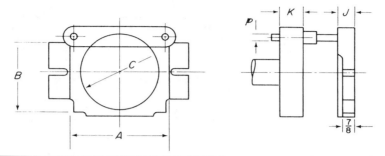

Die Zone (in.)			Thickness (in.)				
Right to Left, A	Front to Back, B	Dia., C	Die Shoe, J		Punch Holder, K		Min. Guide Post Dia., P (in.)
			From	To	From	To	
3	3	3	1	$1\frac{1}{4}$	1		$\frac{3}{4}$
4	6		$1\frac{1}{2}$	$2\frac{3}{4}$	$1\frac{1}{4}$	$2\frac{1}{4}$	1
5	5	5	$1\frac{1}{2}$	2	$1\frac{1}{4}$	$1\frac{3}{4}$	1
6	4	5	$1\frac{1}{2}$	$2\frac{3}{4}$	$1\frac{1}{4}$	$2\frac{1}{4}$	1
	6	$6\frac{1}{2}$	$1\frac{1}{2}$	$2\frac{1}{2}$	$1\frac{1}{4}$	$2\frac{1}{4}$	1
7	5	$5\frac{3}{4}$	$1\frac{1}{2}$	3	$1\frac{1}{4}$	$2\frac{1}{4}$	1
	7	$7\frac{1}{2}$	$1\frac{1}{2}$	$2\frac{1}{2}$	$1\frac{1}{4}$	$2\frac{1}{4}$	1
8	6	7	$1\frac{1}{2}$	3	$1\frac{1}{4}$	$2\frac{1}{4}$	1
	8	$8\frac{1}{2}$	$1\frac{1}{2}$	$2\frac{1}{2}$	$1\frac{1}{4}$	$2\frac{1}{4}$	1
9	12		$1\frac{3}{4}$	$3\frac{1}{2}$	$1\frac{1}{2}$	$2\frac{1}{4}$	$1\frac{1}{2}$
10	5		$1\frac{1}{2}$	$2\frac{1}{2}$	$1\frac{1}{4}$	$1\frac{3}{4}$	$1\frac{1}{4}$
	10	10	$1\frac{5}{8}$	$2\frac{3}{4}$	$1\frac{3}{8}$	$2\frac{1}{4}$	$1\frac{1}{4}$
11	9	10	$1\frac{3}{4}$	$3\frac{1}{2}$	$1\frac{1}{2}$	$2\frac{1}{4}$	$1\frac{1}{4}$
12	12	$12\frac{1}{2}$	$1\frac{3}{4}$	$3\frac{1}{2}$	$1\frac{3}{4}$	$2\frac{1}{4}$	$1\frac{1}{2}$
14	10	$11\frac{1}{4}$	$1\frac{3}{4}$	$3\frac{1}{4}$	$1\frac{5}{8}$	$2\frac{3}{4}$	$1\frac{1}{2}$
18	14	15	2	3	$1\frac{3}{4}$	$2\frac{1}{4}$	$1\frac{1}{2}$
20	5		$1\frac{3}{4}$	$2\frac{1}{2}$	$1\frac{1}{2}$	2	$1\frac{1}{2}$
25	14		$1\frac{3}{4}$	3	$1\frac{1}{2}$	$2\frac{1}{4}$	$1\frac{1}{2}$

material has three components: (1) plastic deformation, (2) shear, and (3) break. The control of these three components determines the quality of the workpiece produced.

Plastic deformation (Fig. 15.4b).   Very soon after the punch has contacted the work, the continued downward movement of the punch exerts forces on the work material. Once the elastic limit of the material has been exceeded, plastic deformation takes place. The combination of elastic-plastic deformation results in an upper radius band on the scrap strip and a lower radius band on the workpiece.

Shear (Fig. 15.4c).   The slug is pushed farther into the die opening by the punch, but at this point in the operation, the material has begun to separate as a result

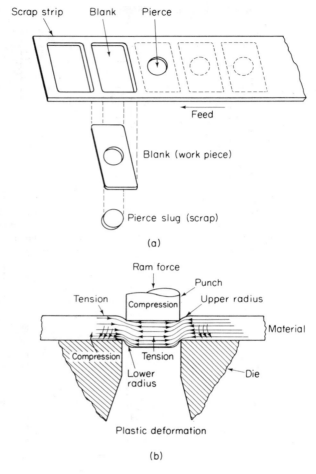

Figure 15.4(a,b)

of being cut. The material is able to resist fracture at the cutting edges of the punch and the die, and therefore the punch cuts (shears) the material at the cutting edges of the punch. At the same time the slug is pushed farther into the die opening. Again, the material resists fracture and is sheared at the cutting edge of the die opening. So far the two operations result in a radius band and a shear band about one-third down from the top of the hole and one-third up from the bottom edge of the slug. These actions are shown in Fig. 15.4c.

Break (Fig. 15.4d).   The cutting operation is completed when the material can no longer resist the forces of cutting. It fractures as the punch pushes farther into the material. Because there is clearance between the cutting edge of the punch and the die, the forces on the material are at an angle and the fracture band takes place as

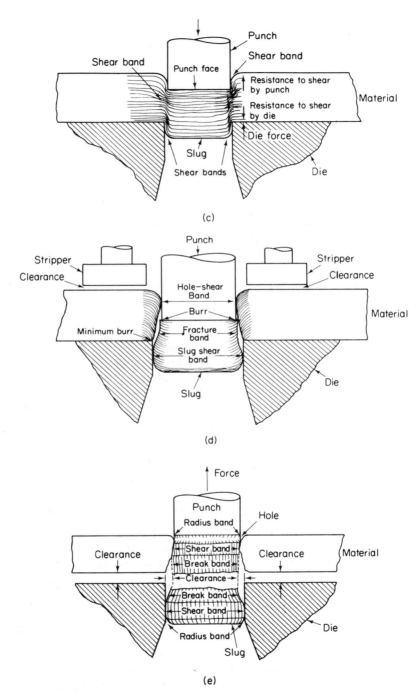

Figure 15.4(c,d,e)

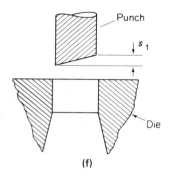

Figure 15.4(f)

shown in Fig. 15.4d. A slight burr (jagged sharp edge) is generally left at the bottom of the hole and the top of the slug.

The slug is pushed farther into the die opening. The slug shear band expands and is held in the die opening as shown in Fig. 15.4e. The hole shear band contracts and clings to the punch. As the punch retracts, the material moves up and closes the clearance, Fig. 15.4d, between the obstruction built into the die and the material. This obstruction is called a *stripper*. It, in fact, strips the stock from the punch as the punch continues to move up.

The tonnage required for shearing is a function of the length of the cut edge being sheared at any instant and the shearing strength of the material. Table 15.3 shows the shearing strength of some common materials.

The equation for determining the force required to cut through material is given by

$$F = \frac{LtS_s}{2000}$$

L = length of cut edge, in.
t = thickness of material, in.
S_s = shear strength, psi
F = force, tons
F_s = force of sheared punch, tons

If shear is applied to the punch, Fig. 15.4f, the force necessary to cut through the material is reduced according to the penetration before break occurs (see Fig. 15.4d). The ratio of the penetration to the shear applied to the face of the punch (or die) gives the reduction factor. Thus the force equation becomes

$$F_s = KF \qquad \text{where} \qquad K = \frac{tp}{s}$$

K = constant
p = % penetration
s = shear, in.
t = thickness, in.

The value of p is shown in Table 15.4.

It should be noted that s should always be greater than tp; otherwise, F_s will be greater than F and there would be no purpose in applying shear to the punch (or die).

TABLE 15.3 Strength of Materials

Material	Shear (× 10³ psi)	Ultimate (× 10³ psi)	Yield (× 10³ psi)
Aluminum (anneal)			
2SO	9.0	13.0	5.0
17SO	18.0	26.0	10.0
24SO	18.0	25.0	11.0
Alclad	16.0	32.0	14.0
Asbestos	5.0	—	—
Brass (anneal)			
Gilding metal	30.0	38.0	14.0
Comm. bronze	32.0	41.0	15.0
Red	34.0	46.0	18.0
Cartridge	35.0	54.0	22.0
Yellow	35.0	54.0	22.0
Bronze (anneal) high silicon (3%)	45.0	63.0	30.0
Cardboard			
Soft	4.0	—	—
Hard	8.0	—	—
Cellulose acetate	10.0	—	—
Cloth	8.0	—	—
Copper (anneal) electrolytic	22.0	32.0	10.0
Fiber			
Hard	28.0	—	—
Phenol	25.0	—	—
Inconel (anneal)			
Hot-rolled	65.0	95.0	45.0
Cold-rolled	63.0	90.0	37.0
High T.S.	85.0	156.0	108.0
Lead, commercial	2.0	—	—
Leather			
Soft	9.0	—	—
Hard	14.0	—	—
Magnesium (annealed)	21.0	36.0	22.0
Hard rolled	20.0	37.0	27.0

Material	Shear (× 10³ psi)	Ultimate (× 10³ psi)	Yield (× 10³ psi)
Mica	10.0	—	—
Monel			
Soft	50.0	75.0	35.0
Hard	65.0	110.0	100.0
Monel-K			
Annealed	75.0	120.0	60.0
Cold-worked	100.0	160.0	120.0
Nickel, rolled	52.0	63.0	20.0
Nickel-silver			
Annealed	56.0	60.0	30.0
Hard	80.0	85.0	74.0
Paper			
Bristol	6.5	—	—
Soft	4.0	—	—
Hard	5.0	—	—
Rubber, hard	20.0	—	—
Steel (annealed)			
0.10C	35.0	56.0	—
0.20C	42.0	69.0	—
0.30C	52.0	85.0	—
0.40C	70.0	100.0	—
0.50C	80.0	110.0	—
1.00C	110.0	125.0	—
SAE			
2320	98.0	160.0	—
2340	125.0	—	—
3130	110.0	—	—
3140	130.0	—	—
3250	165.0	—	—
4130	115.0	—	—
Soft	55.0	90.0	—
Hard	90.0	12.5	—
Stainless steel, No. 302	56.0	74.0	30.0

TABLE 15.4   % PENETRATION

Material	p (%)	Material	p (%)
Carbon steel		Aluminum alloy	60
0.10 anneal	50		
0.10 CRS	38	Brass	50
0.20 anneal	40		
0.20 CRS	28	Bronze	25
0.30 anneal	33		
0.30 CRS	22	Copper	55
Silicon steel	30	Nickel alloy	55
		Zinc alloy	50

The stripping force, Fig. 15.4e, is given by the equation

$$f_s = 3500Lt \qquad f_s = \text{stripping force, lb.}$$

Example 15.1

A washer is 0.060 in. thick and has a $\frac{3}{8}$-in. hole and an outside diameter of $\frac{3}{4}$ in. as shown in Fig. 15.5. The material is SAE 1010 steel, ultimate shearing strength of 35,000 psi. Calculate (a) the force needed to do the job if both punches operate at the same time with no shear and (b) the force if the punches are staggered and only one operates at a time. (c) Assume 60% penetration when the punchs have 0.050 in. shear. What force is needed if the punches are staggered? (d) What stripping force is required to strip both punches?

Solution

(a) The force of both punches acting at the same time with no shear is

$$F = \frac{LtS_s}{2000} = \frac{\pi(D + d)tS_s}{2000} = \frac{\pi(0.750 + 0.375)(0.060)(35,000)}{2000}$$

$$= 3.7 \text{ tons}$$

$t = 0.060$ in.
$S_s = 35{,}000$ psi
$L = $ length cut, in.
$D = $ large diameter, in.
$d = $ small diameter, in.

(b) If the punches are staggered, the punch taking the longest cut will determine the tonnage. In this case it is the $\frac{3}{4}$ in. punch. Thus

$$F = \frac{\pi DtS_s}{2000} = \frac{\pi(0.750)(0.060)(35,000)}{2000}$$

$$= 2.47 \text{ tons}$$

This is more than enough tonnage to take care of the small punch, since it does not begin to operate until the large punch finishes cutting.

(c) If 0.050-in. shear is added to the staggered punches, the force is

$$F = KF = \frac{tp}{s}(F) = \frac{0.060 \times 0.60}{0.050}(2.47)$$

$$= 1.78 \text{ tons}$$

$t = 0.060$ in.
$p = 60\%$
$s = 0.050$ in.
$F = 2.47$ tons

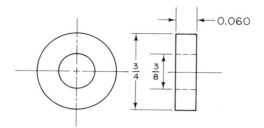

Figure 15.5

(d) The stripping force is

$$f = 3500 \, Lt = 3500\pi(0.750 + 0.375)(0.060)$$
$$= 742 \text{ lb}$$

15.4 PUNCH AND DIE CLEARANCE

If the *clearance* between the punch and the die opening is *correct,* the fracture lines will start (ideally) at the cutting edge of the punch and also at the die as shown in Fig. 15.6a. They will proceed toward each other until they meet as shown in Fig. 15.6b. The edge of the blank will appear as shown in insert, Fig. 15.6c. The radius and shear band widths will be about a third of the thickness of the material with a minimum burr as shown.

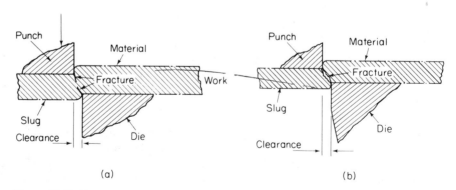

Figure 15.6(a,b)

If the *clearance* between the punch and die is *excessive,* Fig. 15.6d, plastic deformation will continue for a longer period of time than as shown in Fig. 15.6c. The radii that result in the material and the slug will increase because of the additional plastic flow. The punch will in effect draw the material into the clearance space between the punch and the die. As the process proceeds, the material develops an increased resistance to plastic flow and shear takes place followed by fracture. The

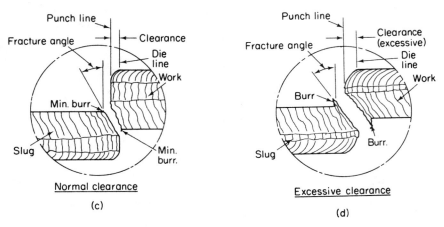

Figure 15.6(c,d)

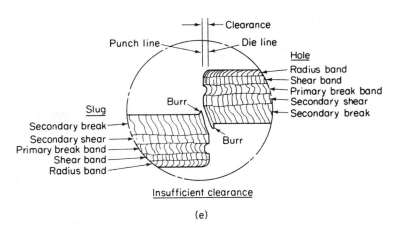

Figure 15.6(e)

results are large radii bands, short shear bands, increased angles of fracture, and burrs on the underside of the hole and upper side of the slug.

If the *clearance* between the punch and die is *insufficient,* Fig. 15.6e, the plastic flow time period is shortened and small radii are produced. The shear bandwidths are also somewhat narrower. However, small fracture lines develop at the punch and die, but because these fracture lines do not meet, the resistance to fracture builds up momentarily and secondary shear bands appear with resulting secondary fracture bands. The high compressive forces operating, because of insufficient clearance between the punch and die, will also cause burrs to develop.

If the punch and die are *misaligned,* an objectionable combination of all three conditions are present. One side of the punch–die combination has too much clearance, the opposite side too little clearance, whereas the center may be exactly right. The hole in the material and the slug exhibit all the telltale conditions just discussed.

TABLE 15.5

Material	Clearance Allowance (%)
Aluminum	6.0
Brass	3.0
Copper	3.0
Soft steel	3.0
Medium steel	3.5
Hard steel	4.0

Clearance, by definition, applies to the space between the punch and the die. Thus, for a round punch and die, the clearance in Table 15.5 refers to the difference between the radii of the punch and the die.

The general rule for the application of clearances is as follows:

The size of the punch determines the size of the pierced hole. The size of the die opening determines the size of the blanked part.

Application of the rule can best be explained with an illustration. However, before considering the illustration, one other condition must be taken into account. That is the condition of "recovery" that takes place within the structure of material. It has already been pointed out that the material being worked will cling to the piercing punch. Once this material is stripped off the punch, the material recovers and the hole size decreases so that the hole is actually smaller than the punch which produced it. The same thing happens when a part is blanked in a press. The blanked part is larger than the die opening which produced it.

The next question is: How much smaller is a pierced hole than the punch that produced it, or how much larger is the blanked part than the die opening that produced it? The answer depends on many variables. Usually, 0.002 in. is arbitrarily taken as an allowance for a round punch and die. If the punch is not round, 0.001 in. may be used. Thus, to produce a 1-in. hole, the punch must be made 1.002 in. in diameter. To produce a 1-in. blank, the die opening should be 0.998 in. in diameter.

Example 15.2

The washer in Figure 15.7 is to be punched. Calculate: (a) the clearance, (b) the size of the piercing punch, (c) the size of the piercing die, (d) the size of the blanking die, (e) the size of the blanking punch.

(a) The clearance for soft steel is

$$c = at = 0.03(0.060)$$
$$= 0.0018 \text{ in.}$$

c = clearance (one side)
t = thickness 0.060 in.
a = allowance, 3%

(b) The size of the hole is determined by the punch size. The hole will shrink approximaely 0.002 in. Therefore, the punch size is

$$P_p = 0.375 + 0.002$$
$$= 0.377 \text{ in.}$$

P_p = piecing punch

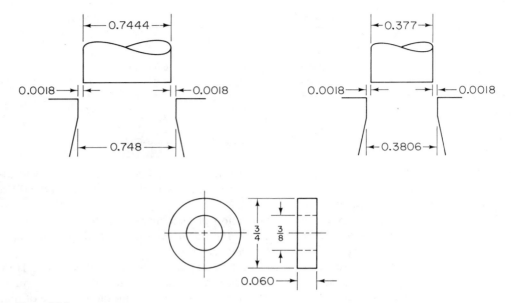

Figure 15.7

(c) The piercing die opening will need to be larger than the punch. Therefore,

$$D_p = P_p + 2c = 0.377 + 2(0.0018) \qquad D_p = \text{piercing die opening}$$
$$= 0.3806$$

(d) The size of the blank (0.750 in. dimension) is determined by the blanking die opening. The washer will expand approximately 0.0092 in. Therefore, the die opening must be

$$D_b = 0.750 - 0.002 \qquad D_b = \text{blanking die opening}$$
$$= 0.748 \text{ in.}$$

(e) The clearance must be subtracted from the blanking punch size:

$$P_b = D - 2c = 0.748 - 2(0.0018) \qquad P_b = \text{blanking punch}$$
$$= 0.7444 \text{ in.}$$

All dimensions are shown in Fig. 15.7.

15.5 SCRAP STRIP LAYOUTS

Because of the capability of high production, the major portion of the cost for producing a stamping is the material. The layout of the scrap strip is therefore of utmost importance from the cost standpoint. In addition, the scrap strip layout is a good place to start when designing a die.

Strips of stock are fed into the die from the right, left, or front to rear, depend-

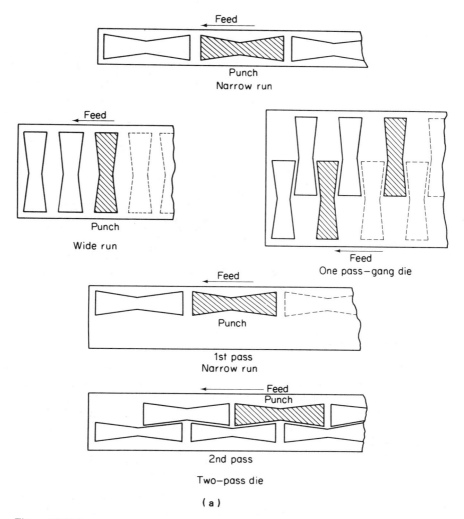

Figure 15.8(a)

ing on the method selected. Strips may be sheared to desired widths at the mill, or they may be purchased as rolled stock.

The grain direction generally runs lengthwise of the strip for rolled or sheet stock. If the grain direction must run perpendicular to the length of the strip, it can be ordered sheared from sheet stock in that direction.

The position of the workpiece relative to the direction of feed of the strip is an important consideration. Generally speaking, wide run layouts are more economical than narrow run layouts, and side-by-side layouts will also save material and handling time. Figure 15.8a also shows a one-pass gang die strip layout with two displaced punches operating on the same stroke, and a two-pass strip layout where one punch is operating. After the strip is run through once it is reversed and run through again.

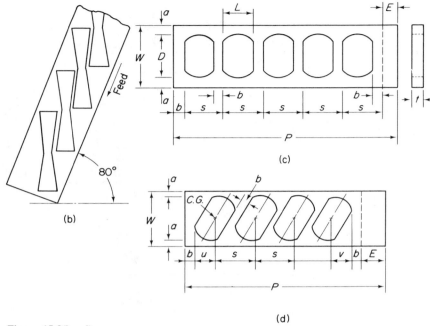

Figure 15.8(b,c,d)

As indicated in Fig. 15.8a, many patterns of scrap strip layouts are possible. If it should become necessary to lay out a scrap strip at an angle, it is good practice to simply draw the shape of the scrap as shown in Fig. 15.8b and then draw the strip at an angle. Assume a straightforward *single-pass* layout as shown in Fig. 15.8c. Before the material requirements can be calculated, the values of *a* and *b* must be determined.

When the thickness of the stock is *greater* than 0.025 in. the value of *a* may be calculated from the formula

$$a = t + 0.015D$$

a = edge of blank to side of strip
t = thickness of stock
D = width of blank

The value of *b* is found from Table 15.6. If the thickness of the material is *less* than 0.025 in. the values of *a* and *b* may be taken from Table 15.7. For a *double-pass* layout Table 15.8 should be used when the thickness of the material is less than 0.025 in.

The number of blanks that can be produced from one length of stock is given by

$$N = \frac{P - b}{s}$$

P = length of stock
N = number of blanks

The scrap at the end of the strip is

$$E = P - (Ns + b)$$

TABLE 15.6   WHERE $t > 0.025$

Material Thickness, t (in.)	Stock Allowance, b (in.)
0.025–0.03125	0.03125
0.03125–0.1875	t
Over 0.125	0.125

TABLE 15.7   SINGLE-PASS WHERE $t < 0.025$

Strip Width, W (in.)	Dimensions a and b
0–3	0.0312
3–6	0.0625
6–12	0.09375
Over 12	0.125

TABLE 15.8   DOUBLE-PASS WHERE $t < 0.025$

Strip Width, W (in.)	Dimensions a and b
0–3	0.0625
3–6	0.09375
6–12	0.125
Over 12	0.15625

Where the workpiece is produced at a slant as shown in Fig. 15.8d, the equation that will give the number of blanks is given by

$$N = \frac{P - (u + v + 2b) + s}{s} = \frac{P - (u + v + 2b)}{s} + 1$$

If the scrap strip requires a double pass, the number of blanks may be found from

$$N = \frac{2[P - (u + v + 2b)]}{s} + 1$$

Example 15.3

Assume the blank in Fig. 15.8c to be $D = 3$ in., $L = \frac{5}{8}$ in., and $t = \frac{3}{16}$ in. Calculate (a) the value of a, (b) the value of b, (c) the width of the strip W, (d) the length of one piece of stock needed to complete one part, (e) the number of parts that can be produced from an 8-ft length, and (f) the scrap that remains at the end of the run.

Solution

(a) The value of a is

$$a = t + 0.015D = 0.1875 + 0.015(3)$$
$$= 0.2325 \text{ in.}$$

$t = 0.1875$ in.
$D = 3$ in.
$P = 8$ ft. $= 96$ in.
$L = 0.625$ in.

(b) The value of b is

$$b = 0.1875 \text{ in. (from Table 15.6)}$$

(c) The width of the strip is

$$W = D + 2a = 3.000 + 2(0.2325)$$
$$= 3.465 \sim 4.000 \text{ in.}$$
$$= 4.000 \text{ in.}$$

(d) The length of one piece is

$$s = L + b = 0.625 + 0.1875$$
$$= 0.8125 \text{ in.}$$

(e) The number of blanks from an 8-ft strip is

$$N = \frac{P - b}{s} = \frac{96 - 0.1875}{0.8125}$$
$$= 117.92 \sim 117 \text{ blanks}$$

(f) The scrap at the end of the run is

$$E = P - (Ns + b) = 96 - [117(0.8125) + 0.1875]$$
$$= 0.750 \text{ in.}$$

15.6 TYPES OF DIES

The components generally incorporated in a piercing or blanking die are shown in Fig. 15.9a. This figure shows the die in the conventional closed position. The die set is made up of the punch holder, which is fastened to the ram of the punch press, and the die shoe, which fastens to the bolster plate of the punch areas.

As indicated previously, the *conventional* position of the *die* set is to have the punch fastened to the punch holder and aligned with the opening in the die block. Figure 15.9a shows one type of stripper and push-off pins. The stripper holds the scrap strip so that the punch may pull out of the hole. The push-off pins are needed to free the blank in instances where the material clings to the bottom of the punch. This may be necessary for thin material, or where lubricants are used on the material.

Sometimes it becomes necessary to interchange the punch and the die positions. This may become necessary when the opening in the bolster plate is too small to permit the finished product to pass through the bolster opening. Figure 15.9b shows such a die.

Inverted dies are designed with the die block fastened to the punch holder and the punch fastened to the die shoe. As the ram descends, the blank is sheared from the strip. The blank and shedder are forced back into the die opening, which loads a compression spring in the die opening. At the same time, the punch is forced through the scrap strip and a spring attached to the stripper is compressed and loaded. On the upstroke of the ram, the shedder pushes the blank out of the die opening and the stripper forces the scrap strip off the punch. The finished part falls, or is blown, out the rear of the press Fig. 15.9b.

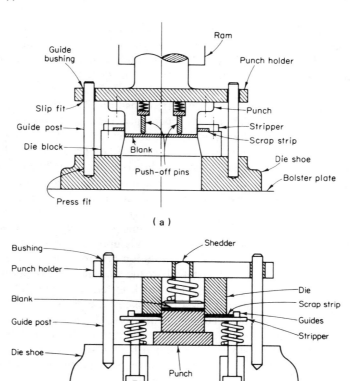

Figure 15.9(a,b)

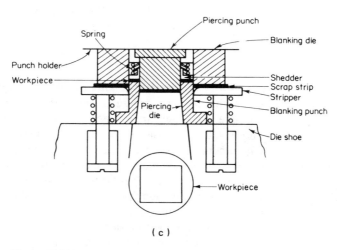

Figure 15.9(c)

Compound dies, Fig. 15.9c, combine the principles of the conventional and inserted dies in one station. This type of die may produce a workpiece which is pierced and blanked in one station and in one operation. The piercing punch is fastened in the conventional position to the punch holder. Its matching die opening for piercing is machined into the blanking punch. The blanking punch and blanking die opening are mounted in an inverted position. The blanking punch is fastened to the die shoe and the blanking die opening is fastened to the punch holder.

Progressive dies, Fig. 15.10, are made with two or more stations. Each station performs an operation on the workpiece, or provides an idler station, so that the workpiece is completed when the last operation has been accomplished. Thereafter each stroke of the ram produces a finished part. Thus after the fourth stroke of a four-station die, each successive stroke will produce a finished part. Operations which may be included are piercing, blanking, forming, drawing, cut-off, and so on. The list of possible operations is long. The number and types of operations which may be performed in a progressive die depends upon the ingenuity of the designer.

Figure 15.10f shows a four-station progressive die. The die block is machined out of four pieces and fastened to the die shoe. This permits the replacement of broken or worn die blocks. The stock is fed from the right and registers against a finger stop (not shown). The first stroke of the press, Fig. 15.10a, produces a square hole and two notches. These notches form the left end of the first piece.

The ram retracts, and the stock moves to the next station against a finger stop (not shown). The stock is positioned for the second stroke. The second station is an idler, Fig. 15.10b. The right end of the first piece, the left end of the second piece, and a second square hole are pierced.

The ram retracts and the scrap strip moves to the third station against an automatic stop, Fig. 15.10c. This stop picks up the notched V and positions the scrap strip. The third stroke of the ram pierces the four holes as shown in Fig. 15.10c. The fourth stroke, Fig. 15.10d, cuts off and forms the radii at the ends of the finished pieces. Thereafter every stroke produces a finished part, Fig. 15.10e.

15.7 PUNCHES

As indicated, punches may be termed *blanking* or *piercing.* The main concerns when designing punches are that they are designed so that they do not deflect, that the heat treatment is appropriate, that they are strong enough to withstand stripping forces, and that they do not rotate as a result of the cutting action.

Deflection of punches may be avoided by making the body diameter of the punch larger than the cutting diameter. A large radius is the bridge between the required operating diameter and the larger body diameter. Figure 15.11 shows several methods for mounting blanking punches on a punch holder. The rectangular and flanged punch in Fig. 15.12 shows the socket head screws and locator dowel pins used to hold the punch to the punch holder.

Sometimes it is more economical to use a machine steel spacer and a tool steel plate for a punch, than to make a large blanking punch out of one piece of tool steel.

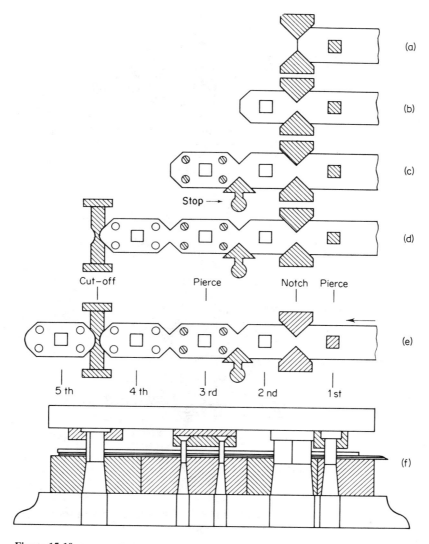

Figure 15.10

Figure 15.13a shows one such setup. Figure 15.13b shows another material-saving punch construction. The ring is made from tool steel and set in a machine steel spacer.

Sometimes it is more economical and easier to make punches out of sections and fit them into the desired pattern than it is to attempt to make the punch from one piece. This may be desirable when the punches need to be large or irregularly shaped. Another advantage is that a section may be removed and replaced if it becomes worn or broken. One such composite punch is shown in Fig. 15.14. Sometimes large

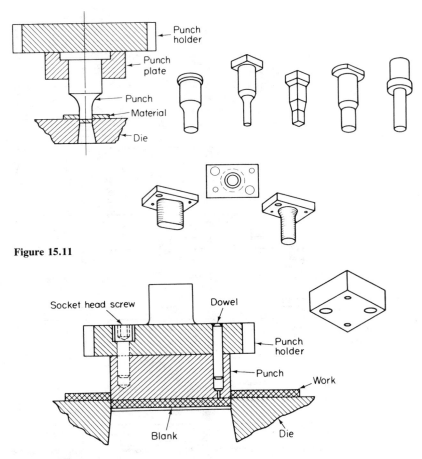

Figure 15.11

Figure 15.12

composite punches are welded from several smaller pieces of tool steel. In some cases the shapes are flame cut from larger tool steel blocks or sheets and then welded or bolted to machine steel spacers. Shear may be applied to the faces of punches as shown in Fig. 15.15.

Piercing punches are smaller and generally longer than blanking punches. They must be designed to withstand shock, deflection, and buckling. Because of the high probability of damage, they must be designed so that they can be easily removed and replaced.

Small-diameter punches must be supported for most of their length. It may be necessary to guide them through the stripper plate if the punch diameters are less than $\frac{1}{4}$-in. In some instances it is necessary to back up small diameter punches with hardened plates so that they do not loosen as they are being used. When long thin punches are used, it is a very difficult task to machine heads on them. The procedure

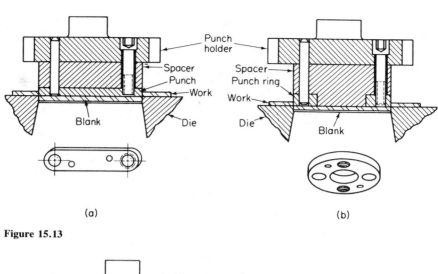

(a)

(b)

Figure 15.13

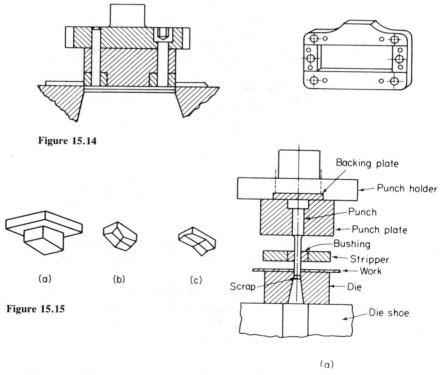

Figure 15.14

(a)   (b)   (c)

Figure 15.15

(a)

Figure 15.16(a)

used is to peen a head at one end of the punch, machine on 82° taper to match a countersink hole in the punch plate, and then to grind the punch plate flush with the punch head.

Figure 15.16a shows a long thin punch guided through a bushing in the stripper

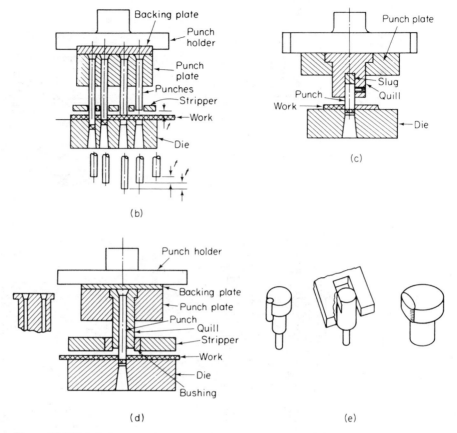

Figure 15.16(b,c,d,e)

plate. Headless bushings may be used for this purpose. The hardened backing plate should be press fitted into the punch holder. Provisions must be made for removing this plate.

Figure 15.16b shows a series of small piercing punches staggered to reduce the force required to shear through metal, and also to prevent punch breakage due to metal displacement or crowding. The offset of one row of punches over another is generally from 60% to one full thickness of the metal. The insert shows another pattern for arranging offset.

Figure 15.16c shows a quill used to support a short punch. This avoids the need for long unsupported punches. Figure 15.16d shows another method for supporting a long punch in a quill. Sometimes two and three punches are supported this way. Dowels, flats, keys, and pins may be used as shown in Fig. 15.16e to keep punches from turning in the punch plates.

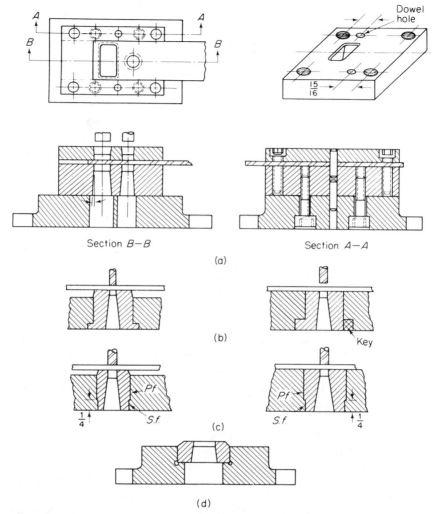

Figure 15.17(a,b,c,d)

15.8 DIE BLOCKS

Die blocks are the construction components that house the opening to receive punches. These die openings may be machined into a solid block of tool steel. Intricately shaped die openings or large die blocks will generally be sectioned. These blocks are predrilled, tapped, and reamed before being fastened to the die shoe. Provision must be made for a stripper plate and a back gage against which the raw stock registers. Figure 15.17a shows one method for fastening a die block to a shoe.

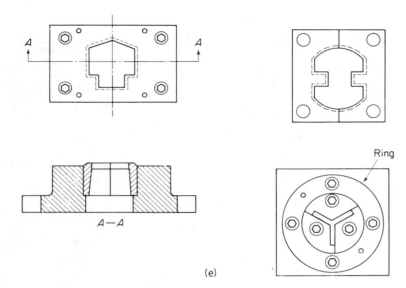

(e)

Figure 15.17(e)

Socket head screws are inserted from the bottom of the die shoe into threaded holes in the die block. Dowels are used to prevent a shift in position of the block.

At times the die opening is made from a bushing and inserted into a machine steel retainer. If the bushing has a shoulder, it is held in the retainer as shown in Fig. 15.17b. If it has no shoulder, it is pressed into the retainer as shown in Fig. 15.17c. The lower end of the bushing has a $\frac{1}{4}$-in. length reduced diameter to ensure alignment when pressed into the retainer.

At times a ring is used as the die button, Fig. 15.17d. The purpose is to save the cost of large amounts of tool steel. In addition, provision is usually made to replace the die ring if it is worn excessively or damaged.

If the die opening is complicated, or if it becomes difficult to machine contours, the die may be sectioned. Several types of sectioned dies are shown in Fig. 15.17e. The Y-section is made up of three pieces fitted into a ring which is then used as a die ring insert in the die shoe. This technique may be used to produce small projections on the workpiece which are usually difficult to machine into a solid piece of tool steel.

It is also possible to apply shear to the die face. Figure 15.18 shows shear applied to the face of a die. Figure 15.18a shows convex shear and Fig. 15.18b shows concave shear applied to a die opening.

In Fig. 15.18b the flats provide a clean first entrance into the material by the punch. The shear is generally equal to the stock thickness. Sometimes the die surface is scalloped as shown in Fig. 15.18c. It should be remembered that shear will distort the slug which is forced into the die opening. If this slug is the workpiece and not the scrap, the distortion may be objectionable.

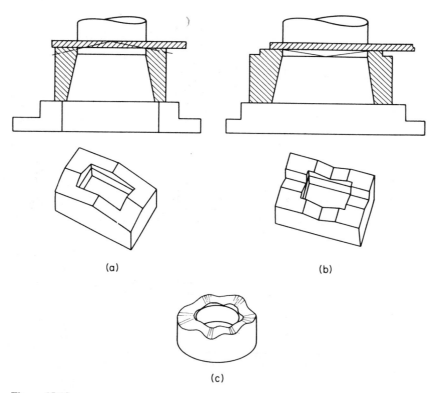

Figure 15.18

15.9 PILOTS

Pilots are used in progressive dies to locate the work strip so that the relationships between stations may be maintained, Fig. 15.19a. They are generally made from tool steel, hardened for toughness and polished. They may be made from drill rod.

If the operation is slow, the pilots may be press fitted into a hold in the center of the punch. If there is any danger of the pilot dropping out, it should be fastened. They may have a radius end which terminates in the diameter of the pilot, or the end of the pilot may be blunted as shown in Table 15.9. Some variations of pilot nose design are shown in Fig. 15.19b. The pilot should fit the work hole with a tolerance of from 0.0005 in. for precision work to 0.005 in. for average stamping operations. It should be remembered that pierced holes shrink.

Pilot lengths should be enough so that they register the material and position it before any punches, strippers, or pressure pads contact the work. There must, therefore, be a clearance hole in the die block to permit the pilot to continue to move down as the punches operate. Pilot clearance in the die opening should be slightly larger than punch clearance. Not over two to three times more clearance should be applied to the die opening for receiving the pilot. It should also be tapered and guided through the stripper as shown in Fig. 15.20.

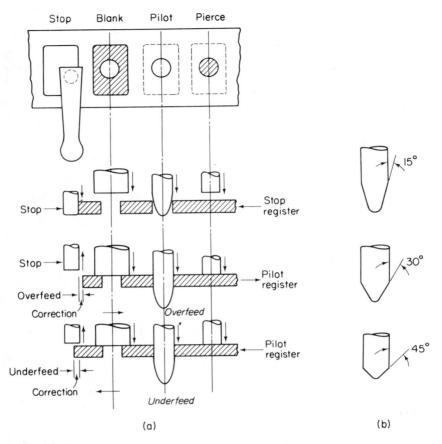

(a)

(b)

Figure 15.19

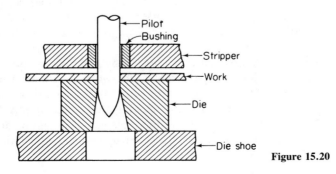

Figure 15.20

TABLE 15.9 PILOTS[a]

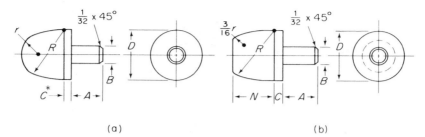

(a)   (b)

(a) Radius End—Drawing

D	B	A	R	r
$\frac{1}{8}$	$\frac{3}{32}$	$\frac{1}{4}$	$\frac{1}{8}$	$\frac{1}{32}$
$\frac{3}{16}$	$\frac{1}{8}$	$\frac{5}{16}$	$\frac{3}{16}$	$\frac{3}{64}$
$\frac{1}{4}$	$\frac{3}{16}$	$\frac{7}{16}$	$\frac{1}{4}$	$\frac{1}{16}$
$\frac{5}{16}$	$\frac{1}{4}$	$\frac{1}{2}$	$\frac{5}{16}$	$\frac{5}{64}$
$\frac{3}{8}$	$\frac{1}{4}$	$\frac{1}{2}$	$\frac{3}{8}$	$\frac{3}{32}$
$\frac{7}{16}$	$\frac{5}{16}$	$\frac{1}{2}$	$\frac{7}{16}$	$\frac{7}{32}$
$\frac{1}{2}$	$\frac{5}{16}$	$\frac{1}{2}$	$\frac{1}{2}$	$\frac{1}{8}$
$\frac{5}{8}$	$\frac{3}{8}$	$\frac{9}{16}$	$\frac{5}{8}$	$\frac{5}{32}$
$\frac{11}{16}$	$\frac{3}{8}$	$\frac{9}{16}$	$\frac{11}{16}$	$\frac{11}{64}$

(b) Flat End—Drawing

D	B	A	R	N
$\frac{3}{4}$	$\frac{3}{8}$	$\frac{5}{8}$	$\frac{3}{4}$	$\frac{3}{8}$
$\frac{7}{8}$	$\frac{3}{8}$	$\frac{5}{8}$	$\frac{7}{8}$	$\frac{7}{16}$
1	$\frac{7}{16}$	$\frac{11}{16}$	1	$\frac{1}{2}$
$1\frac{1}{4}$	$\frac{1}{2}$	$\frac{3}{4}$	$1\frac{1}{4}$	$\frac{5}{8}$
$1\frac{3}{8}$	$\frac{1}{2}$	$\frac{3}{4}$	$1\frac{3}{8}$	$\frac{5}{8}$
$1\frac{1}{2}$	$\frac{5}{8}$	1	$1\frac{1}{2}$	$\frac{3}{4}$
$1\frac{3}{4}$	$\frac{3}{4}$	$1\frac{1}{4}$	$1\frac{3}{4}$	$\frac{3}{4}$
2	$\frac{3}{4}$	$1\frac{1}{2}$	2	1

*C equals the thickness of the stock if greater than $\frac{1}{16}$ in.

Figure 15.21a shows various methods for mounting pilots in punches. It should be noted that the reamed hole which receives the pilot should be extended through the punch so that the pilot may be removed if it breaks.

Retractable pilots are made so that the shank of the pilot and the hole into which it fits is a slip fit. These pilots are generally backed up with heavy springs which are strong enough to cause the pilot to pierce the stock in case of a misfeed. Sometimes the spring will compress, the shank of the pilot will bottom, and the pilot will pierce the stock. Figure 15.21b shows a design of a spring pilot.

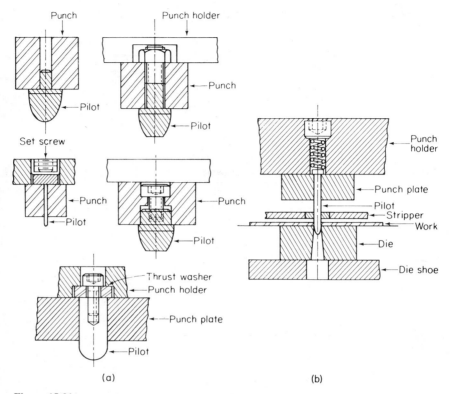

Figure 15.21

15.10 KNOCKOUTS AND SHEDDERS

Knockouts are used to remove workpieces that adhere to the die opening as shown in Fig. 15.22. The upper end of the knockout registers against the knockout bar of the press, which pushes the shedder down so that the workpiece is pushed back into the scrap strip, or out of the die opening so that it may be ejected. The pinned collar keeps the knockout from dropping out of the punch.

Sometimes a hole is provided in the punch so that air may be directed through it to act as a shedder. An intermittent or continuous stream of air may be used to prevent slugs from sticking to the bottom of the punch.

15.11 STRIPPER PLATES

Earlier it was stated that when a punch shears its way through a workpiece, the material contracts around the punch to the degree that it takes a substantial force to

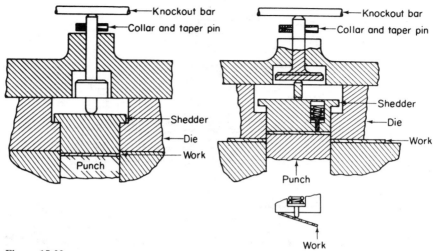

Figure 15.22

withdraw the punch from the material. The equation used to calculate the force required to strip the material from the punch was given in Sec. 15.2.

The cheapest and simplest design for a stripper is shown in Fig. 15.23a. A spacer is provided to raise the stripper so that there is clearance between the top of the strip and the stripper plate.

Two additional methods for mounting solid stripper plates are shown in Fig. 15.23b and c. Figure 15.23b is a flat piece of stock that has had a channel milled into its surface. Figure 15.23c uses the same idea except that two spacers are used under the stripper plate.

The tunnel dimension $(w + c)$ should allow about $\frac{1}{32}$ in. per foot of stripper length to permit free passage for the width of the stock. The tunnel height $(t + G)$ provides clearance in the height direction. This clearance (G) should not be less than $0.5t$ but may be up to $5t$, depending on the conditions of the operation. These allowances are shown in Fig. 15.23d.

Example 15.4

The length of a channel stripper, Fig. 15.23d, is 16 in. The strip stock used is $3 \times \frac{3}{16}$ in. Calculate (a) the median clearance G, (b) the tunnel height, and (c) the allowable tunnel width.

(a) The median clearance is

$$G = \frac{5t - 0.5t}{2} = 2.25t = 2.25(0.1875)$$

$$= 0.422 \text{ in.}$$

$$w = 3.000 \text{ in.}$$
$$t = 0.1875 \text{ in.}$$
$$l = 16 \text{ in.}$$

(b) The height is

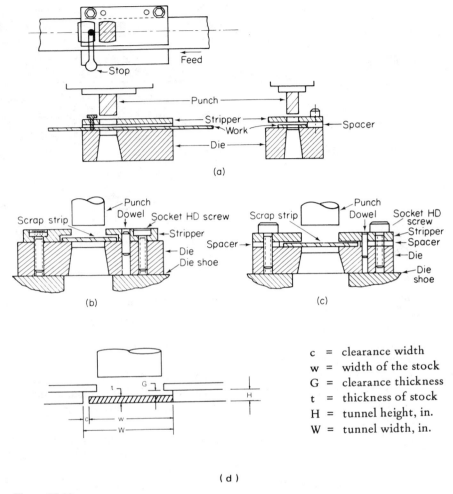

Figure 15.23

$$H = t + G = 0.1875 + 0.422$$
$$= 0.610 \text{ in.} = \tfrac{39}{64} \text{ in.}$$

(c) The tunnel width is

$$W = w + c = w + (\tfrac{1}{32})I = 3.000 + (\tfrac{1}{32})(16)$$
$$= 3.500 \text{ in.}$$

A solid stripper plate must have clearance between the stripper and the scrap strip so that the latter may move freely at the end of each up stroke of the ram. Sometimes it is desirable to hold the scrap strip in a flat position before the punch

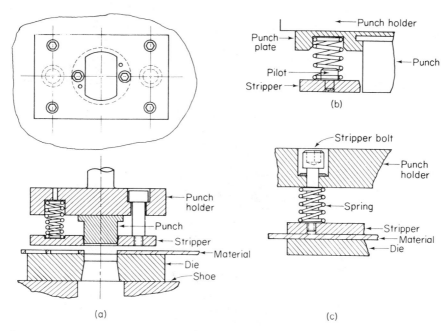

Figure 15.24

contacts the work. This is necessary where very accurate blanks are needed, when very thin material is punched, or when thin punches are used. This class of strippers is spring loaded.

This stripper is generally made from a plate that provides the desired configuration. It is generally suspended from the punch holder with stripper bolts and compression springs, equally spaced about the center of pressure of the punches. One such stripper plate is shown in Fig. 15.24a. Figure 15.24 shows several ways of retaining springs. In Fig. 15.24a both the punch holder and the stripper plate are counterbored. A counterbore in the stripper plate is not often possible. Under circumstances where this is not desirable, a spring pilot Fig. 15.24b, may be used. Also shown is the counterbore in the punch plate, rather than in the punch holder.

Figure 15.24c shows the spring guided by the stripper bolt. The spacer, the head of the stripper bolt, is used to position the bottom face of the stripper plate correctly relative to the length of the punch.

In some instances small hydraulic cylinders or rubber pads are used to produce the force necessary to strip the scrap strip from the punch.

Punches must go through the stripper plate. The hole through the stripper may be a simple chamfer and a $\frac{1}{8}$- to $\frac{1}{4}$-in. straight section, Fig. 15.25a; a counterbored hole and a straight section, Fig. 15.25b; or hardened guide bushing as shown in Fig. 15.25c and d.

When strippers need to be guided, this can be done by pressing one end of a hardened and ground guide pin into the punch holder, Fig. 15.26a, and allowing the

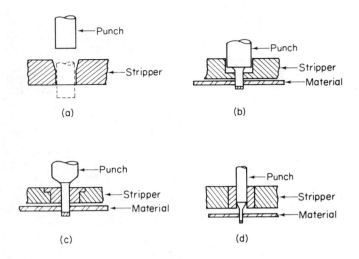

Figure 15.25

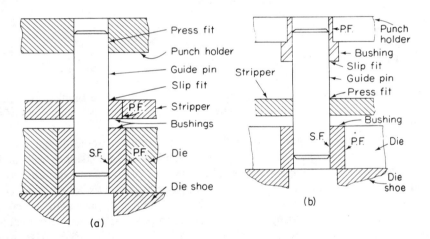

Figure 15.26

guide pin to slide in the stripper and the die block. Another method for guiding the stripper is to press it on the guide pin and allow both ends of the pin to slide in the punch holder and die block as shown in Fig. 15.26b. Under certain conditions it is possible to press fit the guide pin into the die shoe.

15.12 DIE STOPS

Die stops are used to either *stop* the material strip or *register* it. When a die stop halts the strip, the process is said to *stop* the strip. After stopping the strip, if a pilot

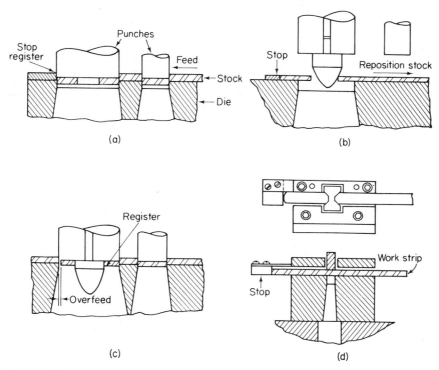

Figure 15.27

repositions the strip so that it may be punched to desired dimensional locations, the pilot is said to *register* the stock strip. In some instances a stop block may perform both functions as shown in Fig. 15.27a. In Fig. 15.27b the stock has overfed and stopped against the stop block. The pilot has picked up the right side of the pierced hole and is about to move the stock back to its proper position. The pilot should be long enough to position and hold the stock before the punches begin to operate. Figure 15.27c shows the stock strip in the proper position and the punches cutting.

The amount of overfeed permitted is a function of the pilot diameter and the thickness of the material. Of course, the stock material must be strong enough so that the forces of repositioning do not collapse the strip. The larger the diameter of the pilot relative to the thickness of the material strip, the greater the permissible overfeed. The overfeed ranges from 0.002 to 0.050 in. for pilots with a range of diameters of $\frac{1}{8}$ through 1 in. and stock thicknesses of 0.005 through 0.125 in. Sometimes the back gage is extended and used to support a stop. This is especially useful where the workpiece is long. Such a stop is shown in Fig. 15.27d.

Solid pins, with or without heads, may be used as stops. The pin should be a light press fit into the die shoe. It should extend above the die block face about equal to or $1\frac{1}{2}$ times the thickness of the stock. A clearance hole should be provided in the stripper plate and under the pin in the die shoe in case it needs to be removed. Figure 15.28 shows several solid pin type stops.

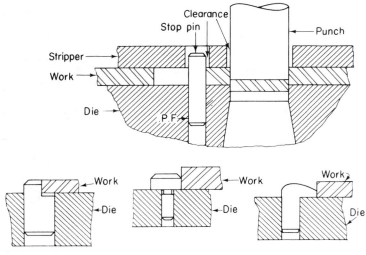

Figure 15.28

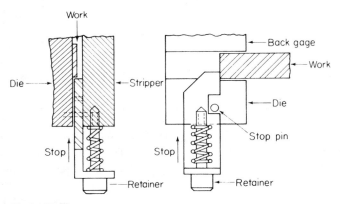

Figure 15.29

Finger stops are used to start new strips in the proper location in a die. They are operated with the finger by pushing them into the stock channel until they seat. The press is tripped, the stop is released and returns to its out position. It is not used again until a new strip is started. One such stop is shown in Fig. 15.29.

The automatic stop shown in Fig. 15.30 is a very common construction. The press ram has an adjustable trigger which contacts the pad on the stop and pushes the opposite end of the stop up so that the stock may pass under it on the upstroke. Note that the small side movement of the stop is enough to allow it to come down on top of the stock strip while the punches are still in the holes which they have cut. As the punches lose contact with the holes, the feed moves the scrap strip forward. The stop is now sliding on top of the scrap strip. It falls into the hole, registers against the right

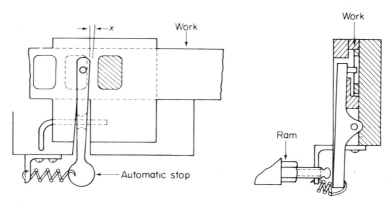

Figure 15.30

side of the hole and is forced all the way to the left. On the next downstroke of the ram the operation repeats.

The clearance between the slot in the stripper and the sides of the pilot arm should be held to a minimum. As little as 0.002 will give good results. The movement, x, should be no more than $\frac{1}{16}$ in.

15.13 DESIGNING A DIE

The die designer must draw from material presented in this chapter and from his field experience while he is given a workpiece for which he is to produce a die. The following problem is worked out with the intention of illustrating the reasoning processes and the step-by-step procedure used when designing a die. If it is desired the dimensions of the punch, die, and die set can be inserted after the process, explained below, is finished.

Figure 15.31a shows the workpiece that is to be produced. This blank is used to develop the scrap strip, Fig. 15.31b. At this stage of development, the workpiece should be positioned in various orientations to the direction of flow of the scrap strip and the grain direction. Calculations should be done for the distances between the end (*b*) and sides (*a*) of the workpiece and the scrap strip and the distance between the centerlines of the stations. Calculations should also be done to support the orientation finally selected in terms of conservation of handling time, operation, material, and economy.

At least three alternatives to the scrap strip chosen are shown in Fig. 15.31b. They are: a two-station pierce blank; a two-station pierce form cutoff; a compound die. The scrap strip used throughout this section makes the development of the die somewhat easier than had some other scrap strip been chosen.

The next step is to match the die block to the scrap strip which is finally chosen. This is done in Fig. 15.31c. Three views of the die block are developed. The top view is drawn to match the scrap strip. The block is made large enough so that screws and

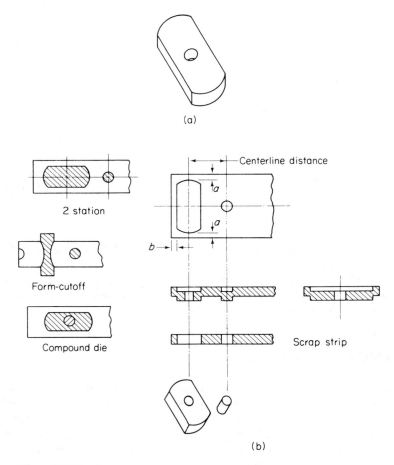

2 station

Form-cutoff

Compound die

Centerline distance

Scrap strip

Figure 15.31(a,b)

dowels needed to fasten the block to the die set will have enough room. The front and side views are to be sectional views. Note that the side sectional view is taken through the large punch. The pictorial view is drawn here to clarify the development for the student.

The next step, Fig. 15.31d, is to design the punches, their orientation to each other, and the pilot, if one is used. The distance between the centerlines of the two punches has been established in Fig. 15.31b. The shape of the punch heads must therefore be designed to accommodate this distance. It should be noted that the convention for drawing the views is followed here. The upper right view shows the punches inverted and rotated so that the view is looking up at the punches from the bottom, but rolled 180° as if the cover of this book were being opened. The upper left view is that of the die openings looking down after the punch holder has been removed. The large punch flange is shown here with enough allowance for socket

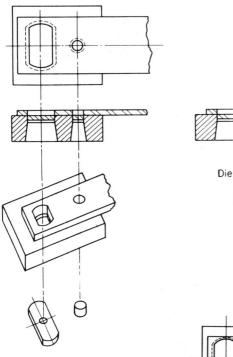

Figure 15.31(c)

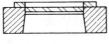

Die block

(c)

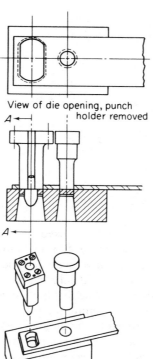

View of die opening, punch
holder removed

$A \leftarrow$

$A \leftarrow$

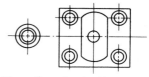

View of punches looking up

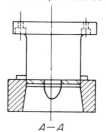

$A-A$

Punches and pilot

Figure 15.31(d)   (d)

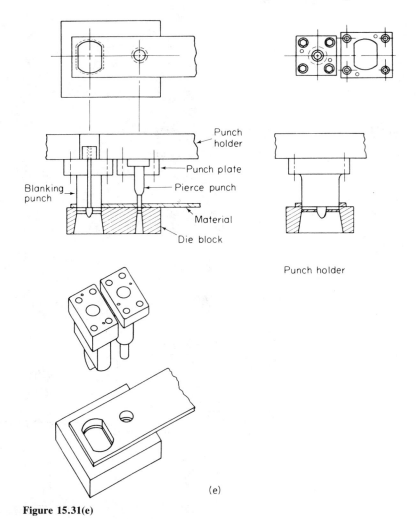

Punch
holder

Punch plate

Blanking
punch

Pierce punch

Material

Die block

Punch holder

(e)

Figure 15.31(e)

head screws and dowels. The pilot is used to establish the relationship of the previously pierced hole to the second station blanking operation.

Figure 15.31e shows the punch plate designed to match the flange of the larger
blanking punch. It must not only fit into the space provided, but it must allow for
socket head screws and dowels. Note that the punch holder has a through hole drilled
to receive the special nut for fastening the pilot. Figure 15.31f shows the die block
fastened to the die shoe. Added to the die block is a back gage, a spacer strip in the
front of the die block to complete the scrap strip channel and two stops. The finger
stop is used at the beginning of each run to position the first stroke accurately.
Thereafter the automatic stop takes over.

The die assembly is now ready for the stripper plate, which mounts on the back

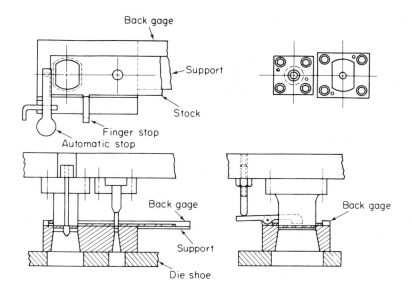

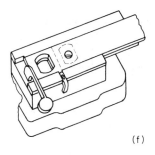

(f)

Figure 15.31(f)

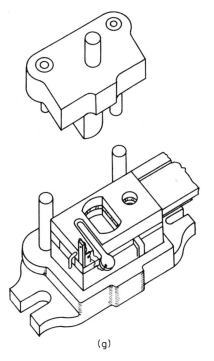

Figure 15.31(g)   (g)

gage and spacer strips as shown in Fig. 15.31g. The entire assembly is now ready for mounting in the die set.

PROBLEMS

15.1. **(a)** Make a sketch of an OBI press and label all parts.
  (b) How do they differ from straight-sided presses?

15.2. **(a)** Can the length of the stroke of a press be changed?
  (b) Can the position of the stroke be changed? Explain.

15.3. How is the shut height of a press determined?

15.4. Describe
  (a) a single-action press,
  (b) a double-action press, and
  (c) a triple-action press?

15.5. Explain each elevation in Fig. 15.2a through d.

15.6. Explain the use of the technique of splitting a die down the vertical centerline (Fig. 15.2e).

15.7. A square washer measures $3 \times 4 \times \frac{1}{8}$ in. and has a 2-in.-diameter hole. It is made from yellow brass. Calculate
  (a) the needed tonnage to complete the washer if both punches operate at the same time and neither has shear,
  (b) the force if the punches are staggered, and
  (c) the punches are staggered and 0.090 in. shear is applied to the faces of both punches, assuming 40% penetration.
  (d) What stripping force is required in part (a)?

15.8. A washer is 0.050 in. thick, has a $\frac{1}{2}$-in. hole, and an outside diameter of 1 in. as shown in Fig. 15.32. The material is SAE 1020 steel, ultimate shearing strength of 40,000 psi. Calculate
  (a) the force needed to do the job if both punches operate at the same time with no shear and
  (b) the force if the punches are staggered and only one operates at a time.
  (c) Assume 60% penetration when the punch has 0.040 in. shear. What force is needed if the punches are staggered?
  (d) What stripping force is required to strip both punches?

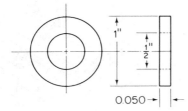

Figure 15.32

15.9. Calculate
  (a) the tonnage required to cut the workpiece down in Fig. 15.33a if the material is SAE 1010 steel and

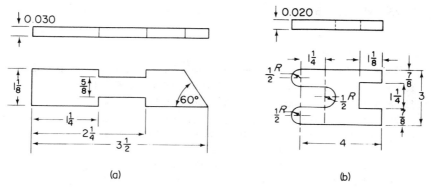

Figure 15.33

   (b) the stripping force.

15.10. **(a)** Calculate the tonnage required to blank the workpiece shown in Fig. 15.33b from 0.020-in.-thick electrolytic copper.
   (b) What is the stripping pressure? Assume the punch as 0.20 in. shear and the penetration as 50%.

15.11. The washer in Fig. 15.32 is to be punched. Calculate
   (a) the clearance,
   (b) the size of the piercing punch,
   (c) the size of the piercing die,
   (d) the size of the blanking die, and
   (e) the size of the blanking punch.

15.12. Define
   (a) *blanking* and
   (b) *piercing*.

15.13. What are the three components of the operation when a punch cuts through the scrap strip?

15.14. Discuss elastic-plastic deformation as the punch forces increase on the work strip. Make a drawing and show all forces.

15.15. Explain the mechanism of shear, after initial deformation, as the punch proceeds to push into the material. Support your explanation with a sketch.

15.16. The last stage of the cutting operation by the punch and die is that of break. Explain this component of the mechanism of shearing through a strip of material. Show all radii, straight, and angular configurations on the cut edge of the blank and scrap strip.

15.17. **(a)** Why is it necessary to strip material off punches?
   (b) What happens to the material that is forced into the die opening?

15.18. Draw views of the edges of the scrap strip and the slug when the clearance is correct for a shearing operation. Label all parts and explain how it progresses.

15.19. Repeat Prob. 15.18 when the clearance is excessive.

15.20. Repeat Prob. 15.18 when the clearance is not enough.

15.21. Repeat Prob. 15.18 when the punch and die are misaligned.

15.22. Define *clearance* as it applies to die work.

15.23. What determines the application of clearance to
   (a) the punch and
   (b) the die?
   (c) How is "recovery" handled?

15.24. Calculate the punch and die dimensions for the square washer in Prob. 15.7.

15.25. Calculate the dimensions for the punch and die for Fig. 15.33a.

15.26. Calculate the punch and die dimensions for Fig. 15.33b.

15.27. If the configuration in Fig. 15.33a were a *hole* in a blank instead of the outer contour of a workpiece, would this affect the sizes of the punch and the die in Prob. 15.25? Explain and show all calculations.

15.28. Why is the layout of the scrap strip critical to the die designer?

15.29. What determines the grain direction in raw stock? Explain.

15.30. Draw at least two scrap strip layouts for Fig. 15.33a.

15.31. Draw at least three scrap strip layouts for Fig. 15.33b.

15.32. Assume that the workpiece in Fig. 15.33a were to be positioned on a scrap strip at 60° with the length of the strip. How would you do this?

15.33. Given a scrap strip as shown in Fig. 15.34a, calculate
   (a) the value of *a*,
   (b) the value of *b*,
   (c) the width of the strip *W*,
   (d) the length of one piece of stock needed to produce one part,
   (e) the number of blanks from one 8-ft strip, and
   (f) the scrap that remains at the end of each run.

15.34. Repeat Prob. 15.33 for Fig. 15.34b.

15.35. Repeat Prob. 15.33 for Fig. 15.34c for a double-pass for the 3-in. equilateral triangle.

15.36. Assume the blank in Fig. 15.8c to be $D = 4$ in.; $L = \frac{3}{4}$ in., and $t = 0.125$ in. Calculate
   (a) the value of *a*,
   (b) the value of *b*,
   (c) the width of the strip *W*,
   (d) the length of one piece of stock needed to produce one part,
   (e) the number of parts that can be produced from an 8-ft length, and
   (f) the scrap that remains at the end of the run.

15.37. When is it advisable to build in push-off pins into the punches?

15.38. Make a drawing of an inverted die and label all parts. Explain its operation.

15.39. Make a drawing of a compound die and label all parts. Explain its operation.

15.40. Discuss the operation of a progressive die.

15.41. List the four concerns of a designer when he is asked to design punches. Discuss each concern.

15.42. Discuss the piercing punches in Fig. 15.11 in light of Prob. 15.41.

15.43. Explain the purpose for the designs in Fig. 15.13 as preferred to Fig. 15.12.

15.44. You are asked to design a large equilateral punch. You have a choice of building it out of smaller pieces as shown in Fig. 15.14, welding several small pieces, or flame cutting the shape from the tool steel plate. Explain
   (a) the advantages and

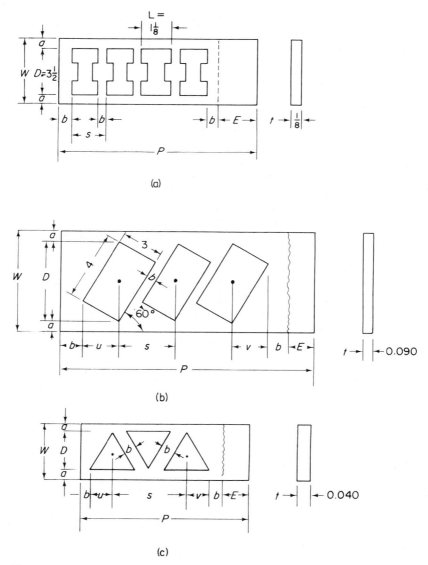

Figure 15.34

    (b) the disadvantages of each process.

    (c) Sketch the punches for each method.

15.45. What precautions must be observed when designing small-diameter piercing punches?

15.46. (a) What is the purpose of a backing plate when used with piercing punches?

    (b) How are countersunk heads put on piercing punches?

15.47. How should a large punch be guided through a stripper plate? Illustrate.

15.48. (a) Draw two methods for offsetting a series of piercing punches.
   (b) Why should this be done?

15.49. What is the reason for using a quill with long punches? Illustrate.

15.50. How may punches be prevented from rotating? Illustrate.

15.51. Discuss, and support with sketches, the use of hardened bushings as die openings.

15.52. Illustrate and discuss the use of die buttons.

15.53. Assume that you are asked to produce a sectioned die opening for producing the part in Fig. 15.33a. Make a sketch of the section and explain why you chose your design.

15.54. Repeat Prob. 15.53 for Fig. 15.33b.

15.55. (a) Describe the three types of shear which may be applied to a die block.
   (b) Repeat for the shear when applied to a punch face.

15.56. (a) Sketch five methods for fastening pilots in punches.
   (b) Sketch two additional designs.

15.57. (a) List some of the considerations which govern pilot length.
   (b) How much clearance should exist between the punch diameter and the diameters of the die opening?

15.58. (a) How strong should the compression spring be which backs up a spring pilot?
   (b) Make a sketch of another type of spring pilot other than the pilot shown in Fig. 15.21b.

15.59. Explain the operation of the knockout and shedder in Fig. 15.22.

15.60. Figures 15.23a through c show three designs for solid strippers. Compare all three designs. Explain them.

15.61. The strip stock is $3 \times \frac{1}{16}$ in. A channel stripper is 5 in. long. Calculate
   (a) the tunnel width and
   (b) the tunnel height if $1.5t$ is used for clearance.

15.62. Assume that the length of a channel stripper, Fig. 15.23d, is 12 in. The strip stock used is $2\frac{1}{2} \times \frac{1}{8}$ in. Calculate
   (a) the median clearance G,
   (b) the tunnel height, and
   (c) the allowable tunnel width.

15.63. When is it appropriate to use spring-loaded strippers?

15.64. What determines the positioning of springs and stripper bolts used with a stripper plate? Explain.

15.65. Discuss the three methods used to retain springs in Fig. 15.24. Make sketches of at least two additional methods of retaining springs and discuss them with your instructor.

15.66. Discuss the methods for guiding punches through a stripper plate as shown in Fig. 15.25. Make sketches of at least two additional methods and discuss them with your instructor.

15.67. Discuss the three methods for guiding a stripper plate with guide-pins. Sketch another method for guiding strippers, other than the three above. Discuss it with your instructor.

15.68. (a) Distinguish between the concept of stopping a strip and registering it.
   (b) Can both functions be performed by one obstruction? Explain.

15.69. (a) What is the range of overfeed permitted when stopping a scrap strip?
   (b) What factors control the amount of overfeed allowed?

15.70. Discuss the solid pin stops in Fig. 15.28. Why is the top of the pin extended above the material? How does the material get over the pin?

15.71. (a) Why is a finger stop necessary for most dies?
(b) Sketch another design for a finger stop. Explain its operation.

15.72. Explain the construction and operation of the automatic stop in Fig. 15.30. Check your library for at least one more automatic stop design. Draw a sketch of it, and describe it.

15.73. Given a washer with two holes as shown in Fig. 15.35, design a two-station die, using the procedure of Sec. 15.13. Start with the scrap strip, etc.

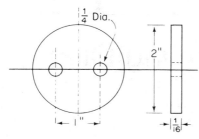

Figure 15.35

15.74. Repeat Prob. 15.73 for Fig. 15.33a.

15.75. Repeat Prob. 15.73 for Fig. 15.33b.

15.76. Repeat Prob. 15.73 for Fig. 15.34c.

Chapter 16

Bending and Forming Dies

16.1 THE THEORY OF METAL FLOW

The mechanism of metal displacement during the bending operation is quite different from the mechanism for drawing a shell. Since this chapter deals with both bending and drawing, the theory of metal flow will be discussed here.

Figure 16.1a shows the bending operation. The material is clamped as shown, so that it cannot escape the clamping device. The unsupported end of the material is forced down over the radius. The only movement of the grains in the material takes place at the point where the bend takes place. The outer fibers in the external radius are stretched because they are under tension. The fibers at the inner radius are under compression. At some point within the thickness of the material the forces are neither compressive nor in tension. This plane of the material is called the *neutral bending line*.

It should be noted that bending should take place at right angles to the grain direction, as shown in Fig. 16.1b. If the bending operation takes place parallel to the grain direction, as shown in Fig. 16.1c, a separation may occur and cracking will develop. Stock may be bent safely at angles up to 45° with the grain direction. If there is any doubt whether a piece will bend without cracking, a test should be run on the material to be used.

Usually, quarter-hard cold rolled sheet can be bent with the grain if a small radius is permissible in the inner corner of the bend. Half-hard steel should not be bent less than 45° with the grain. Hard steel should be bent only across the grain.

The drawing operation, Fig. 16.2, on the other hand, results in a radical displacement of the grains throughout the structure of the material. As the material is pulled and caused to slide over the radius of the die ring from underneath the pressure

462

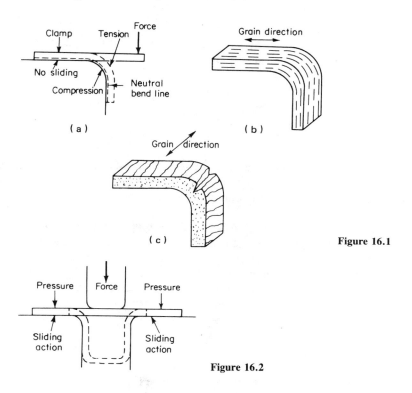

Figure 16.1

Figure 16.2

pad, every crystal in the material is displaced. At the radius the crystal displacement is analogous to water flowing over a dam.

The bottom of the shell is not affected by any of the forces operating during the drawing operation. The sidewalls of the shell are under tension. The larger-diameter blank being pulled and forced into a smaller diameter die opening places the material under compression at the lip of the die ring opening. The material under the pressure pad is under tension. It should be pointed out that at the beginning, when the bottom of the punch first touches the unsupported material, the operation is a bending or forming operation.

If the wall thickness of the shell is to be the same thickness as the original material from which it is to be drawn, the area of the original blank will be approximately the same as the area of the final shell. However, it should be emphasized that except for the bottom of the shell and possibly the lower radius, every crystal will have been rearranged.

If the crystal structure is short (cast iron), the metal crystals will not stretch or slide easily and the metal may separate. Long-crystal planes (lead and copper) permit easy bending. Some metals may prove poor risks for drawing because they do not offer enough resistance to the tensile forces in the sidewalls of the shell. The materials may stretch too much and rupture. Other materials, such as brass or hearth steels, may be drawn or bent rather freely without rupturing.

Figure 16.3

Many materials require heat treating between successive drawing operations to make them more plastic (remove work hardening, stresses, slip planes, etc.) Given enough time, enough operations, and enough heat treatment, almost any metal can be bent or drawn—even cast iron.

An intermediate type of operation is called *forming*. In this operation the shape of the punch and die are reproduced in the part. There is very little of the type of metal flow that is characteristic of the drawing operation. The mechanism of forming is quite similar to bending except that it is generally performed along a curved radius line. Figure 16.3 shows the three types of operations—bending, forming, and drawing—which seem similar but are, in many respects, structurally different.

16.2 DEVELOPED LENGTH

Blueprints show the finished dimensions applied to the finished product. The material before bending is flat stock. The question arises as to how long the flat stock is to be cut so that, once the stock is bent, the finished part will meet the print requirements. It should be noted very carefully in Fig. 16.4a that a piece of material that is cut 4.500 + 3.000 = 7.500 in. long will be too long once bent. If the internal dimension is taken and the strip cut to (4.500 − 0.090) + (3.000 − 0.090) = 7.320 in., the legs will be too short once the bend is completed. The length of the neutral bending line, shown in Fig. 16.4b, will give you the true length of the piece after it has been bent because this line remains unchanged. Its length is the same after bending as before bending.

All bending equations in this book are based on the length of the neutral bending

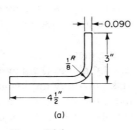
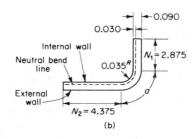

(a)   (b)

Figure 16.4

line and *internal* dimensions. Figure 16.4a with *external* dimensions applied has been converted to Fig. 16.4b showing *internal* dimensions. It is good practice to make a sketch and to convert all external dimensions to internal dimensions before applying the equations which follow.

The general equation for the length of the arc at the neutral bending is given by:

$$a = \frac{\theta\pi}{180}(Kt + r)$$
$$= 0.01745\theta(Kt + r)$$

a = length of arc, in.
r = internal radius, in.
t = thickness of material, in.
θ = bend angle
K = constant

Since the majority of the bend angles that are used are 90°, it is advisable to state the equation for 90° bends. This equation is

$$a = 1.5708(r + Kt) \qquad a = \text{arc length for 90° bends}$$

When the length of the *internal radius* is smaller than the material thickness, $K = \frac{1}{4}$. When the length of the internal radius is equal to or twice the thickness of the material, $K = \frac{1}{3}$. When the length of the internal radius is more than twice the stock thickness, $K = \frac{1}{2}$. To summarize:

If $r < t$ $K = \frac{1}{4}$
If $t \le r \le 2t$* $K = \frac{1}{3}$
If $r > 2t$ $K = \frac{1}{2}$

It is also important to point out that the angle θ is that angle through which the material is bent. Thus in Fig. 16.5 the angle through which the material is bent, is 30°, not 60°.

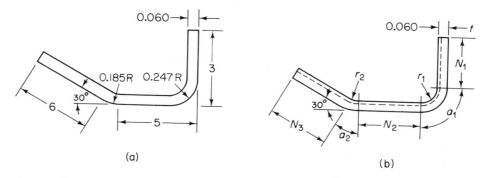

Figure 16.5

*r is between t and $2t$; or r is greater or equal to t and less than or equal to $2t$.

Example 16.1

Calculate the developed length of the part shown in Fig. 16.5a.

Solution The first requirement is to redimension the drawing to internal dimensions. This is shown in Fig. 16.5b.

The internal radius r_1 is

$$r_1 = 0.247 - 0.060$$
$$= 0.187 \text{ in.}$$

The internal radius r_2 is

$$r_2 = 0.185 - 0.060$$
$$= 0.125 \text{ in.}$$

The length of leg N_1 is

$$N_1 = 3.000 - (0.060 + 0.187) = 3.000 - 0.247$$
$$= 2.753 \text{ in.}$$

The length of leg N_2 is

$$N_2 = 5.000 - (0.060 + 0.187) = 5.000 - 0.247$$
$$= 4.753 \text{ in.}$$

The length of leg N_3 is

$$N_3 = 6.000 \text{ in.}$$

Since r_1 is greater than t and greater than 0.120, or $2t$, the value of K is $\frac{1}{2}$. Therefore the length of arc a_1 at the neutral bending line is

$$a_1 = 1.5708 \ (Kt + r_1)$$
$$= 1.5708 \left(\frac{0.060}{2} + 0.187 \right)$$
$$= 0.341 \text{ in.}$$

$$t = 0.060 \text{ in.}$$
$$r_1 = 0.187 \text{ in.}$$
$$K = \tfrac{1}{2}$$

Since 0.125 is greater than t, the value of K is $\frac{1}{2}$. Therefore, the length of the arc at this neutral bending line is

$$a_2 = 0.01745 \times \theta(Kt + r_2)$$
$$= 0.01745(30) \left(\frac{0.060}{2} + 0.125 \right)$$
$$= 0.081 \text{ in.}$$

$$t = 0.060 \text{ in.}$$
$$K = \tfrac{1}{2}$$
$$\theta = 30°$$
$$r_2 = 0.125 \text{ in.}$$

The developed length is the summation of all these lengths along the neutral bending line. It is

$$L = N_1 + N_2 + N_3 + a_1 + a_2$$
$$= 2.753 + 4.753 + 6.000 + 0.341 + 0.081$$
$$= 13.928 \text{ in.}$$

It should be explained that the line where the *internal radii* are tangent to the legs of a bend are called the *bend lines*. They are shown as dashed lines in Fig. 16.6. A line drawn parallel to, but halfway between, both bend lines is the *layout line* at

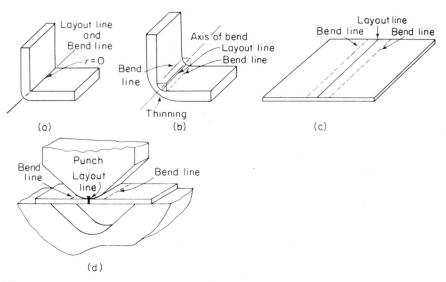

(a)

(b)

(c)

(d)

Figure 16.6

which point the punch contacts the workpiece. In Fig. 16.6a, the internal radius is $0°$, and therefore the layout line and the bend lines coincide. In Fig. 16.6b the axis of bend is the line from which the radius is generated. Note the bend lines and the layout lines. Fig. 16.6c shows the developed length, the two bend lines, and the layout line, which is the contact line for the punch shown in Fig. 16.6d.

Example 16.2

Using the values from Fig. 16.5 and the calculations from Ex. 16.1, calculate the bend lines and the layout lines for the developed length.

Solution   See Fig. 16.7. The right dimension to the layout line is

$$N_1 + \frac{a_1}{2} = 2.753 + \frac{0.341}{2}$$

$$= 2.924 \text{ in.}$$

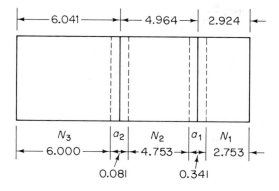

Figure 16.7

The center dimension between the two layout lines is

$$N_2 + \frac{a_1}{2} + \frac{a_2}{2} = 4.753 + \frac{0.341}{2} + \frac{0.081}{2}$$

$$= 4.964 \text{ in.}$$

The left dimension is

$$N_3 + \frac{a_2}{2} = 6.000 + \frac{0.081}{2}$$

$$= 6.040 \text{ in.}$$

Figure 16.8a shows a part which has been bent so that front, rear, left side, and bottom are mutually perpendicular to each other. The rear wall is notched out (dashed lines), and the right wall is bent 30° to the base.

Example 16.3

Calculate the developed length of the multiple bends in the boxlike structure, Fig. 16.8b of (a) L_1, (b) L_2, and (c) L_3.

Solution   All internal radii are 0.250 in.
Since

$$r > t$$

then

$$K = \tfrac{1}{4}$$

(a) The length L_1 is the sum of all the segments, Fig. 16.8a, through section A-A. This is shown in Fig. 16.8c.

$$N_1 = 3.000 - (0.090 + 0.250) \qquad t = 0.090 \text{ in.}$$
$$\qquad\qquad\qquad\qquad\qquad\qquad\quad r = 0.250 \text{ in.}$$
$$= 2.660 \text{ in.} \qquad\qquad\qquad K = \tfrac{1}{4}$$

$$N_2 = 4.500 - (0.090 + 0.250)$$

$$= 4.160 \text{ in.}$$

$$N_3 = 5.000 \text{ in.}$$

$$a = 1.5708(Kt + r) = 1.5708\left(\frac{0.090}{4} + 0.250\right)$$

$$= 0.427 \text{ in.}$$

$$b = 0.017450(Kt + r) = 0.01745(30)\left(\frac{0.090}{4} + 0.250\right)$$

$$= 0.142 \text{ in.}$$

$$L_1 = N_1 + N_2 + N_3 + a + b$$

$$= 2.660 + 4.160 + 5.000 + 0.427 + 0.142$$

$$= 12.390 \text{ in.}$$

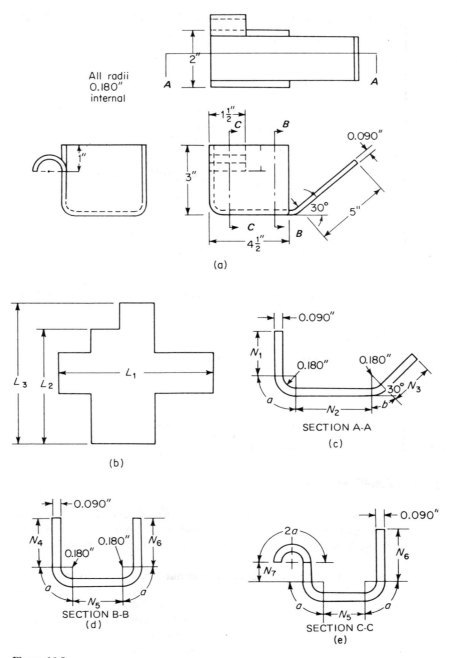

All radii
0.180"
internal

Figure 16.8

(b) The length of L_2 is taken through section B-B, Fig. 16.8a, and is shown in Fig. 16.8d.

$N_4 = 3.000 - (0.090 + 0.250)$

$\quad = 2.660$ in.

$N_5 = 2.000 - 2(0.250 + 0.090)$

$\quad = 1.320$ in.

$N_6 = 3.000 - (0.090 + 0.250)$

$\quad = 2.660$ in.

$a = 1.5708 \ (Kt + r) = 1.5708\left(\dfrac{0.090}{4} + 0.250\right)$

$\quad = 0.427$ in.

$L_2 = N_4 + N_5 + N_6 + 2a$

$\quad = 2.660 + 1.320 + 2.660 + 2(0.427)$

$\quad = 7.494$ in.

(c) The length of L_3 is taken through section C-C, Fig. 16.8a, and is shown in Fig. 16.8e.

$N_6 = 2.660$ in.

$N_5 = 1.320$ in.

$N_7 = 3.000 - (1.000 + 0.090 + 0.250)$

$\quad = 1.660$ in.

$L_3 = N_6 + N_5 + N_7 + 4_a$

$\quad = 2.660 + 1.320 + 1.660 + 4(0.427)$

$\quad = 7.348$ in.

16.3 BENDING FORCES

Figure 16.9b, c, and d show the three types of bends encountered most frequently in die work. Figure 16.9b is a span bend, called a *V air bend* when the radius of the forming block does not match the punch radius. Figure 16.9c is a *U-channel bend*. Figure 16.9d is a *cantilever bend*.

The force required to bend material into a V is given by the equation

$$F = \frac{KLSt^2}{W}$$

F = bending force, tons
L = contact length (width of work), in.
S = ultimate tensile strength, tons/in.^{2}
t = material thickness, in.
W = width of V die, in. span of beam
K = die opening factor

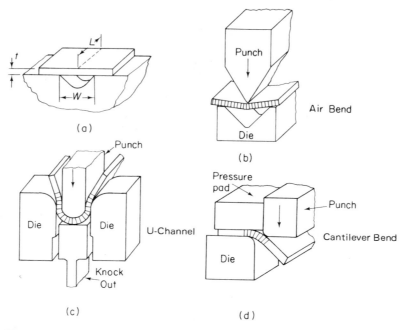

Figure 16.9

It shall be noted that the value L is the width of the strip in contact with the bend points of the die opening width W as shown in Figure 16.9a. The value W is actually the span value of the opening.

For V bends, Fig. 16.9b, K has values of 1.33 when the die opening W is 8 times the material thickness. It has a value of $K = 1.2$, when W equals 16 times the material thickness. As indicated, where the punch does not bottom, Fig. 16.9b, the operation is called *air bending*.

For U and channel bending, Fig. 16.9c, the constant K is double that used for air bending. Thus K will range from 2.67 to 2.4. For edge bending, Fig. 16.9d, K is one-half air bending. The range is therefore from 0.67 to 0.60.

Example 16.4

A part is to be bent into the shape shown in Fig. 16.10a. The material has an ultimate tensile strength of 60,000 psi. It is to be bent into a 90° angle in an air-bending type die as shown. Calculate the tonnage required to produce the part. The bend length is 8 in., the stock thickness is $\frac{1}{4}$ in., and the beam length is 3 in.

Solution   The span length W is 3 in. Since 16 times the material thickness ($16 \times \frac{1}{4} = 4$ in.) is greater than W, K equals 1.33 is appropriate. The force required to bend this work piece is

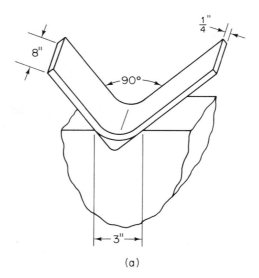

(a)

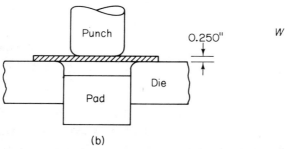

(b)

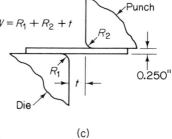

(c)

Figure 16.10

$$F = \frac{KLSt^2}{W} = \frac{1.33 \times 8 \times 30 \times 0.250^2}{3}$$

$$= 6.65 \text{ tons}$$

$W = 3$ in.
$t = \frac{1}{4}$ in.
$L = 8$ in.
$S = 60,000$ psi
$ = 30$ tons
$K = 1.33$

Example 16.5

Assume the same conditions as in Ex. 16.4 except that the bend is a U bend, as shown in Fig. 16.10b. Calculate the force required to make the bend.

Solution   The value of k is 2.67, where W is less than 16 times the thickness. Thus the force is

$$F = \frac{KLSt^2}{W} = \frac{2.67 \times 8 \times 30 \times 0.250^2}{3}$$

$$= 13.35 \text{ tons}$$

$W = 3$ in.
$t = \frac{1}{4}$ in.
$L = 8$ in.
$S = 30$ tons
$K = 2.67$

Example 16.6

Assume the same conditions in Ex. 16.4 except that the bend is cantilevered, as shown in Fig. 16.10c, edge bent. If the radii are $\frac{3}{8}$ in., calculate the force required to make the bend.

Solution   The value of W is

$$W = R_1 + R_2 + t = 0.375 + 0.375 + 0.250$$
$$= 1 \text{ in.}$$

The force is

$$F = \frac{KLSt^2}{W} = \frac{0.67 \times 8 \times 30 \times 0.250^2}{1.0}$$
$$= 10 \text{ tons}$$

$S = 30$ tons/in.^{2}
$t = \frac{1}{4}$ in.
$L = 8$ in.
$K = 0.67$
$R_1 = \frac{3}{8}$ in.
$R_2 = \frac{3}{8}$ in.

16.4 TYPES OF BENDS

As indicated earlier, the amount of bending that a material can take is limited by the properties of the material, the length of the bend, the condition of the cut edge at the ends of the band line, and the direction of orientation of the axis of the bend to the direction in which the material was rolled.

If the length of the bend is *greater than* eight times the metal thickness, the bend radius may be a minimum and is the same for most materials. If the length of the bend is *less than* eight times the thickness, the radii must be increased.

Low-hardness, high-ductility materials can usually be bent 180° with a sharp internal radius without fracture. Bends with smaller radii can usually be made when bending across the grain direction than when made parallel to the grain direction.

Steels that are subject to work hardening may crack while being bent because of burrs or sharp edges in the bent area. It is also true that if sharp edges are kept on the inside of the bend, the part is less likely to crack than if the burr is on the outside of the bend.

Press brakes are machines used to bend long workpieces. The die opening is relatively narrow in comparison to the length of the bend. Most operations which are done on a conventional press may be done on a brake. The setup is very fast and operations on long workpieces that may not be done in the standard press may be quickly done with a brake. Figure 16.11 shows some of the shapes that may be pressed on a brake.

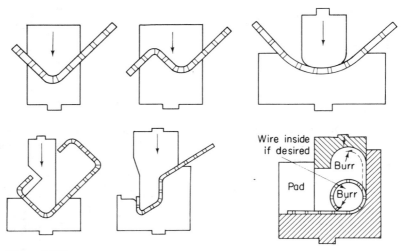

Figure 16.11

16.5 FORMING DIES

Forming is a variation of bending. The part may take the shape of the punch or die. Rarely is the thickness of the metal reduced at the bend line. Generally speaking, the speed of the press should be greater than 50 feet per minute and less than 200 feet per minute for optimum production, quality, and tool life.

Sometimes *stretch strains* occur in materials that are under tension or compression. When low-carbon material is cold worked and is in tension, stretcher lines in the form of valleys appear on the surface of formed parts. If the material is in compression, the stretcher lines take the form of elevations. They can generally be moved by tempering, or by roller leveling the stock before it is used through a machine that reduces the internal strain. The strength of the material is not affected by stretcher lines.

16.6 FORMING

Figure 16.12 shows the essential differences between bending and forming. Forming may be any of the following types: solid form, pressure pad, curling, embossing, coin, swaging, bulging, flanging, extruding, and assembly dies.

Solid forming dies (Fig. 16.13a). Solid forming dies usually have simple construction. The punch is made from a good grade of tool steel and shaped to the desired contour of the workpiece. The die block is shaped to the matching contour of the workpiece. It is generally made from several tool steel pieces (sectioned) so that it can *give* slightly when in use. The sections are sunk into the die shoe and keyed so that they are held in position. Socket head screws and dowels should be large enough

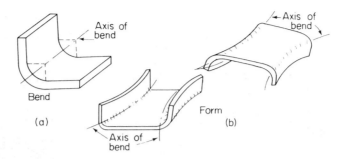

Figure 16.12

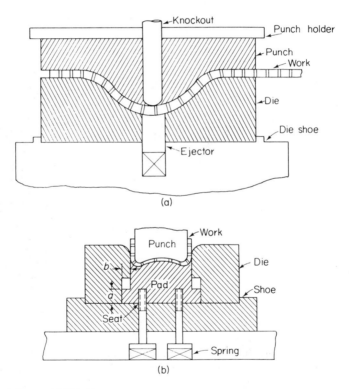

Figure 16.13(a,b)

to take the additional lateral stresses built up during forming. Figure 16.13a shows a solid forming die. Solid-type forming dies are usually added to progressive dies.

Forming dies with pressure pads (Fig. 16.13b).   These types of dies are used where greater accuracy is demanded. The pad holds the blank in position while it is being worked on. It also provides the resistance to the forming operation that is

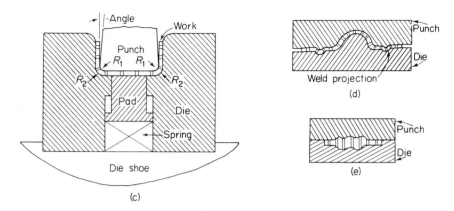

Figure 16.13(c,d,e)

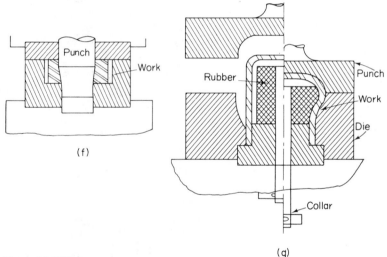

Figure 16.13(f,g)

needed to perform the operation. Springs, air cushions, or hydraulic cylinders are used to provide this resistance. At times the springs are replaced with pneumatic or hydraulic cylinders. Sometimes a combination of both is used and mounted through the hole in the bolster plate.

The lip a on the pressure pad should be about $1\frac{1}{2}$ times the width b, as shown in Fig. 16.13b. The height of the pad should extend slightly above the die block face when in its up position. In the down position the pad must seat so that the punch may "set" the material before it starts up again. If it becomes necessary for the bottom and the sides of the workpiece to be at right angles, the corner radii must be set. This is done by making the corner radii R_1 about 10% smaller than the die radius R_2 and making certain the die block provides the matching radii, as shown in Fig. 16.13c.

Curling (Fig. 16.11).   Curling is the process of causing the edge of a metal part to follow a circular path. The radius of the finished product should not be less than twice the thickness of the metal. The metal needs to be ductile so that it rolls easily. The edge of the piece should have a starting bend and, if possible, the burr should be inside the bend. At times, the curl is caused to roll around a wire to give the curl stiffness. This is the case when curls are made around tops of pails, pans, or cans. It should be noted that the radius causing the curl must be smooth to facilitate rolling.

Embossing (Fig. 16.13d).   This operation is a shallow type of forming. Its purpose is to produce patterns that are reproduced on the punch and die in exact detail. The operation is one of stretching the metal into the forms, rather than causing the metal to flow into the forms. The die set may have as its purpose the raising of ribs as stiffeners around the contour of a sheet metal object, such as a round or rectangular pan. It may also be used as a lettering device, for forming intricate patterns in sheet metal parts such as metal buttons, or for forming small projections used in projection welding. The criterion is the duplicating of the pattern on both sides of the sheet metal without changing the metal thickness. Figure 16.13d shows the cross section of a simple pattern with two buttons to be used in projection welding.

Coining (Fig. 16.13e).   Coining differs from embossing in that the metal is caused to flow into the provided cavity. The result is a workpiece that has a variable thickness throughout. Coins, plates, jewelry, and medals are objects wherein the details have been coined into the faces of the part. High pressures are required for coining because the metal work hardens as it is being worked and because it takes a great deal of pressure to cause metal to flow. As much of the thickness of the metal as possible is confined in the die cavity. The die blocks must be built to withstand the high pressures. Any sharp change in contour, such as a very sharp corner or deep scratch, could act as a line of stress concentration and cause the die to break. The punch and die cavity must have draft to allow the part to be removed after it has been formed. The two halves of the die must be aligned exactly to avoid the possibility of the metal overflowing between the faces of the punch and die. It should be noted that very accurate and intricate patterns may be coined using this process.

Swaging (Fig. 16.13f).   This is another of the forming operations that requires very high pressures. Consequently, die blocks, die sets, and machines must be made stronger than usual. High-strength chrome, or chrome-tungsten steels, should be used in the dies. The contour of the part to be swaged is very often machined into the die block. Draft on the workpieces and relief, where no swaging is to take place, is included in the die cavity. The swaging operation is generally a sizing operation, called *planishing,* which work hardens the swaged surface so that the workpiece surface has greater wear resistance than comparable machined surfaces. This comes under the general heading of *cold workings.* Another class of swaging operations is the upsetting of rivets, pins, and bolts into heads. Figure 16.13f shows a swaging operation which is sizing a bushing.

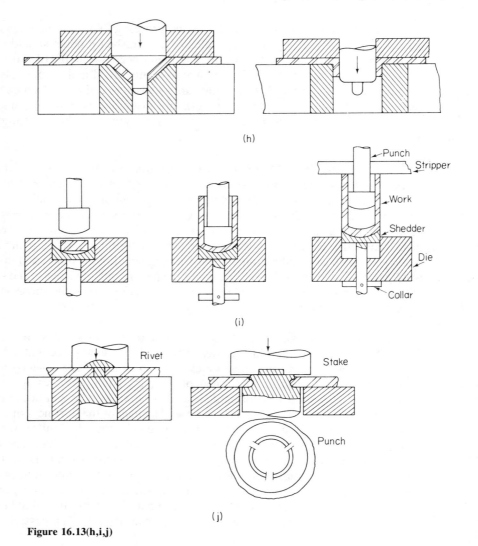

(h)

(i)

(j)

Figure 16.13(h,i,j)

Bulging (Fig. 16.13g).   This operation is generally done by inserting rubber, water, or oil into a preformed shell or tube. The form of the finished shell is machined into a split die so that when pressure is applied to the rubber (or other media) it forces the workpiece into the forming cavity. The die is opened and the rubber is removed. The metal shell usually needs to be annealed for maximum ductility.

Medium-hardness rubber is used and is easy to handle. Water, oil, grease, or any other fluid used requires careful design of the dies so that leaking does not occur. Besides being uncomfortable, leaking reduces the necessary volume of fluid needed for satisfactory performance.

When bulging tubes, one end is closed with a solid plug. The tube is filled with the fluid (water), and the other open end is sealed with the plug, which has the feeder line attached. The assembly is put into a die, which is closed. The ram is lowered to hold the die and the pump is started. The fluid pressure builds up to complete the job.

Flanging (Fig. 16.13h). Hole flanging may be accomplished for a rivet head, countersunk for a flat head screw, or pierced and extruded into a 90° hole flange. The hole may be punched in the same station that the flanging takes place; or it may be done in a two-station die. These punches are shown in Fig. 16.13h.

Extruding (Fig. 16.13i). This is a process wherein the volume of the material needed to form a tube is calculated. A block of this volume of material is placed in a close die cavity. A punch is pushed into the material, which "splashes" up the side of the punch. A knockout pad and stripper are needed to push the finished shell out of the die and off the punch.

Assembly dies (Fig. 16.13j). Assembly dies use operations that have been discussed in this section. Thus dies may be used to perform the operations of staking, curling, riveting, seaming, or press fitting one part to another. Figure 16.13j shows a punch that rolls over (stakes) three prongs so that a threaded stud may be fastened to a sheet metal part.

PROBLEMS

16.1. (a) Discuss the forces present when a strip of sheet steel is bent to a 90° angle.
(b) What is the neutral bending line?

16.2. (a) Discuss the differences between the bending and the drawing operations.
(b) Compare the forces operating in both operations.

16.3. Discuss and compare the three operations of bending, forming, and drawing.

16.4. Calculate the developed length of the part shown in Fig. 16.14a.

16.5. Using the values from Fig. 16.14a and the calculations from Prob. 16.4, calculate the bend lines and the layout lines for the developed length.

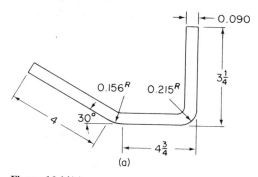

(a)

Figure 16.14(a)

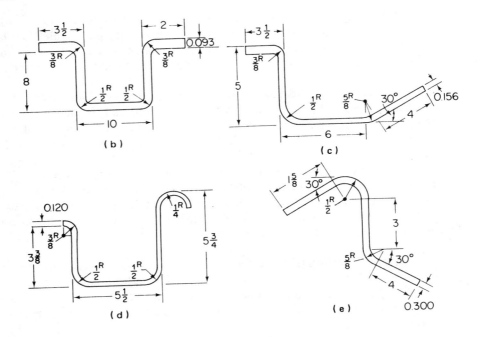

Figure 16.14(b,c,d,e)

16.6. **(a)** Calculate the developed length of the bent-up workpiece in Fig. 16.14b.
   (b) Draw the bend lines (dashed) and the layout lines for your solution.

16.7. **(a)** Calculate the developed length for Fig. 16.14c.
   (b) Lay out the developed length showing the bend and layout lines.

16.8. **(a)** Calculate the developed length of the part in Fig. 16.14d.
   (b) Lay out the developed length and show all layout and bend lines.

16.9. **(a)** Calculate the developed length of Fig. 16.14e.
   (b) Lay out the developed length showing all bend and layout lines.

16.10. Make three sketches of air, U-, and cantilever bending operations.

16.11. An air bend is to be made as shown in Fig. 16.15a if the ultimate tensile strength of the material is 40,000 psi. Calculate the force required.

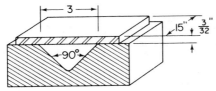

Figure 16.15(a)

16.12. Calculate the force required to bend the material in Prob. 16.11 into a U-bend.

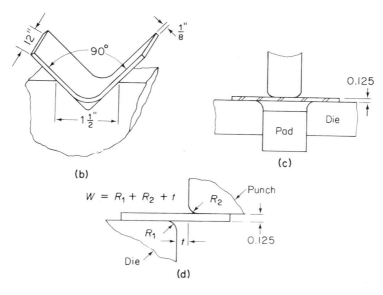

Figure 16.15(b,c,d)

16.13. Calculate the force required to bend the material in Prob. 16.11 for a cantilevered bend if the radius of the die is $\frac{1}{2}$ in. and the radius of the punch is $\frac{3}{8}$ in.

16.14. Assume the bend as shown in Fig. 16.15b. If the material has an ultimate tensile strength of 70,000 psi and is to be sent into a 90° angle in an air-bending type die as shown, calculate the tonnage required. The bend length is 12 in., the stock thickness is 0.125 in., and the beam length is 1.5 in.

16.15. Assume the same conditions as in Prob. 16.14 except that the bend is a U-bend, as shown in Fig. 16.15c. Calculate the force required to make the bend.

16.16. Assume the conditions of Prob. 16.14 except that the bend is cantilevered, as shown in Fig. 16.15d, edge bent. Calculate the force required to make the bend if the radii are $\frac{1}{2}$ in.

16.17. What conditions determine the amount of bending that a material can undergo?

16.18. A workpiece is $\frac{1}{8}$ in. thick by 12 in. long by 2 in. wide. The grain runs parallel to the 12 in. A minimum bend radius of $\frac{1}{4}$ in. is made across the grain.
  (a) Assume the width of the work is increased to 3 in. Can the same bend radius be made?
  (b) Assume the width of the work is reduced to $\frac{3}{4}$ in. Can the same bend radius be maintained? Explain your answers.

16.19. Explain the effects the following conditions have upon the success or failure of a bend:
  (a) material hardness,
  (b) ductility,
  (c) grain direction,
  (d) burrs.

16.20. (a) What is a brake?
  (b) Figure 16.11 shows several shapes that can be formed on a brake. Check your handbook and sketch at least two additional types of bends that can be done on a brake.

16.21. (a) What are stretcher strains?

(b) How do they form?

(c) How can they be removed?

16.22. There are 10 types of forming operations listed in Sec. 16.6. Briefly explain each type.

16.23. (a) Make a sketch of a solid forming die and explain its action.

(b) Why is the die block sometimes sectioned?

16.24. (a) What purpose does a forming pad serve in a pressure pad forming die?

(b) List some of the precautions and characteristics of this type of die.

16.25. (a) Make a drawing of a curling die.

(b) What precautions must be observed for the successful operation of this type of die?

16.26. (a) Explain the operations of coining and embossing.

(b) How do they differ?

(c) What precautions must be observed in both operations?

16.27. (a) Describe the operation of swaging with a die.

(b) What effect does it have on the swaged surface?

16.28. (a) Describe the operation of bulging.

(b) How are long tubes bulged with a fluid?

16.29. Draw the punch contour for flanging a rivet hole and a flathead screw.

16.30. Describe the process of splash extruding a drawn shell.

16.31. (a) Explain how a staking operation resembles a curling operation.

(b) Make sketches of other assemblies that can be done with dies using the operations discussed thus far.

16.32. Examine several objects that could have been or were done using dies. Describe your idea of the process. Make sketches supporting your descriptions.

Chapter 17

Drawing and Progressive Dies

17.1 THE THEORY OF DRAWING (IN GENERAL)

The forces necessary in drawing a shell were discussed in Sec. 16.1. The area of the developed blank before drawing should be the same as the surface area of the shell after drawing, provided that the thickness of the material remains unchanged.

As the metal is pushed through the die opening over a radius, the drawing action causes the shell to hug the punch. This requires a stripping action. One of these devices is shown in Fig. 17.1. There are many more.

The pressure pad, Fig. 17.1, exerts a force on the "lip" of the parially drawn shell. This force holds the lip down with a pressure just strong enough to permit the material to slide out from underneath once the punch starts to descend. Under this type of action, the metal must flow over the die radius without flapping up.

If the metal lips on the partially drawn shell were permitted to flap up, a larger circumference of material would try to crowd into the smaller die opening circumference. Folding would take place at the die radius. As the punch continued down, these folds would be too thick to pass through the punch and die clearance. The folds would be "ironed" out and appear on the finished product as wrinkles. If the metal lips are held too tightly, the punch would either punch the bottom out of the shell, or stretch the material in the walls of the shell.

Based on the statements just made, large die radii or unconfined metal will cause wrinkles to form in the finished shell. These may be ironed out so they are not visible. However, the weakness created is always there no matter how much ironing the shell has undergone. Certainly, if there are to be several drawing operations, defects such as wrinkles, cracks, or thinning should be corrected when they occur. If the radii are too small, breaks will occur near the punch radius soon after the material

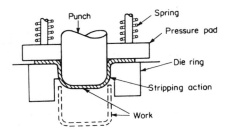

Figure 17.1

enters the die opening. As a general rule, for sheet metal the radius over the die ring should be about four times the thickness of the material. The punch radius should be at least the same size as or larger than the radius in the die ring.

Since the material is pulled over the die-ring radius, friction is a factor that must be considered. The die-ring radius must be highly polished and lapped. This will reduce the tensile forces on the shell and will eliminate the possibility of the radius picking up metal from the shell walls, which will cause scratches in the sides of the shell.

It is also possible to reduce the wall thickness of shells by controlling the clearance between the punch and the die. As a general rule wall thickness for ductile materials may be reduced by 10% of the original thickness of the material. Other factors, such as the type of material, the ductility, and the work hardness that takes place during the operation, must be considered. When the operation reduces the wall thickness of the material, the developed blank will be different from a drawn shell, which has the same material thickness as the operation described just above.

17.2 SHELL-BLANK CALCULATIONS

Before a shell can be drawn, the diameter of the developed blank must be known. The calculations that follow are approximate. They are calculations that yield results when the operation is performed under ideal conditions. If the wall thickness of the material remains the same after drawing as it was before drawing, the surface area of the blank and the subsequent surface area of the drawn shell should remain the same.

If the shell has no corner radius and is plain, as shown in Fig. 17.2a, the formula for calculating the developed blank diameter by the method of areas can be

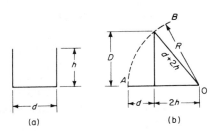

(a)   (b)   **Figure 17.2(a,b)**

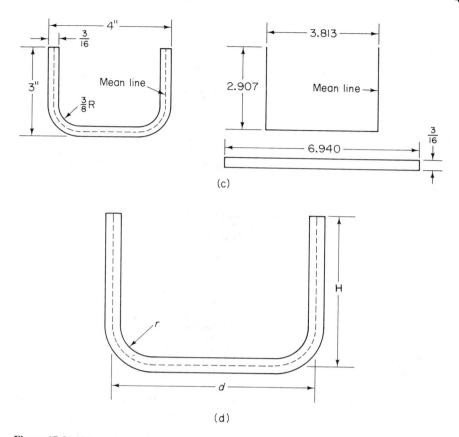

Figure 17.2(c,d)

derived as follows from Fig. 17.2b. It should be noted that the single solid line $d + 2h$ (Fig. 17.2a) is the *mean* line shown in Fig. 17.2c. This line and its associated dimensions should be used when calculating the blank diameter for drawn shells. It will yield values close enough for most design problems.

1. Lay out the mean diameter d of the base of the shell and twice the height $2h$ of the shell on d extended.
2. With O as the center and a radius of $d + 2h$, scribe the arc AB.
3. Erect a perpendicular from d to the arc AB.
4. D is the diameter of the required blank.
5. From the triangle shown,

$$R^2 = D^2 + (2h)^2$$

d = mean diameter, in.
D = diameter of blank, in.
h = mean height of shell, in.

6. Solve for D:

$$D = \sqrt{R^2 - (2h)^2}$$

7. Substitute $d + 2h$ for R:

$$D = \sqrt{(d + 2h)^2 - (2h)^2}$$

8. Evaluate:

$$D = \sqrt{d^2 + 4dh + 4h^2 - 4h^2}$$
$$= \sqrt{d^2 + 4dh}$$

Example 17.1

Calculate the blank diameter in Fig. 17.2c. Use the method of areas.

Solution The mean diamter (Fig. 17.2c) is

$$d = 4.000 - 0.187 = 3.813 \text{ in.}$$

The mean height is

$$h = 3.000 - \frac{0.187}{2} = 2.907 \text{ in.}$$

The blank diameter for a square corner shell is

$$D = \sqrt{d^2 + 4dh} = \sqrt{3.813^2 + 4(3.813)(2.907)}$$
$$= 7.673 \text{ in.}$$

If the shell is plain but has an internal mean radius r, as shown in Fig. 17.2d, the equation for converting the shell to a square shell is

$$h = H - 0.43r \qquad \begin{array}{l} h = \text{mean height zero radius} \\ H = \text{mean height of shell} \\ r = \text{radius at neutral bend line} \end{array}$$

Example 17.2

Assume the same dimensions as in Fig. 17.2c except that the shell has a $\frac{3}{8}$-in. internal radius. (a) Calculate the developed blank for this shell. (b) Use an equation from Fig. 17.3 to check your solution.

Solution
(a) The mean radius (Fig. 17.2c) is

$$r = 0.375 + \frac{0.187}{2} \qquad \begin{array}{l} H = 2.907 \text{ in.} \\ d = 3.813 \text{ in.} \end{array}$$

$$= 0.468 \text{ in.}$$

To convert the shell to a square corner shell, the value of h is

$$h = H - 0.43r = 2.907 - 0.43(0.468)$$
$$= 2.706 \text{ in.}$$

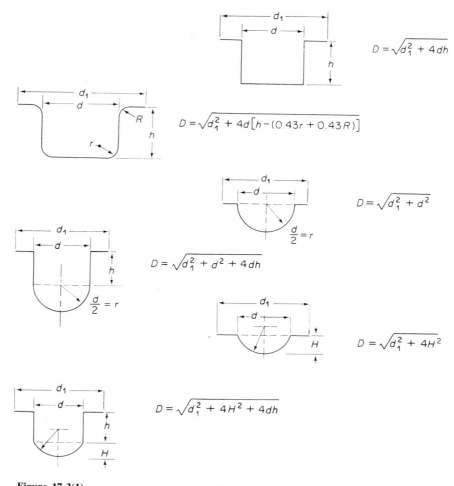

Figure 17.3(1)

The developed blank diameter is

$$D = \sqrt{d^2 + 4dh} = \sqrt{(3.813)^2 + 4(3.813)(2.706)}$$
$$= 7.470 \text{ in.}$$

(b) Using an equation from Fig. 17.3, the solution is

$$D = \sqrt{d^2 + 4dH - 1.72rd}$$
$$= \sqrt{3.813^2 + 4(3.813)(2.907) - 1.72(0.468)(3.813)}$$
$$= 7.470 \text{ in.} \quad \text{check}$$

The formulas for calculating many different types of developed lengths are shown in Fig. 17.3.

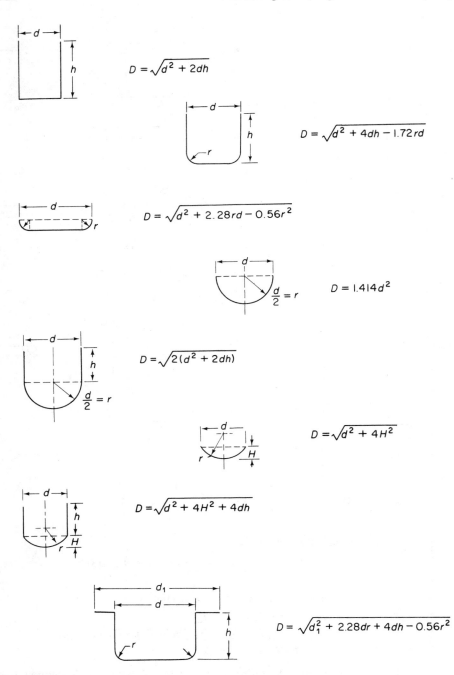

$$D = \sqrt{d^2 + 2dh}$$

$$D = \sqrt{d^2 + 4dh - 1.72\,rd}$$

$$D = \sqrt{d^2 + 2.28rd - 0.56r^2}$$

$$\frac{d}{2} = r \qquad D = 1.414d^2$$

$$D = \sqrt{2(d^2 + 2dh)} \qquad \frac{d}{2} = r$$

$$D = \sqrt{d^2 + 4H^2}$$

$$D = \sqrt{d^2 + 4H^2 + 4dh}$$

$$D = \sqrt{d_1^2 + 2.28dr + 4dh - 0.56r^2}$$

Figure 17.3(2)

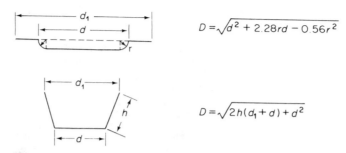

$$D = \sqrt{d^2 + 2.28rd - 0.56r^2}$$

$$D = \sqrt{2h(d_1 + d) + d^2}$$

Figure 17.3(3)

If the shell walls are to be reduced in thickness, the equation for the mean height is a function of the ratio of the shell wall thickness after ironing of the shell wall thickness before ironing. This equation gives a good approximation of the mean height to be used in the general equation. Thus

$$h = \frac{h_i t_i}{t}$$

h = mean height, in.
h_i = iron height, in.
t_i = ironed thickness, in.
t = original thickness, in.

The percent of reduction of a shell may be calculated from

$$\% \text{ reduction} = \frac{(t - t_i)100}{t}$$

In many instances it is necessary to draw a shell several times if the final diameter after drawing is more than a 50% reduction of the diameter before drawing. Reductions of the diameter for the first draw, from the first developed blank to the first shell, may be as much as 50%. If the shell is annealed before the second shell is drawn (first redraw), the reduction of the diameter may be as much as took place during the first drawing. If there is no annealing operation intervening, the percent of reduction of the diameter must be 40% or less. The percent of reduction of the shell diameter must be decreased if there is no intervening annealing operation.

The percent reduction in Ex. 17.1 is about 50% and is acceptable.

$$7.673x = 100(7.673 - 3.813)$$
$$x = 50.3\%$$

The required shell diameter is 3.813 in. The shell in Ex. 17.4 would require several drawing operations.

Example 17.3

A shell is to have its wall thickness reduced from 0.050 in. by 12.5%. The new shell will have a new height of 2.750 in. and a new diameter of 2.000 in. (a) What is the mean height of the shell? (b) What is the shell diameter needed to draw the ironed shell?

Solution

(a) The new wall thickness will be

$$t_i = 0.050 - 0.050(0.125) \qquad t = 0.050 \text{ in.}$$
$$= 0.044 \text{ in.} \qquad h_i = 2.750 \text{ in.}$$

The mean height of the shell is

$$h = \frac{h_i t_i}{t} = \frac{2.750(0.044)}{0.050}$$
$$= 2.420 \text{ in.*}$$

(b) The shell diameter required to draw the ironed shell is

$$D = \sqrt{d^2 + 4dh} = \sqrt{1.750^2 + 4(1.750)(2.42)}$$
$$= 4.472 \text{ in.}$$

The blank diameter may also be found by using the sum of the areas of segments of the shell. The table of area parts Figure 17.4 gives various areas so that a shell may be broken down into matching components and the area of each component calculated. The sum of the area components is inserted into the equation below. The result is the developed blank.

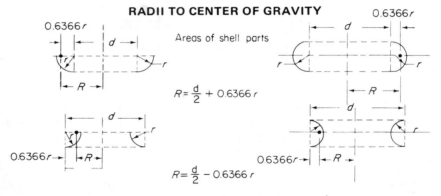

Figure 17.4

The area of a circular blank is

$$A = \frac{\pi D^2}{4}$$

The diameter of the developed blank is

$A =$ sum of the areas
$D =$ diameter of developed blank

$$D = \sqrt{\frac{4A}{\pi}} = 1.128\sqrt{A}$$

Note: A reduction of the diameter of 2.750 in. by 12.5% equals 2.406 in.

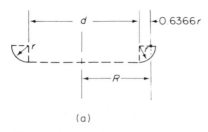

Figure 17.5

The area of a segment is based on the radius from the center of the shell part to the center of gravity of the configuration (see Fig. 17.4). Thus for an exterior radius r, as shown in Fig. 17.5a, the radius R to the center of gravity of the external radius will sweep out an area A, which may then be used in the equation above.

The distance to the center of gravity on the radius r lies on the radius bisector $0.6366r$ from the center of the radius r. The radius R is given as follows:

Exterior radius, Fig. 17.5a:

$$R = \frac{d}{2} + 0.6366r$$

r = radius of arc
R = radius to center of gravity of r
d = diameter to center of generation of radius r
A = area of segment
L = length of arc

Interior radius, Fig. 17.5b:

$$R = \frac{d}{2} - 0.6366r$$

The length of the arc is

$$L = \frac{\pi r}{2}$$

The area in both instances is

$$A = 6.283RL$$

When the thickness of the stock is given, the distance L should be obtained from the equation

$$L = 1.5708\left(\frac{t}{3} + r\right)$$

Example 17.4

Figure 17.6a shows a shell to be drawn from a circular blank. Using the methods of area segments, find the diameter of the blank needed to draw the shell. Neglect the thickness of the material.

Solution   The blank may be divided into five areas. These are shown in Fig. 17.6b, the ring; Fig. 17.6c, the interior arc; Fig. 17.6d, the cylinder; Fig. 17.6e, the exterior arc; and Fig. 17.6f, the disk.

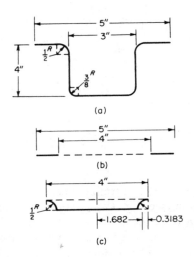

Figure 17.6(a,b,c,d,e,f)

The area of the ring, Fig. 17.6b, is

$$A_b = \frac{\pi}{4}(5^2 - 4^2) = 7.069 \text{ in.}^2$$

For Fig. 17.6c, the interior arc, the radius R is

$$R = \frac{d}{2} - 0.6366r = \frac{4}{2} - 0.6366(0.500) = 1.682 \text{ in.}$$

The area of the segment is

$$A_c = 6.283RL = 6.283(1.682)\frac{2\pi 0.500}{4} = 8.300 \text{ in.}^2$$

The area of the cylinder, Fig. 17.6d, is

$$A_d = \pi dh = \pi(3.000)(3.125) = 29.452 \text{ in.}^2$$

For the interior arc, Fig. 17.6e, the radius R is

$$R = \frac{d}{2} + 0.6366r = \frac{2.250}{2} + 0.6366(0.375) = 1.364 \text{ in.}$$

The area of the segment is

$$A_e = 6.283RL = 6.238(1.364)\frac{2\pi 0.375}{4} = 5.012 \text{ in.}^2$$

The area of the disk, Fig. 17.6f, is

$$A_f = \frac{\pi d^2}{4} = \frac{\pi(2.250)^2}{4} = 3.976 \text{ in.}^2$$

The sum of the areas is

$$A = A_b + A_c + A_d + A_e + A_f$$
$$= 7.069 + 8.300 + 29.452 + 5.012 + 3.976$$
$$= 53.809 \text{ in.}^2$$

The developed blank diameter is

$$D = 1.128 \sqrt{A} = 1.128 \sqrt{53.809}$$
$$= 8.274 \text{ in.}$$

Example 17.5

Assume a shell to be drawn to a diameter of 2.500 in. and a height of 8 in. How many drawing operations would be necessary if there were no annealing operations intervening? Assume reductions of 50, 40, and 30% for each draw without annealing.

Solution   The blank diameter needed is

$$D = \sqrt{d^2 + 4dh} = \sqrt{2.5^2 + 4(2.5)(6)} \qquad d = 2.500 \text{ in.}$$
$$= 8.140 \text{ in.} \qquad\qquad\qquad\qquad\qquad h = 6.000 \text{ in.}$$

A blank diameter, 8.140 in. in diameter, drawn to a shell of 50% reduction, would produce a diameter after the first draw of

$$D_1 = 8.140 - (8.140 \times 0.50)$$
$$= 4.070 \text{ in.}$$

A second draw with a reduction of 40% would produce a shell diameter of

$$D_2 = 4.070 - (4.070 \times 0.40)$$
$$= 2.442 \text{ in.}$$

A third draw may be necessary to bring the shell diameter to 2.500 in.

The height of the shells may be found by considering the vertical height only. That is, if the shell has radii, flanges, or a bottom, the area of these segments should be taken and subtracted from the area of the developed blank. The area remaining is the area of the vertical wall (cylinder). Since the area of a cylinder is the circumference times the height, the height may be found by dividing the net area by the circumference. Thus the area of the blank, base, and vertical cylinder are:

$$\text{area of the blank} = \frac{\pi D^2}{4} = \frac{\pi 8.140^2}{4} = 52.040 \text{ in.}^2$$

$$\text{area of the base} = \frac{\pi 4.070^2}{4} = 13.010 \text{ in.}^2$$

$$\text{area of the vertical cylinder} = 52.040 - 13.010 = 39.030 \text{ in.}^2$$

The height of the first draw is

$$h_1 = \frac{39.030}{\pi D} = \frac{39.030}{\pi 4.070} = 3.052 \text{ in.}$$

The height of the second draw may be found in the same way:

$$\text{area of the base} = \frac{\pi 2.442^2}{4} = 4.684 \text{ in.}^2$$

area of the cylinder $= 52.040 - 4.684 = 47.356$ in.

The height of the second draw is

$$h_2 = \frac{47.356}{\pi 2.442} = 6.173 \text{ in.}$$

To find the height of the third draw:

$$\text{area of the base} = \frac{\pi 2.5^2}{4} = 4.909 \text{ in.}^2$$

$$\text{area of the cylinder} = 52.040 - 4.909 = 47.131 \text{ in.}^2$$

The height of the third draw is

$$h_3 = \frac{47.131}{\pi 2.5} = 6.000 \text{ in.}$$

Therefore, three drawing operations are required to draw an 8.000-in.-high shell.

17.3 DRAWING RADII AND CLEARANCE

Because the material is pulled over the die radius, it is necessary that this radius have an optimum value. The value for the optimum radius on the die ring depends upon the print requirements and the type of material being drawn. The smaller the die ring radius, the greater the needed force to draw the shell. If the radius of the die ring is too large, too much of the material will not be confined as it passes over the radius. The material will thicken, fold, and wrinkle. The radius of the drawing ring may be from four to eight times the thickness of the material. The usual practice is to start with a radius equal to $4t$ and increase its size if necessary. The recommended radius for the die rings is given in Table 17.1.

The radius at the bottom of the shell is shaped by the punch nose radius. If this radius is too small, the bottom radius of the cup may tear out. It may be necessary to make the radius larger than needed, and reduce its size in subsequent drawing operations. As a start, a $4t$ radius-to-stock thickness may be used. If more than one drawing

TABLE 17.1 DIE-RING
DRAWING RADIUS

Stock Thickness (in.)	Drawing Radius (in.)
$\frac{1}{64}$	$\frac{1}{16}$
$\frac{1}{32}$	$\frac{1}{8}$
$\frac{3}{64}$	$\frac{3}{16}$
$\frac{1}{16}$	$\frac{1}{4}$
$\frac{5}{64}$	$\frac{3}{8}$
$\frac{3}{32}$	$\frac{7}{16}$
$\frac{1}{8}$	$\frac{9}{16}$

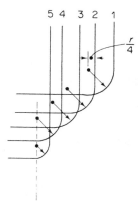

Figure 17.7

operation is needed, each successive radius should be made smaller than the previous one.

If there is to be several redrawing operations, the first redraw should be as shown in Fig. 17.7. The center of the radius of the first draw should lie $r/4$ inside the next drawing operation. The centers of the radii of the last two drawing operations should lie to the same centerline. The last drawing operation (from 4 to 5, Fig. 17.7) generally reduces the radius to its final size.

The clearance between the walls of the punch and the die should be the thickness of the metal plus an allowance of 6 to 20%. As the shearing strength of the metal decreases, the clearance allowance increases. Table 17.2 shows this relationship.

TABLE 17.2 DRAWING-CLEARANCE ALLOWANCE

Thickness, t (in.)	First Draw (%)	Second Draw (%)
Up to 0.015	7–9	8–10
0.016–0.050	8–10	9–12
0.051–0.125	10–12	12–14
Over 0.125	12–14	15–20

Example 17.6

A shell has an outside diameter of 4 in. and is made from soft steel $\frac{3}{16}$ in. thick (see Fig. 17.2c). Calculate the dimensions of the punch and die.

Solution   The punch size is 4 in.

$$d = 4.000 \text{ in.}$$

$$d = 4 \text{ in.}$$
$$t = \tfrac{3}{16} \text{ in.}$$
$$c\% = 12\% \text{ (Table 17.2)}$$

The clearance is

$$c = 0.187 \times 12\% = 0.022 \text{ in.}$$

The die opening size is

$$\text{die size} = d + 2t + 2c = 4.000 + 2(0.1875) + 2(0.022)$$
$$= 4.419 \text{ in.}$$

17.4 DRAWING FORCES

The drawing forces F (in tons) necessary to draw a shell is given by the equation

$$F = \pi dtS\left(\frac{D}{d} - c\right)$$

D = blank diameter, in.
d = outside diameter shell, in.
c = constant (0.6–0.7)
S = yield strength, tons/in.^{2}
t = thickness of material, in.

Example 17.7

A 4-in.-diameter cup is to be drawn from a $\frac{3}{16}$-in.-thick blank 6.940 in. in diameter, see Fig. 17.2. The yield strength of the material is 40,000 psi. Calculate the force needed to draw the shell. Assume a constant of 0.65.

Solution   The force needed to draw the shell is

$$F = \pi dtS\left(\frac{D}{d} - c\right)$$

$$= \pi(4.000)(0.1875)(20)\left(\frac{6.940}{4} - 0.65\right)$$

$$= 51.13 \text{ tons}$$

t = 0.1875 in.
D = 6.940 in.
d = 4.000 in.
c = 0.65
S = 40,000 psi
   = 20 tons/in.^{2}

17.5 TYPES OF DRAWING DIES

Single-action dies.   This type of die uses a simple punch and die ring attached, respectively, to the punch holder and the die shoe. A blank, which has been precut, is inserted into a nest. It is pushed through the die opening and drawn. It is stripped from the punch by the underside of the die. A simple pressure pad is incorporated. One such die is shown in Fig. 17.8a. The punch is vented to aid in stripping the work from the punch. The solid pressure pad may be replaced with spring-type pressure pads.

If the shell needs to be redrawn, thick material can be redrawn without the need of a pressure pad, or ring. Such a redrawing operation is shown in Fig. 17.8b. If the material is thin and the shell is to be redrawn, the shell must be held by a pressure pad, or ring, to prevent wrinkles from forming. This type of redrawing operation is shown in Fig. 17.8c.

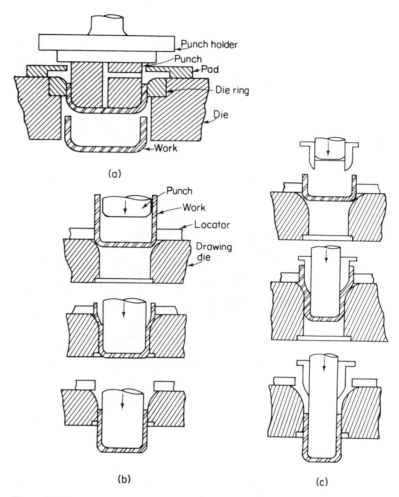

Figure 17.8

Double-action dies.   Figure 17.9a shows one type of double-action die. The die is mounted on the die shoe in the conventional position. The punch is attached to the punch holder, which is itself attached to the inner slide. The pressure pad is attached to the outer slide. The pressure pad puts pressure on the blank so that as the punch pushes the blank through the drawing ring, the work is held with a force great enough to prevent the material from raising and lightly enough so that the material is able to slide from under the pad and move over the die ring radius.

Figure 17.9b shows an inverted die, which is another type of double-action die. However, this die set is used on a single-action press. The cushion acts as a pressure pad. The die ring is mounted on the punch holder and the punch is mounted on the die

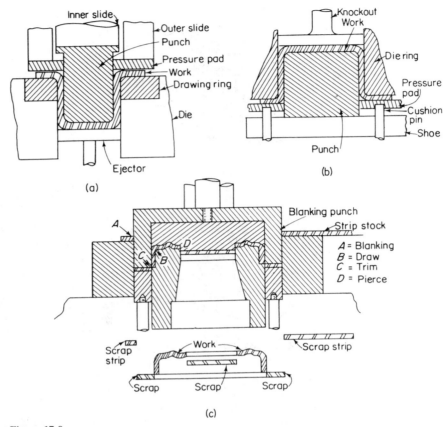

Figure 17.9

shoe. The drawn shell is ejected from the drawing ring when the knockout rod strikes a stationary bar in the press ram.

Compound dies. These drawing dies are designed to blank, draw, pierce, and in some instances trim the edge of a shell for each stroke of the ram. Figure 17.9c shows a die that cuts a blank from a strip of stock at *A;* the drawing operation then takes place at *B.* As the punch descends, the edge of the blank is trimmed at *C.* The last operation is the shearing of the center hole at *D.*

Progressive drawing dies. These dies blank, pierce if necessary, draw, and redraw shells in successive stations. Figure 17.10 shows a scrap strip and the successive shells drawn in a progressive die. The workpiece must be attached to the scrap strip so that it may be carried from one station to the next until it reaches the last station. Therefore, sometimes they are called "cut and carry" dies. One of the difficulties encountered with progressive drawing is that of moving the drawn shell

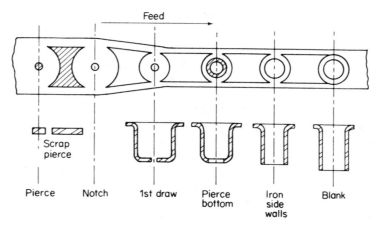

Figure 17.10

from one station to the next. This is especially true if the draw is deep. If the draw is deep, the press must have a long stroke.

Transfer dies.   This process also relies on multiple stations to complete the job. However, instead of relying on the scrap strip to move the shell from one station to the next, it utilizes a transfer mechanism. The shells are, therefore, separated from the scrap strip. Thus precut blanks may be used to draw a shell in the first station of a transfer agreement. Transfer mechanisms are expensive and are used where movement of the work is difficult along a straight line.

Reverse drawing.   The process of inverting a shell and drawing it inside out is called reverse drawing. Figure 17.11 shows a partial reverse draw made to strengthen the bottom of the shell. The operation could have continued until the entire shell had been turned inside out. The process is used to eliminate wrinkles. With this process wrinkles do not form because the forces of compression encountered during the drawing operation are changed to forces of tension. The radii should be not less than four times the material thickness, and the reverse-draw depth should not be greater than $1\frac{1}{2}$ times the diameter.

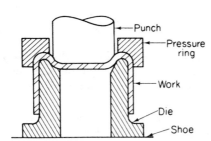

Figure 17.11

17.6 SHELL DEFECTS: THE DRAWING OPERATION

There are many reasons why shells develop defects during the drawing operation. However, tool engineers have had much experience in analyzing these defects and have documented many of the reasons for these defects.

Defects during drawing. If the shell fractures during the drawing operation, the problem may be that the clearance between the punch and the die is incorrect or uneven. This can be a direct result of the punch and die having been designed with incorrect allowance or made incorrectly. It could also result if the thickness of the workpiece is out of tolerance, or nonuniform, or if the punch and die are not properly aligned. If the pressure pad is applying too much force to the blank, the shell will rupture. A small punch nose radius or die ring radius will cause the shell to rupture because of the increased pressure required to draw the cup. Scratches, dirt, or any surface defects of the punch, die ring, or pressure pad will increase the required drawing forces and may cause shell rupture. All polishing lines should be parallel to the direction of flow of the material, and not perpendicular to the flow. As a final check all lubrication should be adequate. Of course, the material will fracture if it does not have the proper ductility.

If the pad pressure exerts too little pressure, Fig. 17.12a, or if the punch or die ring radii are too large, wrinkles will appear at the top flange of the cup. This results because of too little resistance to the drawing operation, or because of unconfined metal. Sometimes wrinkles may be caused by too much metal trying to crowd over the die ring when the clearance between the punch and the die is not enough.

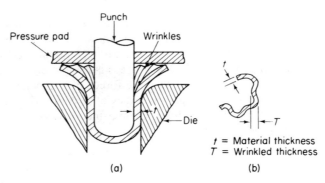

t = Material thickness
T = Wrinkled thickness

(a)   (b)

Figure 17.12

Figure 17.12b shows a broken out section of the buckled material. It shows the material thickness and the punch and die clearance as t and the buckled thickness as T. If T is thin enough to be forced into the space t and ironed, there will be wrinkles in the drawn shell.

If the redrawing operation is too long for the diameter, too much metal will be forced under the flange and try to force its way into the clearance. If wrinkles are formed on one side of the flange only, the trouble may be uneven pressure by the pressure pad. Wrinkles will also appear if the drawing speed is too great.

Die defects which may cause uneven flanges to be drawn include scratches, nicks, burrs, or dirt. If the pressure pad exerts uneven pressure on the blank so that the metal is not drawn evenly into the die, the flange may be drawn uneven. Off-center punches will produce the same effect. Differences of directional properties of the material will cause uneven tops such as cracks or scalloped edges.

Defects during redrawing.   If the material at the flange fractures in the second or third draw, it may be because the punch and die clearance is too little. In some instances the fault may be in the original drawing, or blank. Hard spots, improper grain size, and thinned material are all causes that need not show up in the first draw but may show up in subsequent drawing operations.

Fractures which occur at the bottom of shells that have been redrawn may result (1) from punch and die radii which are too small; (2) where the reduction from the previous shell diameter drawing is too great; (3) from surface defects in the previous drawing operation which create weaknesses in subsequent drawing operations; (4) from nicks, scratches, and any defects that increase the friction; and (5) when there is a thinning of the shell wall near the bottom radius of the shell from a previous operation.

Wall thinning near the bottom of the previously drawn shell results usually when the radius of the punch nose is not blended properly with the sides of the punch. If the reduction in diameter of the shell is too severe or if the die ring radius is too small, the material in the wall may thin out. Not quite as common is the use of material not suited to multiple drawing operations.

If wrinkling occurs on the redrawing operation, the causes may be (1) too large a die ring radius, (2) excessive clearance between the punch and die, or (3) the wrong material thickness. If the sidewalls of a previously drawn shell have been thinned, wrinkles will appear. They will also appear if the redraw operation is too deep or if the previously drawn shell is too shallow. The latter condition may also result in a cracked flange in the redrawn cup.

17.7 PROGRESSIVE DIES

Progressive dies, Fig. 17.13f, may be defined as those dies that have two or more stations. Each station performs an operation on the workpiece so that the workpiece is completed when the last operation has been accomplished. Thus a four-station die produces a finished piece after the fourth stroke and a finished piece with each successive stroke. Operations that may be included are piercing, blanking, forming, drawing, and cutoff. The list is long. The number and types of operations that may be performed in a progressive die depend on the ingenuity of the designer.

Figure 17.13f shows a progressive die with four stations. The die block is machined out of four pieces and fastened to the die shoe. Thus, if one of the die openings wears or is damaged, it can be replaced economically. The stock is fed from

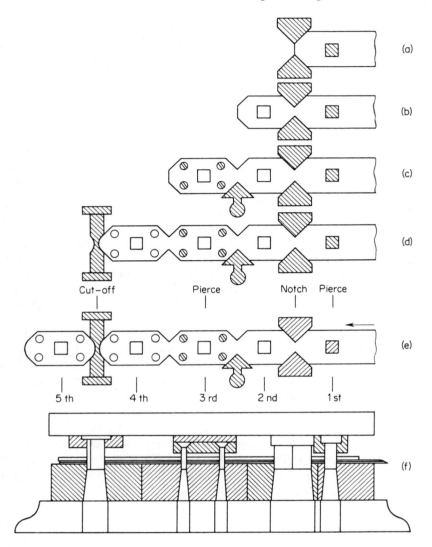

(a)

(b)

(c)

(d)

Cut-off   Pierce   Notch Pierce

(e)

5 th   4 th   3 rd   2 nd   1 st

(f)

Figure 17.13

the right; it registers against a finger stop (not shown). The first stroke, Fig. 17.13a, of the press produces the square hole and two notches. These notches form the left end of the first piece.

The press ram retracts, and the stock moves to a finger stop (not shown) which positions it in preparation for the second press stroke. The second station, Fig. 17.13b, is an idle station. On the second ram stroke the right end of the first piece, the left end of the second piece, and the square hole in the second piece (first station) are pierced.

The ram retracts, and the stock moves to the third station. At this point an automatic stop, Fig. 17.13c, comes into use. This stop picks up the V pierced in the first station. The third stroke of the ram pieces the four holes, Fig. 17.13c. The fourth stroke, Fig. 17.13d, cuts off and forms the radii at the ends of the finished part. Thereafter every stroke will produce a finished part, as shown in Fig. 17.13e.

The methods for developing scrap strip or no-scrap strip dies were discussed in Chap. 15. Several scrap strip developments are shown in Fig. 17.14.

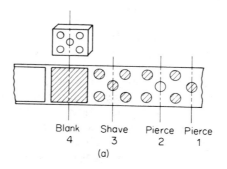

(a)

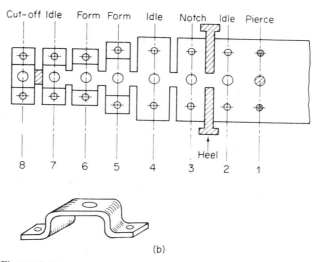

(b)

Figure 17.14

Figure 17.14a shows a single four-station progressive die strip where the center hole is pierced and used as a pilot hole thereafter. The second station pierces the four peripheral holes which are shaved to size in station 3.

Figure 17.14b is an eight-station progressive die strip. Three holes are pierced in the first station, and two notches pierced in the third station, with an idler station in between these two. The fourth station is also an idler station which separates the first forming operation from the notching station. The final forming operation and the parting operation conclude the sequence. Figure 17.10 is a progressive strip that illustrates the inclusion of the drawing operation.

17.8 PROGRESSIVE DIES: DESIGN CONSIDERATIONS

The design of a scrap strip and of the progressive die to match it is a difficult process. With this in mind, several precautions must be observed.

The developed shape of the blank should be studied and divided into simple shapes. These shapes should be chosen so that the punches and the matching die openings are easy to machine. If not, there is a possibility that the shapes may need to be broken down further; or sectional construction may need to be used.

If the operation requires a drawing station, the design of the position of the shell—upright or inverted—may need to be studied carefully. Should it occur at the beginning, middle, or end of the progression? The decision made here may determine when to cut off the part from the scrap strip.

If the workpiece has bent sections, the grain direction of the strip stock should be studied carefully. If lugs are bent up or down, the designer must make sure that they will clear everything in subsequent stations, or if they are to be ejected, they must be ejected with no interference.

In almost all instances pierced holes should be done as soon as possible and used for piloting or stops. If this is not feasible, it may be that an edge notch can be used for positioning or guiding the strip.

Another consideration is the position of the pierced holes, notches, or blanking operations. If they are too close together, the die block may be weakened. Sometimes these punches need to be separated and included in subsequent stations. In other instances weak thin walls in the die block can be averted by including an idler station after the punches have been separated.

No-scrap strip dies should be considered first. If the shape of the workpiece makes the use of a no-scrap die not feasible, a carrier strip should be designed so that the scrap remaining is a minimum. If possible, the arrangement of the punches in a progressive die should be such that the force loading on the die is distributed evenly.

In addition, some specifics need to be considered when designing progressive dies. They are:

1. Pilots should be removable in case they are broken, or if the punch faces need to be ground for sharpening. If spring pilots are used, are they strong enough to cope with missed feeds? The piloting conditions should be examined very carefully. The diameters of the pilots should be designed so that they do not stick. Neither should they fit too loosely. The clearance through-holes should be adequate, smooth, and tapered in case of missed feeds.

2. Small-diameter piercing punches should be guided in the stripper plate. They should be backed up by hardened steel plates, which serve two purposes. The first is to keep the punch head from sinking into the punch holder, and the

second is to provide some means of maintaining the relative height positions of the punch faces to the faces of the other punches.

      If punches are too close to permit hardened bushings in the stripper, it may be useful to machine accurately spaced holes for the punches in a block that is hardened and inserted into the stripper plate. Wire punches that have 60° peened heads should be not less than two times the diameter of the punch. Where slugs stick to the face of small punches and it is not possible to use push-off pins, the clearance between the punch and the die opening may be reduced so that the slug is held more firmly in the die opening. Sometimes pointing the punch slightly will prevent sticking.

3. Punch plates should be thick enough to avoid collapse. They should be made from a very fine grade of tool steel even though they need not be hardened. At times the punch holder will support a large quill into which is inserted a small diameter, short piercing punch. The quill may be guided in the stripper plate. This adds strength and life to the punch quill.

4. Springs behind pilots, push-off pins, shedders, and guide post springs should be designed to avoid failure. If springs are designed to compress less than 30% their free length, they will generally not fail.

5. The stock guides, finger and automatic stops, and pilot registering should all be selected and designed to meet the needs of the job requirements.

17.9 PROGRESSIVE DIES: DESIGN PROCEDURES

A recommended procedural sequence in the design of progressive dies is discussed next. The blank should be developed so that all the dimensions are included. Templates may then be developed. Using as many templates as necessary, the sequence and number of operations are established. Idler stations are included wherever they are needed. A check is made of the required accuracies, interferences, and distance tolerance. Thus the edge of a hole should be at least 1.5 times the stock thickness from the center of the radius before the bend in Fig. 17.15a is made. The distance from the center of a pierced hole to the edge of the blank(s) should not be held closer than 0.005 in. Thin, weak sections in the die openings and punches should be avoided even if it means dividing the operations into two or more stations.

      When laying out the scrap strip, bends should be between 45 and 90° to the direction of the grain, as shown in Fig. 17.15b. The economics of the cost of the material is a big factor in determining the orientation of the workpieces and must be considered. The location and use of pilot holes should be determined at this point.

      Once the scrap strip has been finalized and its drawing made, the bolster plate location is drawn so that disposal of scrap slugs or finished parts may be ejected without interference. Pierced holes that are to be used as pilots should be punched in the first station and piloted in the second station. Notching punches are usually included in the early stations and backed up with heels if needed (see Fig. 17.14b).

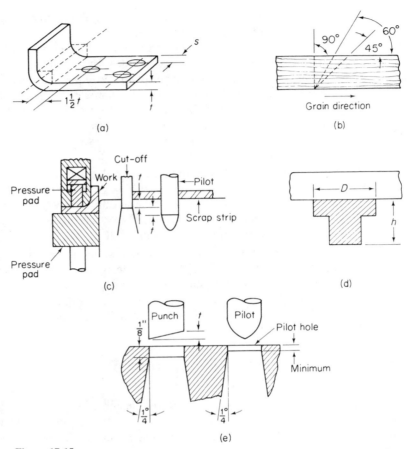

Figure 17.15

Drawing operations are usually done in the first or second station; whereas bending or forming operations are usually done in the last operation in cooperation with the cutoff operation. Thus the full diameter of the pilot must have already registered the work before the cutoff tool operates. The full diameter must extend at least one material thickness t below the face of the cutoff punch, as shown in Fig. 17.15c. As a matter of fact, the pilot must have completely registered the scrap strip before any of the punches begin to operate; otherwise, the strip will bind and prevent the pilot from doing its job. Also shown is the face of the cutoff punch at least one thickness below the bending punch. Sometimes a push-off pin is provided in the bending punch. A pressure pad is also shown at the bending station. Sometimes ears are bent down instead of up as shown. The question should be asked whether or not a scrap cutter is needed.

The method of controlling the location of the scrap strip should be considered carefully. The guide length before the first station is reached should extend at least two strip widths in front of the first station. In conjunction with the latter the type of stripper to be used should be considered. The stripper thickness, based on the stripping force, should be calculated and the channel clearance should be adequate to allow the strip to move freely. Next locate the fingers and automatic stops.

Design the die block, making it long enough and wide enough so that the location of the holes in it will be at least $1\frac{1}{2}$ times the thickness of the block, away from the edges. Select the screws, dowels, and the number of sections needed for grinding, replacement, and so on. Set dowels in nonsymmetrical locations so that sections or parts may be mounted in one position only.

The counterbores in the die block, the tapped holes in the die shoe, and the reamed dowel holes in the die shoe must all be made $\frac{1}{4}$ to $\frac{3}{8}$ in. deeper than needed to allow for grinding of the die block. Choose the die set so that when the die block is mounted it can be ground without having to remove it from the die shoe.

As indicated earlier, small fragile punches should be guided in the stripper plate. If a small punch is set next to a large punch, it should be shortened to prevent defection caused by the action of the large punch. A series of small punches should be staggered. If the punch diameter is twice the material thickness it need not be guided. If the punch produces a hole 40% deeper than its own diameter, the shank diameter must be at least twice the punch size. Sixty-degree heads are used on punches where their diameter is at least twice the material thickness. If a punch diameter is too small for push-off pins, the clearance between the punch and die should be reduced, and the sidewalls of the die should be straight for at least two material thicknesses. Sometimes the end of the punch can be rounded slightly to prevent sticking of the slug on the face of the punch. Hardened backing plates should be used when small-diameter punches are used or when the run is to be long. If the punch heads are small, they should also be used.

Punches should be a tap fit in punch plates. For purposes of stability the flange (D) of a punch should never be smaller in diameter than the length (h) of the punch as shown in Fig. 17.15d. Punches should be secured with pins, flats, keys, and so on, against rotating in the punch holders. The need for air vents in larger punches should be considered.

If a hole is to be pierced, shear should be ground on the punch face. If the operation is blanking, the shear should be applied to the die. The difference between the high and low points of the shear should not exceed one material thickness, Fig. 17.15e.

The taper scrap hole should be reamed to $\frac{1}{2}°$ included angle and an approximately $\frac{1}{8}$-in. straight section left for grinding, as shown in Fig. 17.15e. Pilot holes should be reamed with a $\frac{1}{2}°$ included angle to the face of the die.

PROBLEMS

17.1. Using Fig. 17.3, calculate the diameter of the developed blank in Fig. 17.16a.

17.2. Calculate the blank diameter in Fig. 17.16b. Use the method of areas.

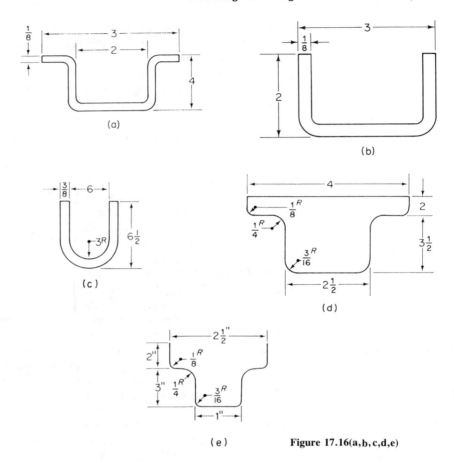

Figure 17.16(a,b,c,d,e)

17.3. Using the method of areas, calculate the blank diameter in Fig. 17.16a.

17.4. Using the method of areas, calculate the blank diameter for drawing the shell in Fig. 17.16c.

17.5. Repeat Prob. 17.4 using Fig. 17.3.

17.6. Assume the same dimensions as in Fig. 17.16b except that the shell has a $\frac{1}{2}$-in. internal radius.
(a) Calculate the developed blank for this shell.
(b) Using an equation from Fig. 17.3 check your results.

17.7. Calculate the blank diameter for Fig. 17.16d using the method of areas.

17.8. A shell wall is reduced from 0.120 in. to 0.108 in. What is the percent of reduction of the shell thickness?

17.9. A shell is to have its wall thickness reduced from 0.064 in. by 12.5%. The new shell will have a new height of 3.430 in. and a new diameter of 2.000 in.
(a) What is the mean height of the shell?
(b) What is the shell diameter needed to draw the ironed shell?

17.10. A shell has a wall thickness after ironing of 0.060 in. This represents a 15% reduction of wall thickness. The height of the shell after ironing is 4.750 in. and the diameter is 1.500 in.
(a) What was the original wall thickness?
(b) What should the blank diameter be to draw the ironed shell?
(c) What should the blank diameter be to draw the ironed shell?
(d) What should the diameter of the blank be if there had been no ironing?
(e) What would the height of the blank be without ironing?

17.11. Using the method of segments, find the diameter of the developed blank for the shell shown in Fig. 17.16e.

17.12. Given a shell 2.000 in. diameter and $6\frac{1}{2}$ in. deep, assume possible successive reductions of 50, $33\frac{1}{3}$, 20, and 15% when the shell is redrawn without annealing. How many drawing operations are needed?

17.13. Find the height of all the shells drawn in Prob. 17.12. Check the last draw.

17.14. Assume a shell to be drawn to a diameter of 3.000 in. and a height of 8.000 in. How many drawing operations would be necessary if there were no annealing operations intervening? Assume reductions of 50%, 40%, 30%, and so on, for each draw without annealing.

17.15. A shell has an outside diameter of 3 in. and is made from soft steel $\frac{1}{8}$ in. thick (see Fig. 17.16b). Calculate the dimensions of the punch and die.

17.16. Discuss the design of the die ring radius and its effect on the drawing operation.

17.17. Repeat Prob. 17.16 for the punch nose radius.

17.18. What should the clearance between the punch and the die be for a drawing operation?

17.19. A shell has an outside diameter of 6 in. and is drawn from $\frac{3}{32}$-in.-thick material. Calculate the dimensions of the punch and the die.

17.20. A shell has an outside diameter of 3 in. and is made from soft steel $\frac{1}{8}$ in. thick (see Fig. 17.16b). Calculate the dimensions of the punch and die.

17.21. Calculate the force necessary to draw the shell in Fig. 17.6a. The material is $\frac{3}{32}$ in. thick and has a yield strength of 18,000 psi. Assume a constant of 0.7.

17.22. A 3-in. diameter cup is to be drawn from a $\frac{1}{8}$-in.-thick blank 5.528 in. in diameter (see Fig. 17.16b). The yield strength of the material is 30,000 psi. Calculate the force needed to draw the shell. Assume a constant of 0.65.

17.23. Make a drawing of a single-action die to punch the shell in Fig. 17.16b. Incorporate all the necessary components in the views you select.

17.24. Repeat Prob. 17.23 for Fig. 17.6a.

17.25. Explain the differences in the operations between Fig. 17.8b and c.

17.26. Describe the action of a pressure pad in a double-action die.

17.27. Describe the operation of an inverted double-action drawing die.

17.28. Describe the action of the compound drawing die, Fig. 17.9c.

17.29. (a) How does a transfer die differ from a progressive die?
(b) Explain and illustrate when it would be used.

17.30. When and why is it desirable to reverse draw a shell?

17.31. There are many causes for shell rupture during drawing. List them and indicate how the cause may be corrected.

17.32. Discuss the formation of wrinkles during the drawing operation.

17.33. What causes uneven flanges to be formed during the drawing operation?

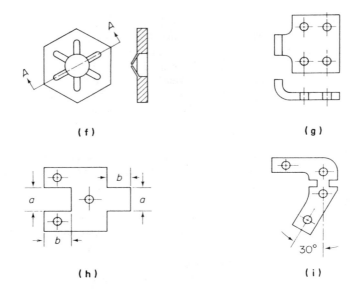

Figure 17.16(f,g,h,i)

17.34. List the causes for shell rupture during the redrawing operations.

17.35. List the causes for wall thinning near the bottom of the shell radius.

17.36. What are the causes for wrinkles which appear during the redrawing operations?

17.37. Draw the scrap strip for the progressive die to make the part in Fig. 17.16f. Crosshatch the punched and mark the stations. Study Secs. 17.7 and 17.8 before attempting these problems.

17.38. Repeat Prob. 17.37 for Fig. 17.16g.

17.39. Repeat Prob. 17.37 for Fig. 17.16h.

17.40. Repeat Prob. 17.37 for Fig. 17.16i.

17.41. When studying a part in preparation for develping a scrap strip for a progressive die, it is important to divide the part into simple shapes.
   (a) Why?
   (b) What precautions must be observed?
   (c) What can be done if it is difficult to machine a die opening or punch?

17.42. **(a)** When should a shell be drawn in a progressive die?
   (b) What precautions need to be observed?

17.43. What precautions must be observed if a bent lug is needed on a work piece in a progressive die?

17.44. There is a best station position in a progressive die for piercing holes, bending lugs, drawing shells. If possible, when should these be done in a progressive sequence? Why?

17.45. If punches are too close together, what can the designer do when developing a progressive die?

17.46. Why is a no-scrap strip progressive die preferable to a cut-and-carry scrap strip? Explain.

17.47. What are some of the precautions to observe when designing pilots?

17.48. Discuss
  (a) the design of small piercing punches,
  (b) when they should be staggered,
  (c) the type of heads used,
  (d) the use of backup plates,
  (e) the sticking of the slugs to the punch face, and
  (f) use of the stripper plate as a guide.

17.49. **(a)** Design at least two quills for holding very small piercing punches.
  (b) If they need to be guided, show how you would do this.

17.50. **(a)** What is the procedural sequence recommended for the designer when designing a progressive die? List them, writing at least a sentence about each step.
  (b) What precaution must be observed for each of the steps listed in part (a)?

Chapter 18

Electric Discharge Machining and Numerical Control

18.1 ELECTRIC DISCHARGE MACHINING

Electric discharge machining (EDM) removes metal by the process of controlled spark erosion. The workpiece must be made from a conductive material. The work and tool are submerged in a dielectric fluid, which serves the dual purpose of controlling the arc and carrying away the waste products of erosion.

A power supply controls the spark frequency, energy, and voltage. The workpiece is the anode and the tool is the cathode. It should be noted that in some instances reverse polarity is advisable. The conventional polarity is shown in Fig. 18.1a. Low frequency coupled with high energy (amperage) is used for rough cutting and high metal removal. High frequencies and low energies are used for slow metal removal, high finish, and dimensional accuracies. It should also be noted that a nonrotating tool is used. The gap between the tool and the work is maintained with a servo system.

The type of tool (electrode) material used is also important for optimum performance. The tool is subject to wear in the same manner as a lathe tool or any other cutting tool. Since the tool and workpiece both erode, the rate of erosion depends on the electrode material, the workpiece material, and the polarity of both. Another factor in electrode wear is the energy concentration at a sharp corner. This energy concentration wears the sharp corners off the electrode, as shown in Fig. 18.1b.

The tool material should have a high melting point and be a good conductor of heat and electricity. As with other tool materials a compromise between wear rate and cost of fabrication of the tool must be made. As noted in Fig. 18.1b, the manner in which the wear takes place is an important factor in the selection of the tool material to be used.

One standard used to determine wear rate is the volumetric wear ratio. Howev-

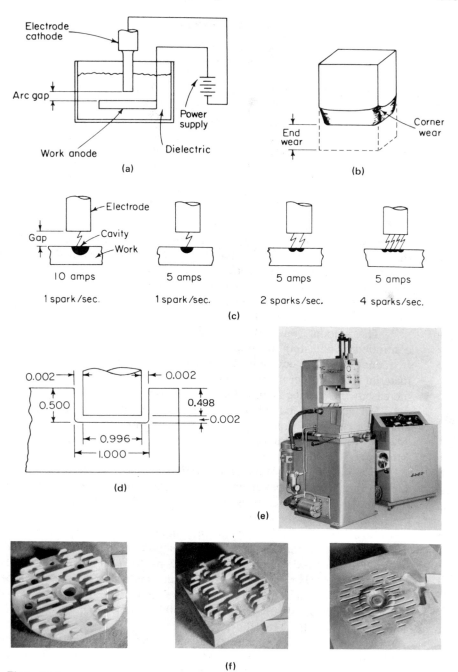

Figure 18.1

TABLE 18.1   WEAR RATIOS

Material	Electrode-to-Workpiece	
Brass	1.5	1 steel
Copper	1.0	1 steel
Graphite	0.4	1 steel
Copper tungsten	0.4	1 steel
Silver tungsten	0.3	1 steel
Brass	4.0	1 carbide
Carbide	1.5	1 carbide
Copper tungsten	0.67	1 carbide
Brass	0.2	1 brass
Brass	7.0	1 tungsten

er, the useful life of a tool is usually determined by its life relative to the machining of sharp corners or contours or to the length of the tool consumed. A useful ratio is one based on the ability of the electrode material to produce a 90° corner in steel. The 90° corner wear ratios of some materials are shown in Table 18.1. Thus 0.400 in. of a copper tungsten electrode will be consumed in the machining of a 1.000-in. length of steel block to a sharp corner.

The amount (rate) of metal removed from the surface of the workpiece is controlled by adjusting the amperage. Assume the workpiece material, tool, gap, and frequency are held constant. If the energy (amperage) is doubled, the volume of material removed per spark is doubled. However, if the metal removed per spark is increased, the surface roughness of the workpiece is also increased.

Surface finish may also be controlled by varying the frequency and holding the amperage constant. Figure 18.1c shows the relationship between amperage and frequency. In this illustration the length of the gap between the electrode and the workpiece is held constant.

If a fixed volume of material is removed when 5 amperes of current is used, the surface finish will be better when a frequency of 4 sparks/sec is used than when a frequency of 2 sparks/sec is used. Thus the higher the frequency used with amperage constant, the better the surface finish is.

Size control is a function of the amperage, frequency, and capacitance used. These quantities create an electronic envelope about the electrode. That is, the space between the electrode and the workpiece constitutes an overcut. If a hole 1,000 in. in diameter by 0.500 in. deep is to be machined and an electronic envelope of 0.002 in. is present, the diameter of the electrode used is 1.000 − 0.004 = 0.996 in. The travel of the electrode needed to produce a 0.500-in. hole is 0.500 − 0.002 = 0.498 in. This is shown in Fig. 18.1d.

The electrolyte (coolant) acts as an insulator, or barrier, to current flow. Once the breakdown voltage of the coolant is reached, a spark is generated by the flow of amperage across this gap. This electrolyte also acts as a coolant for the small vapor particles separated from the workpiece during the machining operation. The coolant prevents the heat generated in the vapor particles from being transferred to the work and electrode. Such transfer of heat could destroy both the workpiece and the electrode. Forcing the coolant through the gap between the workpiece and the electrode removes the particles, which are then filtered out before the coolant is reused.

The development of EDM has been rapid and provides another method for machining metals in the toolroom. It can be used to machine cavities or intricate shapes in very hard materials. Thus it becomes especially useful for machining blind cavities in dies used in die casting, stamping, forging, and injection molding. It makes possible the machining of intricate cavities and holes in hardened steel or special alloy materials without the need of sectioning the cavity.

Figure 18.1e shows an electric discharge machine. Figure 18.1f shows a die-cast cylinder head, mold, and electrode used in the EDM process.

18.2 NUMERICAL CONTROL ECONOMICS

Numerical control (NC) is the operation of machine tools by coded instructions. This means that machine operating commands in the form of symbolic codes are interpreted in such a manner that they control specific operations on a machine tool. These instructions are intended to control every phase of a project from beginning to end. It may position the spindle, direct the movement of the machine, change the rpm of the cutting tool, turn on a coolant, and so on. It will not do anything that is not coded; nor will it deviate from the instructions which it receives. As a matter of fact, it will not do anything that is outside the parameters of the machine tool, even if it is coded to this end. For example, if a program tells the machine table to move 12 in. when its travel is only 8 in., it will not carry out the command.

The process is simply one of taking a dimensioned drawing, Fig. 18.2, and

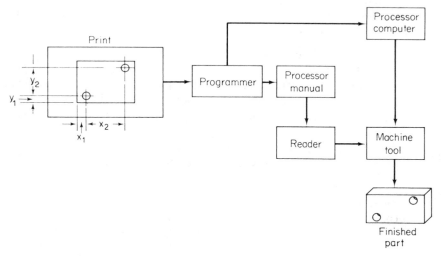

Figure 18.2

having the programmer convert the print into a coded series of instructions. This program controls all movements of the machine from the time the start button is pushed until the cutter returns to the start position. The operator's job reduces itself to loading and unloading the machine, and monitoring it for tool failure or machine breakdown. His judgment in terms of tool motion and other factors is no longer a factor in the success of the operations.

The next step in the sequence is the *machine control unit* (MCU), which is a program reader. Most units are electronic readers. They read the program and generate signals that control the drive mechanism of the machine tool. The program that the MCU must read is stored on tapes, tabulated cards, or magnetic tapes, or the programmer may feed the code directly into a computer storage tape or disk. When the command is called for, it is fed into the drive units of the machine tool.

Reading of the tape is done by a sensing device. This device may be a beam of light that passes through the punched holes and energizes a photoelectric cell. It may also use the vacuum principle to actuate the program. In some instances, brushes sweep the tape and close contacts when they make connection with a terminal at the other side of the hole. Magnetic tape is very common. The tape is coated with ferrous oxide particles which are magnetized so that impulses are created when they pass over a reading head. This principle is similar to the home tape recorder.

The purpose of any automated system is to increase production at a lower cost without decreasing the precision. Usually, as the sophistication of the operation increases, the cost of production increases. A greater number of operations means a greater number of tools, machines, gages, fixtures, and operators. As the precision increases the cost of the fixtures, tools, and machine increases. With numerical control the precision is built into the machine, the control is left to the numerical controls rather than to the operator, and the need for fixtures reduces itself to support blocks, clamps, and a means for locating the work relative to the zero starting point of the setup.

The savings, as a result of numerically over conventionally controlled machine tools, is realized because of several factors:

1. Single-piece machining is feasible because of the nature of the control system and the fact that expensive fixtures are not necessary. The program is written, recorded on a punch tape, and checked by playing it back to an automatic typewriter which prints out the program. The printout is checked against a master tape, blueprint, or a program. Errors may be corrected by typing a new tape and correcting the portion that needs changing; or the portion that needs to be corrected can be typed on a length of tape and spliced into the original tape. This also makes it possible to punch out design changes on a tape and splice them into the old tape.

2. Fixture cost is considerably lower for numerically controlled machines over conventionally controlled machines. The accuracy built into the machine eliminates the need for expensive fixtures. It is true that the accuracy of the operation will be no more accurate than the accuracy of the program. However, once programmed, the positioning accuracy is well within ± 0.001 in. from point to point; and the machine will repeat time and again to well within ± 0.0005 in. Furthermore, the point-to-point errors are not cumulative.

With two datum planes (x and y) and the table acting as the vertical reference plane (z), the need for accurate location of the workpiece is reduced to two banking surfaces and conventional clamps. The reference coordinate point is programmed into the tape. This reference zero point may be on or off the work.

3. Because of the accuracy inherent in numerical control machining, the cost of fixture design, setup and tear-down time, and lead time is drastically reduced. As a corollary changeover from one setup to another is simplified because, once the job has been planned and programmed, all that is required is changing the tape, a new set of tools, and the establishing of a new datum, or reference point.

4. In conventional machining when few workpieces are required, it becomes necessary to lay out at least the first part to make sure hole locations are proper, and so on. This occurs when the production quantity does not warrant the expenditure of money for a fixture or jig. Because the accuracy is programmed into the tape, shop layout has been almost completely eliminated. Programming and tape writing take much less time than building jigs and fixtures and locating gages. Also the tape lasts longer and requires much less storage space than do jigs and fixtures. It also takes much less time to duplicate tape than to build a new fixture.

In addition, modern machines have a built-in check system. If workpieces are to be duplicated, the control will go back to zero upon completion of the cycle. If the control does not return to zero, a warning system alerts the operator. He can then make the appropriate adjustment and proceed. Closed systems, discussed later, keep a continuous check on the commands.

5. Whereas other automated equipment does not lend itself to easy changover, numerical control machines may be programmed for single-piece machining, short runs, or production runs. Once programmed and set up, the operation of the machine does not require a skilled operator. Except for positioning the workpiece on the table, the human element is completely eliminated.

6. Numerically controlled tools force operation and operation sequencing to be planned prior to releasing the job to the shop. Conventional machining places most of the judgment decisions for deciding setups, operational sequencing, tool selection, and feeds and speeds determinations into the hands of the operator. Planning a machining operation for NC requires the planner to study the machining requirements and to produce a document of the required feeds and speeds, operation sequences, inspection tools, cutting tools, holding devices, and a program consistent with the machine tools and employee skills available.

7. Inherent in the advantages of NC is the need for fewer machine tools. In conventional machining, it often becomes necessary to use more setups and machines because of the limited number of operations that can be done with a conventional setup. With NC many different types of operations can be done with one setup.

Machining centers are a good example of the large number of operations that can be done with one setup and one machine tool.

8. Because setup time and lead time are much shorter with NC than with conventional tools, forecasting is much easier and more accurate. In addition, NC machine tools are accurate to 0.0001 in., which places the precision in the machine tool rather than the operator. These items affect how the work force is used, the number of operations used, which machine tool will be used, machine availability for other jobs, and reliability for service to the customer.

It is to be noted that the initial cost of numerical control machines is high. For one thing, they must be more rigidly built than conventional machines to withstand the accelerations dictated by servo controls. Other costs are the electronic control sophistication. To reduce downtime, a skilled electronics technician must be employed to service the electronic controls, even though the reliability of these machines is high.

18.3 NUMERICAL CONTROL SYSTEMS AND COMMANDS

The two basic systems of control are the open loop and the closed loop. The commands are picked off the tape (or card) by a tape reader which sends the signals to servo motors. These signals control the movement of the table or spindle along the x-, y-, or z-axis in a three-axis machine. Two-axis machines have only the x and y axes controlled. Four- and five-axis machines have one or two rotary degrees of freedom controlled as well as the x-, y-, and z-axis control.

In the open type of control system, Fig. 18.3a, the command is fed into the machine but no feedback takes place. Thus there is no check system built into the machine to determine whether or not the command has been carried out. Thus in Fig. 18.3a the operator control console is used to set the initial conditions of the zero position, start the machine, and so on. The tape reader picks off the discrete bits of information from the tape and sends these bits to the distribution control unit which directs the information to the servo control unit for a particular axis control. The command is then transferred to the servomotor, which carries out the command. In Fig. 18.3a the x- and y-axes are controlled. Also included is a z-axis control and a spindle-speed control. This is a three-axis system.

Figure 18.3b shows a closed-loop system which has a transducer attached to each axis. The transducer converts the machine, as directed, from linear (or rotary) motion to an electrical impulse. This impulse is fed back into the control unit as a check on the input signal. If the input signal and the feedback signal match, the servo stops and the desired position is obtained. If the input signal and the feedback signal do not match, the error signal, the difference between input and feedback signal, takes over as the input signal. It remains the control signal until a match is achieved. Thus the correction is made, and the operation continues.

The transducer may be either digital or analog. The digital signal is a pulse signal proportional to the unit motion of the tool or work table. This may be accom-

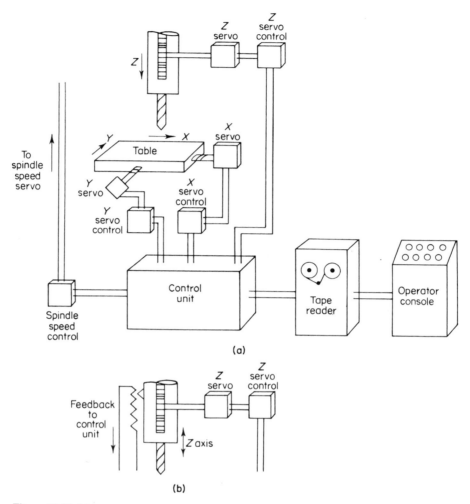

(a)

(b)

Figure 18.3(a,b)

plished by counting the number of pulses needed for a given movement or by matching the pulses to the input signal.

The analog signal is continuous and proportional to the continuous movement of the table, rotation of the spindle, or movement of the spindle. As is the case for a digital transducer, the continuous input voltage is matched to the transducer output voltage to achieve accurate positioning.

The control interfacing, Fig. 18.2c, is basic to computer numerical control. Essentially, the control, Fig. 18.3d, consists of five functional units. They control the flow of information from the operator–machine interface input to the operator–machine interface output.

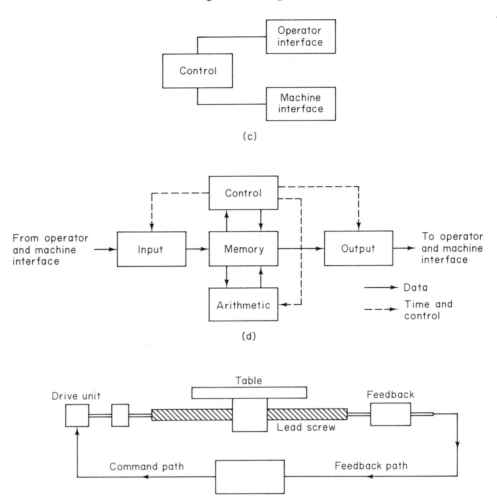

Figure 18.3(c,d,e)

Input devices such as software, programs, and so on, are transmitted to the memory. The memory may be RAM (random access) or ROM (read only). RAM may be accessed immediately, or changed if need be. Part programs are stored in RAM. ROM information cannot be accessed or altered. It contains the systems operating information.

Figure 18.3e shows a simplified diagram of the table and mechanical components of a computer numerical control system. After the signal leaves the control, it is fed into a servomechanism motor that drives the table, or spindle, of the machine. This signal is monitored by a feedback device to the control unit.

These types of commands are used to write programs. The simplest type of command is the *point-by-point system*. This command system requires no extensive

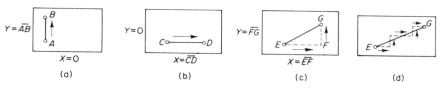

Figure 18.4

knowledge of mathematics. If we assume a two-axis machine (x and y), the movement of the table may be controlled so that the table will move along the y-axis from point A to B, as shown in Fig. 18.4a. In this instance the movement along the x-axis is 0. If the movement is along the x axis, as shown in Fig. 18.4b, then the command along the y-axis is 0. In Fig. 18.4c, the movement is along the x-axis from E to F and then along the y-axis from F to G. Many machines now in existence can also mill a 45° straight line.

It should be noted that all points along a straight line need not be programmed. The coordinates of the endpoints and the feed rate are programmed. In an x, y, z machine, the z-axis would also be programmed. The machine will move from the start point to the finish point. If the feed rate is slow and the cutter is in contact with the work, it will machine the workpiece. If the cutter is raised (table lowered) and a rapid feed rate is programmed, the operation is one of rapid traverse, rather than machining.

If the line EG is to be machined, the x-y movements are reduced so that the movement of the tool approximates the straight line movement EG. This is shown in Fig. 18.4d. It should be noted that the greater the number of x-y movements, the more nearly does the tool approach the straight line EG. This is one of the disadvantages of the point-to-point system. However, the greater the number of x-y movements, the greater is the accuracy achieved. In contour machining, the required number of points may become excessive. In spite of this, many straight-line operations may be done with the point-to-point system.

In *continuous-path systems* the tool is caused to approximate the curved surface which it is generating through the calculation of a great many x-y movements, called linear interpolation; circular movements, called circular interpolation; or parabolic movements, called parabolic interpolation.

In *linear interpolation* the tool follows the desired shape because the commands are far more descriptive than for a point-to-point system. Linear interpolation is capable of causing the cutter to trace out a curve by developing a large number of x and y straight-line movements where x is some function of y, designated $x = f(y)$. Thus, in Fig. 18.5a the cutter would move from A to B while cutting the flat surface. To get to D the cutter would be programmed to cut along the vertical y-axis to C and then along the positive x-axis to D. From D to F, the cutter would need to go *up* to E and then to the right to F. This could be continued until the operation reached G. It is readily apparent that the resultant cut is not a very good trace of the desired curve. The greater the number of x-y movements, the closer the desired curved surface is approximated. A computer can calculate a great number of these movements. The

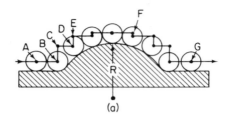

(a)

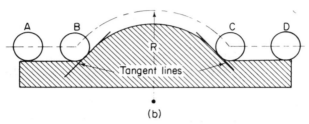

(b)

Figure 18.5

program would be very long. It should be noted that these contouring interpolaters hold both endpoints in their logic until the endpoint of each x-y movement is complete.

In *circular interpolation* the coordinate endpoints of the arc, the coordinates of the center of the circle, the length of the radius of the circle, and the direction of movement of cutter are all that is needed. However, the trace of the curve is still a series of a great number of linear x-y movements of about 0.0002-in. increments. The difference between linear interpolation and circular interpolation is that the former requires that a great number of x-y points be programmed into a long tape, whereas the latter increments are computed inside the computer, after which an output signal moves the cutter.

The program will give the endpoints A and B in Fig. 18.5b. At B the control system will get the signal of the tangent points B and C, the coordinates of the radius R, and the direction of travel of the cutter. The circular interpolator will calculate many x-y coordinates to trace the curve. At C it will receive the endpoints C and D. Compensation must be made for cutter offset as the curve is traced.

Parabolic interpolation is a much more complex process. It progresses between two end points, through a third point which is not on a straight line between these two endpoints. The center point lies on a curve which approximates a part of a parabola. This system is used in free-form design, such as mold and other types of cavity machining.

18.4 CODES AND TAPE PREPARATION

There are three types of coding systems in use today. The *binary-coded decimal system* (BCD), shown in Fig. 18.6, is the one most commonly used today. It may be

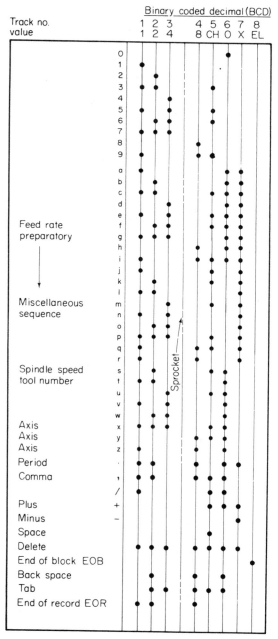

Figure 18.6

used with point-to-point command, or with continuous-path systems. A 1-in. eight-channel tape is used. Besides the eight channels, there is a channel of holes used for the drive sprocket. This ensures positive drive for the tape. This code was developed by the Electrical Industries Association (EIA).

In this system, channels 1, 2, 3, and 4 are used to represent the mathematical powers of 2. Thus channel 1 is the number 1; channel 2, the number 2; channel 3, the number 4; and channel 4, the number 8 from

$$2^0 = 1, \qquad 2^1 = 2, \qquad 2^2 = 4, \qquad 2^3 = 8$$

Table 18.2 shows how the numbers 0 through 9 have been punched into the tape, in Fig. 18.6.

TABLE 18.2   PUNCHED-HOLE CODE—EIGHT-CHANNEL TAPE

Number	Summation	Column
0	0	6
1	1	1
2	2	2
3	1 + 2	1, 2, parity
4	4	3
5	1 + 4	1, 3, parity
6	2 + 4	2, 3, parity
7	1 + 2 + 4	1, 2, 3
8	8	4
9	1 + 8	1, 4, parity

The design of the tape reader dictates that only an odd number of holes punched in a row can be read. Thus if the data require that an even number of holes be punched into a row, the odd-parity hole must be included into row 5. Channel 6 is reserved for zero commands. Channels 7 and 8 are miscellaneous channels used for word commands.

A second system is the *straight binary system*. Multiple simultaneous motions are programmed into one row using the binary number system (base 2). The length of the tape is much shorter than for the BCD system but all dimensions must be binary coded before entry. This system is more widely used when a computer is used in conjunction with numerical control. This system is not used nearly as much as the BCD system.

A third code, now in wide use, is one developed by an industrial committee and the American National Standards Institute, and is called the *American Standard Code for Information Interchange (ASCII)*. It is a system developed with the idea of standardizing the coding systems into one system. It uses the 10-digit system plus two additional tracks for identification symbols. It codes both capital and lowercase letters, whereas the BCD system codes only lowercase letters.

The *parity check* in the BCD system allows for only an odd number of holes to be checked, whereas the ASCII system uses an even-parity check.

The *tab* code is used to arrange information into a column.

The *end of block* (EOB) means end of line, or end of block, which tells the machine that it has received all the information needed to perform the operation and that it is to perform them immediately.

End of record (EOR) provides that after a program has been run the tape will rewind and position itself so that it is ready to repeat the program from the beginning.

Figure 18.6 shows a series of letters that have been labeled word commands. Thus x, y, and z deal with the x, y, z axes of the coordinate system. *Plus* and *minus* signs indicate the standard conventions for rectangular coordinates. The letter n means sequence number, or tab sequence. It is usually coded $n001$, $n002$, and so on. The letter g means that the machine is to do something specific, such as to start, go through a machining cycle, and stop. Thus a $g81$ may direct the machine to do more to the programmed x-y position and drill a hole to the programmed z depth. It then retracts and waits for the next command. The m code is a miscellaneous code that may direct the machine to stop, or instruct the machine to do something specific.

The *feed rate* is given by the *f number* and is usually given as inches per minute. The *s number* is the spindle speed in revolutions per minute. There are many methods for coding feeds and speeds. Below is one method.

One method recommended by the Electronic Industries Association* is that (1) "the feed or speed be expressed as a three-digit coded number plus a letter code, if a word address is used. (2) The first digit of the coded number is to be a decimal multiplier, three greater than the number of digits to the left of the decimal point."

The procedure is to round off the feed or speed rates to two digits, add 3 to the first digit, and let it precede the rounded digits. Some examples will illustrate this process.

Example 18.1

Code the following spindle speeds: (a) 1236, (b) 148.4, and (c) 35.4 rpm.

Solution

(a) Code 1236 rpm.

1. The total number of digits to the *left* of the decimal point is 4. Thus

$$3 + 4 = 7$$

2. The number 1236 is rounded to two digits:

$$12$$

3. The speed number is coded as

$$s712$$

(b) Code 148.4 rpm.

1. There are three digits to the *left* of the decimal point. Therefore,

$$3 + 3 = 6$$

*Standards RS-273 and RS274-A.

2. The number 148.4 is rounded to two digits:

$$15$$

3. The speed number is coded as

$$s615$$

(c) Codes 35.4 rpm.

1. There are two digits to the *left* of the decimal point. Therefore,

$$3 + 2 = 5$$

2. The number 35.4 is rounded to two digits:

$$35$$

3. The speed number is coded as

$$s535$$

Example 18.2

Code the following feeds: (a) 3.62, (b) 0.0586, and (c) 0.000853 ipm.

Solution

(a) Code 3.62 ipm.

1. There is one digit to the *left* of the decimal point. Therefore,

$$3 + 1 = 4$$

2. The number 3.62 rounds to two digits:

$$36$$

3. Thus the code is

$$f436$$

(b) Code 0.0586 ipm.

1. There is one zero to the *right* of the decimal. Thus

$$3 - 1 = 2$$

2. The number 0.0586 is rounded to two digits:

$$59$$

3. The coded number is

$$f259$$

(c) Code 0.000853 ipm.

1. There are three zeros to the *right* of the decimal. Therefore,

$$3 - 3 = 0$$

2. The number 0.000853 rounds to two digits:

$$85$$

3. The coded number is

$$f085$$

All commands are picked off the tape (or cards) by high-speed tape readers. Also important is the zero position in numerical control. This may be either *fixed* or *floating*. If the zero is fixed on the machine when it is manufactured, all movements along the *x*- and *y*-axes must be referenced to that point. In Fig. 18.7 if the fixed zero is at *A* on the table, the *x* dimension will be 6 in. and the *y* dimension will be 5 in.

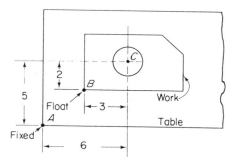

Figure 18.7

If the machine is manufactured with a floating zero, the starting point can be anywhere it is convenient for the designer to put the zero. Thus in Fig. 18.7, if the floating zero is placed at the corner of the workpiece *B*, the *x* dimension to the hole is 3 in. and the *y* dimension is 2 in.

Another area of importance is the method of listing coordinate positions. They may be *absolute* or *incremental*. The absolute system is more widely used. It provides that all measurements are made from a specific zero point. The incremental mode dimensions a point with reference to the preceding point. In this system the + and − signs must be used.

Figure 18.8 illustrates both systems. Some machines are designed to permit switching from one system to the other. An example of the use of the absolute system is shown in Fig. 18.8d.

The commands are shown in Table 18.3 and are correlated at the right of the

TABLE 18.3   ABSOLUTE SYSTEM COMMANDS[a]

Seq. No., *n*	Tab EOB	*x*	Tab EOB	*y*	Tab EOB	*z*	Tab EOB	Misc. *f*	Misc. *n*	Misc. *s*	EOB
001	T	06000	T	05000	T	02500	T	540	T	615	EB
002	T	11000	T	—	T	—	T	—	T	—	EB
003	T	—	T	08000	T	—	T	—	T	—	EOR

[a]Feed of 40 ipm = *f* 540; speed of 150 rpm = s615.

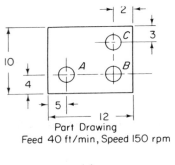

Part Drawing
Feed 40 ft/min, Speed 150 rpm

(a)

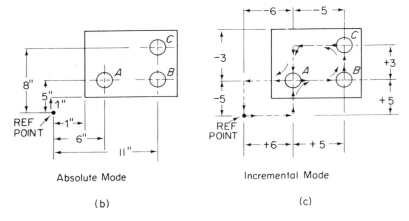

Absolute Mode

(b)

Incremental Mode

(c)

Figure 18.8(a,b,c)

tape. The student should study this tape very carefully. In Fig. 18.8b, *absolute mode,* all movements are from the reference zero point and therefore the plus and minus signs may be omitted. The sequence in which the holes are taken does not matter. In Fig. 18.8c, *incremental mode,* all movement is from the previous hole and the plus and minus signs should be included as shown in Table 18.4.

TABLE 18.4   INCREMENTAL MODE COMMANDS

Seq. No., n	Tab EOB	$\pm x$	Tab EOB	$\pm y$	Tab EOB	$\pm z$	Tab EOB	Misc. f	s	EOB
001	T	+06000	T	+05000	T	+02500	T	540	615	EOB
002	T	+05000	T	—	T	—	T	—	—	EOB
003	T	—	T	+03000	T	—	T	—	—	EOR

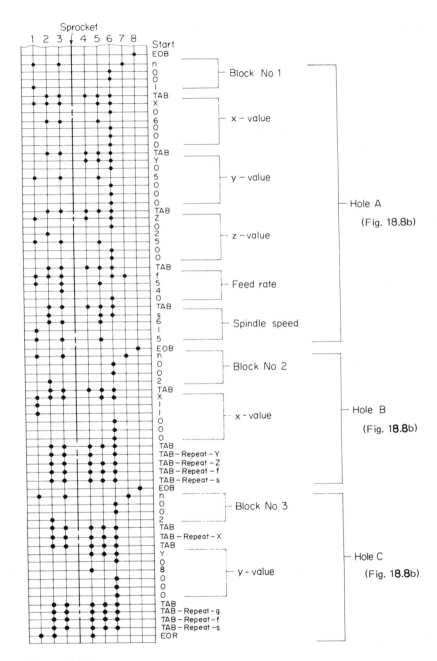

Figure 18.8(d)

18.5 NUMERICAL CONTROL MACHINES

Numerical control is a method for collecting and sequencing a series of operating instructions so that it can be used repetitively. Because of the need for greater discipline when preparing to write a program for numerical machining, a great deal of rethinking has been done about the role of the machine tool, operator, etc. Evolving from all of this is the *machining center,* which is capable of doing a great many more operations repetitively and accurately. As many as 60 tools can be stored in the tool changer, which can be used to machine parts on all sides and at almost any angle of the workpiece. This is accomplished through the use of multiple axes on the machine.

In general, the x axis represents motion parallel to the longest dimension; the y axis, motion primarily parallel to the shortest motion; and the z axis, motion that advances or retracts the tool. The letters A, B, C represent circular motion around the x-, y-, z-axes, respectively. The characters u, v, w are used to identify parallel motions to x, y, z. The letters P, Q, R represent rapid traverse.

The simplest of all the numerical control machines is the drill press, Fig. 18.9a. The workpiece can be moved side to side (x-axis), rear to front (z-axis), or the spindle can be moved up or down (y-axis).

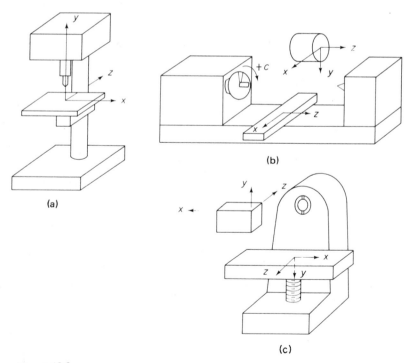

(a)

(b)

(c)

Figure 18.9

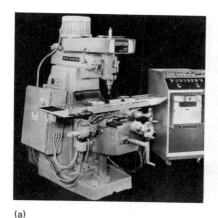

(a)

(b)

Figure 18.10(a,b)

Figure 18.9b shows linear as well as rotary axes on a lathe. Figure 18.9c shows the axes related to a milling machine.

The milling machine shown in Fig. 18.10a is a two-axis (three-axis optional) machine with a preset spindle depth control. The down feed movement of the spindle is programmed into the tape for control of the z-axis. In the three-axis machine the x-, y-, and z-axes are controlled through the use of servomotors which actuate hydraulic cylinders. The tape reader is a photoelectric type reader.

Other features indicated for this machine are emergency manual stops, manual and tape coolant on and off controls, manual or tape spindle start and stop controls, slide hold and reset, and manual full-floating zero for positioning. Features incorporated in the control unit are indicators for normal operation, sequence numbers, tape error, servo error, end of block, and cycle start. Still other features control mode selection and execution.

Optional features may be included in the machine and the controls for greater versatility. Such items are fine feed control, high-speed input, command multipliers, and additional functions which give a wider range of on–off control of auxiliary functions.

Figure 18.10b shows a face cam being milled on a numerically controlled milling machine. Figure 18.10c shows a drill press with numerical control. The down feed controls and tape reader are shown in Fig. 18.10d and e, respectively. A turret for additional tool operations is also shown. Different companies use different methods or tool changers for bringing different tools into position both manually and automatically.

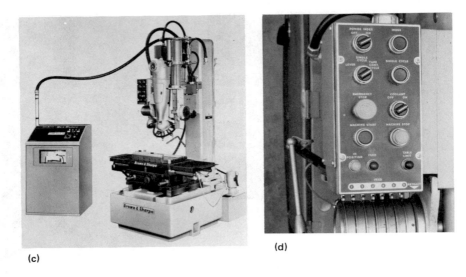

(c)

(d)

(e)

Figure 18.10(c,d,e)

18.6 TOOL DESIGNING FOR NC

Because of the accuracy built into an NC machine, the need for very expensive jigs or fixtures in negligible. Jigs and fixtures may be clamps and support blocks in most instances. The need for locating the workpiece relative to the zero point selected is important and must be dealt with when planning even simple holding devices. Furthermore, if the programmer calls for a 3-in.-long cutter to mill a $\frac{1}{2}$-in.-deep slot and the machinist uses a 2-in.-long cutter, it will go through the cycle without touching the work. Using smaller-diameter end mills than specified will remove too much metal and produce scrap.

Numerical control machines do not think. If a clamp is in the way of the cutter, it will machine the clamp. If the machine table is in the way, it will cut the table. Therefore, the tooling problem becomes one of analyzing the job, using great care in programming and setting up the machine.

It should be noted that one or two piece jobs may sometimes be done more economically on a standard machine than by numerical control. This is especially true if it is not necessary to make expensive jigs or fixtures for the conventional operation. If a production run is large, it is usually better to spend the money for jigs, fixtures, and gages than to resort to NC. This is especially true if the jigs, fixtures, and gages already exist.

One of the major advantages of numerical control over conventional machining with fixtures is the fact that changes which need to be made can be done quickly by correcting the tape. This is a great deal easier and less expensive than reworking a fixture, or scrapping a fixture and making a new one. For example, it is a simpler matter to drill a hole larger or smaller in diameter using numerical control, than it is to rework a drill jig. Incidentally, it takes less space to store tapes than it takes to store jigs, fixtures, and gages for the same job.

The human error is reduced with any numerical control operation over conventional machining. It is almost completely removed when the entire job is automatically controlled. The greater the reliance upon numerical control, the less chance that fatigue, movement of the workpiece, interruption, errors of judgment, and other factors will affect the outcome of the machining operation.

18.7 FIXTURE DESIGN

Numerical control machines use fixtures to hold and locate the workpiece so that the tool can repeat the machining operations accurately. Because tool changers are programmed and used, many operations can be done with one setting of the workpiece. Rigidity is important because of the multiple cutting forces used during the performance of multiple operations. It is therefore important that the designer have a thorough knowledge and understanding of basic machining practices.

It is also necessary that the fixture designer coordinate his efforts with the programmer and the production processer. In a small production plant the tool and fixture designer must also be able to write process sheets and programs for jobs. He must therefore know the limitations and capabilities of the machine tools in the plant. This requires a thorough knowledge of conventional machine tools. It is also advisable for tool and fixture designers to have had prior hands-on machining experience.

When designing fixtures there are considerations that must be taken into account. Reduced setup time, reduced loading and unloading time of parts, accuracy of setup, and reduced secondary activities to machining time are but a few of the items that lead to profit. It has been estimated that about one-fourth of the total time of production is associated with actual machining.

There are four basic elements to fixture design: accurate programs, workpiece processing, rigid holding devices, and accurate location of the workpiece so that the start point is the same for each workpiece.

Because part programs are so easy to incorporate into the memory system of an NC machine, it should be evident that some form of standard procedure for holding and locating the workpiece should be used. One such device is a combination base plate and angle plate that has many tapped holes and dowel holes. Other combinations with the base plate are possible. The base plate is located and fastened to the table of

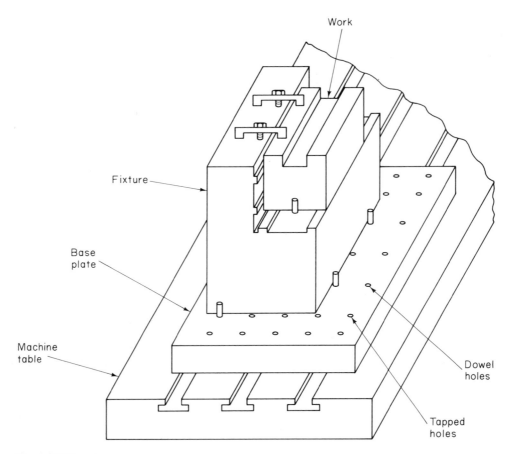

Figure 18.11

the machine. With the base plate fastened, it becomes possible to locate and attach the fixture directly to it, or to attach another holding device to the base plate that in turn mounts the fixture. The fixture locates and holds the workpiece. The advantage of mounting the fixture to the base plate rather than directly to the machine table, is to make it possible to remove the fixture and relocate it accurately if necessary. Such a setup is shown in Fig. 18.11.

Locator dowels and bolt holes should be provided in the base plate so that a wide number of uses are available for multiple setups. The base plate should be fastened to the machine table. Fixtures should be designed to conform to the base plate so that workpieces may be located accurately and clamped to the base plate. Using the base plate/fixture concept, the machine-zero point should always be the same relative to the part-zero point, no matter how many times the fixture is removed and remounted. Of course, removing the base plate defeats the purpose for its use.

The 3-2-1 rule, (Sec. 9.2), applies to NC fixture design. Once the support

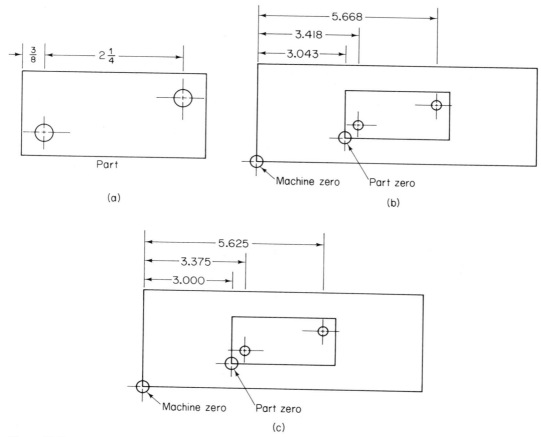

Figure 18.12

planes are determined, the designer must be sure that the program does not run the tool into the clamps or into the fixture.

Floating-zero machines permit location of the workpiece (fixture) anywhere within the reach of the tools that are to operate. If the NC machine has a fixed zero, the location of the machine-to-part zero must be programmed for maximum efficiency and economy.

If a part-zero point is positioned on a machine table so that the dimension is not easy to work with—say, an arbitrary 3.043 in.—the program will reflect the 0.043 throughout. It would be better to locate the machine-to-work dimension with a round number such as 3.000 in. The 3.000 in. is easily identified throughout the program.

The workpiece to be programmed is shown in Fig. 18.12a. In Fig. 18.12b, using an arbitrary 3.043 machine-to-part zero, the 0.043 in. is reflected in all x-axes program numbers. This is not the case in Fig. 18.12c. The possibility for error is reduced.

When designing a fixture there are other considerations besides those already mentioned. One of the first items that needs to be considered is the choice of a reference plane, or hole, that can be used to repeat accurately the location of the workpiece in the fixture relative to the machine zero. Casting and forgings, because of their rough surfaces, do not offer the designer a smooth and flat reference plane. Therefore, a surface or hole(s) must be machined to provide a flat and accurate reference surface. This reference is used to position the workpiece accurately for subsequent operations. Pins, or blocks, may be used to register the initial machined surface so that subsequent operations may be machined to print requirements. As many required operations as possible should be done using the initial reference plane or hole. This makes it easier to maintain the required relationships between machined surfaces. Each time a reference surface, or hole, is changed, inaccuracies result. Furthermore, inaccuracies are cumulative.

PROBLEMS

18.1. Explain the EDM process of machining metals.

18.2. What role does
  (a) the frequency,
  (b) the amperage, and
  (c) the voltage play in EDM machining?

18.3. (a) Define and explain tool wear in EDM machining.
  (b) How is tool wear classified?
  (c) How is it controlled?

18.4. Upon what factors does surface finish depend in EDM machining?

18.5. Explain the term *overcut* as applied to EDM machining.

18.6. Compare the machine operator's role on a numerical control machine and a conventional machine tool.

18.7. (a) What is the function of MCU?
  (b) How does it read a perforated tape?
  (c) How many other ways are there to store the program besides tape?

18.8. Discuss the advantages of machining single pieces with numerically controlled machine tools. What are the advantages?

18.9. Discuss the accuracy factor of numerical control as related to fixture cost on conventional machine tools.

18.10. Repeat Prob. 18.9 as it relates to changeover from one part to another.

18.11. Discuss the decision making process as it relates to conventional versus CNC machining.

18.12. Discuss the advantages of the use of NC machine tools versus conventional machine tools.

18.13. Discuss the statement in Sec. 18.2: "NC places the precision into the machine tool rather than the operator." Use examples.

18.14. What effect has the advent of numerical control had on the need to lay out the first piece for conventional machining? Explain.

18.15. What are some of the costs that make numerical control machining expensive?

18.16. (a) Define open and closed loops in NC machining.
   (b) Explain the operation of each.

18.17. Define
   (a) digital signals,
   (b) analog signals,
   (c) RAM, and
   (d) ROM.

18.18. Explain and illustrate what is meant by a point-to-point system.

18.19. Repeat Prob. 18.18 for a continuous-path system.

18.20. In point-to-point systems does the program need to identify all points on a line? Explain.

18.21. (a) Illustrate graphically how a circle would be machined with a point-to-point system.
   (b) What effect does the number of straight-line movements have on the outcome?

18.22. Explain
   (a) linear interpolation,
   (b) circular interpolation, and
   (c) parabolic interpolation. Illustrate each.

18.23. Describe the system known as
   (a) BCD,
   (b) straight binary, and
   (c) ASCII.

18.24. (a) Draw an eight-channel tape; insert the BCD numbers from 0 through 9.
   (b) How many parity holes are needed?

18.25. What is the purpose of
   (a) the tab code,
   (b) the EOB code, and
   (c) the EOR code?

18.26 (a) Define an *n* word command,
   (b) a *g* word command, and
   (c) an *m* word command.

18.27. Code the following spindle speeds:
   (a) 1027,
   (b) 24,
   (c) 15.8, and
   (d) 176.83 rpm.

18.28. Code the following feeds:
   (a) 0.007,
   (b) 0.015,
   (c) 16.4, and
   (d) 1.2 ipm.

18.29. Describe and illustrate the concept of
   (a) fixed zero and
   (b) floating zero.

18.30. Define and illustrate the method for listing coordinate points using
   (a) the absolute system and
   (b) the incremental system.

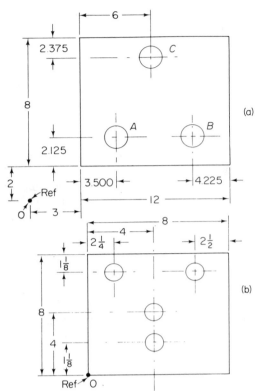

Figure 18.13

18.31. Redraw Fig. 18.13a and dimension it in
   (a) absolute mode and
   (b) incremental mode. Do this from point 0.

18.32. Repeat Prob. 18.31 for Fig. 18.13b.

18.33. Tabulate your results for Fig. 18.13a using tables such as Tables 18.3 and 18.4. Assume a feed of 15.6 ipm and a spindle speed of 485 rpm.

18.34. Repeat Prob. 18.33 for Fig. 18.13b using a feed of 22.86 ipm and a spindle speed of 225 rpm.

18.35. Draw *two* eight-channel tapes and insert the programs in Prob. 18.33 and Fig. 18.13b.

18.36. Draw *two* eight-channel tapes and insert the programs in Prob. 18.34 and Fig. 18.13b.

18.37. **(a)** What is a numerical control machining center?
   (b) What will it do?

18.38. Why is it important that a machinist use the specified cutting tools called for by the programmer? Give at least two examples of when the programmer's recommendation is not followed.

18.39. "Numerically controlled machines do not think." Explain this statement and give examples of what could happen.

18.40. When is it more economical to use conventional machines over NC machines?

18.41. What factors are eliminated when NC is used to its fullest?

18.42. Why is it essential that a tool and fixture designer have a thorough knowledge of, and hands-on experience with, conventional machine tools?

18.43. What are the three basic elements to fixture design? State and discuss each element.

18.44. What is the purpose of
   (a) dowel holes and
   (b) tapped holes in the base plate? Explain.

18.45. Why is it important to use a base plate in CNC? State all your reasons for its use.

18.46. State the 3-2-1 rule in fixture design. How would you apply it to the part in Fig. 18.11?

18.47. Define
   (a) machine zero and
   (b) part zero?

18.48. Discuss the use of a floating zero and a fixed zero. Give examples of each.

18.49. Why is the assigning of arbitrary dimensions to a part-zero point bad practice? Illustrate.

18.50. Discuss the selection of reference planes when designing fixtures.

18.51. Why must a reference surface be machined when designing fixtures for castings and forgings? Can you think of a situation when a casting would not need a reference surface machined as a first operation?

18.52. What is the effect of changing a reference surface before a part is finished? Explain your answer.

Index